JOHN CURR

The Man Who Revolutionised Mining

by Charles P Dixon

Copyright © 2025 by Charles P. Dixon
All rights reserved.

ISBN 978-1905315-88-8

No part of this book may be reproduced in any form or by any electronic or mechanical means, including information storage and retrieval systems, without written permission from the author, except for the use of brief quotations in a book review.

Every effort has been made to trace copyright holders and to obtain their permission for the copyright material. In some instances, through failure to establish contact, permission to use quoted material has not been gained. The author apologises for any errors or omissions and would be grateful to be notified of any corrections that should be incorporated in future reprints or editions of this book. Any errors or omissions are the author's.

Project management and editorial work: Julie Pickard, London
Design & layout: Andrew Chapman
Indexing: Tanya Izzard

A Heritage Hunter publication
www.heritagehunter.co.uk

This biography of John Curr, engineer and servant to Charles Howard, 11th Duke of Norfolk, is dedicated, with permission, to Edward Fitzalan-Howard, 18th Duke.

CONTENTS

	Acknowledgements	vi
	Preface	vii
1	Introduction	1
2	Family	5
3	Education	11
4	Character	23
5	Wealth	25
6	Belle Vue House	27
7	Worship	33
8	Early career	41
9	After dismissal	109
10	Children	143
	Appendices	171
	Glossary	201
	Bibliography	209
	Curr family	228
	Working relationships	229
	Endnotes	231
	Index	255

ACKNOWLEDGEMENTS

This is my first book. My inexperience was the cause of much angst and hand-holding for Julie Pickard, my long-suffering editor.

Particular thanks go to John Hunter for his help with research and to John Taylor for correcting my grammar. Special thanks for help with research are also due to Geoffrey White, Nicola Johnson and Jennifer Hillyard of the Common Room. My thanks go to the other members of the production team: Andrew Chapman, designer, and Tanya Izzard, indexer.

Mentions for their assistance and support are due to Clare Dixon, Frederica Constable, Tom Constable, Henry Dixon, Andrew Garrad, Angus O'Neill, Tim Wainwright, Bob Rae, Lisa Evans of the Stanley Library, Linda James of the Durham Record Office and Henry Woudhuysen, on whose suggestion and encouragement this book was written.

Thanks go to the following for permission to reproduce and quote from material in their possession: Sheffield City Archives, The Common Room (North of England Institute of Mining and Mechanical Engineers); Durham Record Office; Stanley Library; the Newcomen Society; John Hunter; Durham County Local History Society and Ordnance Survey Ltd.

Charles P. Dixon

PREFACE

Samuel Johnson took a great interest in how things – anvils, poetry, plates of iron, bars, spectacles, all sorts of machinery – were made and worked. 'His knowledge in manufactures was extensive,' a friend said soon after his death in 1784, 'and his comprehension relative to mechanical contrivances was still more extraordinary'.[1] Compared to the violence of warfare, it was 'inoffensive industry', Johnson argued, that made people and countries richer.[2] If he had met John Curr, 'the man who revolutionised mining', the two would, no doubt, have had much to say to each other.

Charles Dixon's biography of Curr, the coal viewer and engineer, and his account of the man and his family is a remarkable contribution to our constantly developing knowledge of the industrial and mechanical processes that lay behind the industrial revolution. It is based on a great deal of original research and an impressive understanding of Curr's interests and of his family and its circumstances. There is also an element of what the *ODNB* still calls 'personal knowledge' and 'private information', since the origins of this book lie in its author's interest in his family's history.

Without formal schooling, Curr learned his profession from more experienced men. His high intelligence and powers of observation, his flexibility and pragmatism, are characteristic of the best sorts of practical (and academic) engineers. He was an innovator in his use of the flanged cast-iron rail and developed new trucks for the railroads, revolutionising the transport of coal, and so making mining more efficient. His interests also embraced steam engines and the ancient business of rope making. As well as these subjects, the book introduces the reader to the language that was used to describe the different parts of his various activities, to corves, hurriers, putters, staiths, tipplers and winnings. There is much here that will also be of interest to historians of Roman Catholicism in the North East. The servant of the Duke of Norfolk, John Curr and his wife Hannah had nine children, one of whom became a priest and writer, while two became nuns. There is also an Australian aspect to the story since Curr's eldest son, John Curr III, and his younger brother, Edward, settled in Sydney and Van Diemen's Land, respectively.

John Curr: The Man who Revolutionised Mining tells an extraordinary series of stories about working lives two hundred and more years ago. It is, in all senses, a well-made and well-illustrated book that will add greatly to our understanding of the industries that made the country wealthy.

H.R. Woudhuysen FBA
Lincoln College, Oxford
February 2025

One

INTRODUCTION

Life

It is generally accepted that John Curr was born in County Durham in 1756. His name is little known today. Indeed, the author, a direct descendant, never heard his name mentioned by anyone in our family. The author came across the name some ten years ago, when reading manuscript notes made by Cousin Dorothy Gurner, one of Curr's great great granddaughters.

As early as 1931 Fred Bland, in 'John Curr, originator of iron tram roads', wrote:

> *John Curr was an important man in Sheffield over a hundred and fifty years ago; ... But such is fame – he is almost forgotten and his name is unknown at the Norfolk Estate Office!*[3]

Curr's career divides into two: 1777 to 1801 when he worked at the Duke of Norfolk's Sheffield collieries; and 1801 until retirement in 1820.

It was during the years 1781 to 1801, when Curr was 'Superintendent of the Coal Works of his Grace the Duke of Norfolk', that his ingenuity and inventiveness contributed so greatly to the advancement of technique in mining.

After his dismissal by the duke in 1801 he devoted his time to the promotion and sale of rope, for which he took out nine patents between 1792 and 1813, from his rope-walk, and the cast-iron merchandise from his foundry.

Curr died in Sheffield in 1823.

Recognition

Curr's innovations went a long way towards solving the problem of the increasing cost of moving coal as mines became deeper. That his innovations were so widely adopted is testament to his ability.

Curr's contribution to the development of mining has been acknowledged by many:

Matthias Dunn, a Government mine inspector, wrote in 1855:

> *The most prominent of the advances of mining science within the last hundred years may be enumerated* [there are 6 altogether; Curr

was responsible for nos. 3 & 5] *as follows: -3. The improvements in rope-making, beginning with the flat hempen ropes of Mr. Curr, and ending in the present improved state of wire ropes, which improvements have been the means of extensively conveying below ground, as well as of producing immense savings in the raising of coal. -5. The recent improvements arising from the substitution of wooden conductors, tubs, and cages, for conveying the coals up in shafts, instead of baskets, without the collision attendant upon the old system, has led to immense saving.*

Dunn continued:

These various improvements may be said to be the result of mere practice, since they have all (with the exception of the safety-lamp) been conceived and carried out by the ordinary managers of mines and manufactures.[4]

R. L. Galloway put Curr's abilities thus:

... he introduced a series of improvements [into the Sheffield collieries] *in the mechanical arrangements which exhibited a remarkable degree of ingenuity and originality.*[5]

T. S. Ashton and J. Sykes noted:

In the working out of the many problems to which the new form of transport gave rise, many engineers played their part, but among them the outstanding figure was John Curr ...[6]

G. G. Hopkinson has it that:

... Curr showed a boldness of conception, an ability to weld a number of technical ideas into a harmonious whole, a capacity for large scale organisation and an executive ability which stamp him as by far the greatest mining engineer of his generation in the Sheffield region.[7]

Charles E. Lee,[8] who wrote the Introduction to the 2nd edition of Curr's great book, *The Coal Viewer and Engine Builder's Practical Companion*, noted:

[It is] 'the earliest known engineering work on the railway track'[9] and, 'gave very detailed drawings of turnouts, of curves, and the like; for a book of 1797 it was extraordinarily well-produced and detailed'.[10]

Barrie Trinder considered that:

John Curr [was] the most talented of the 18th century mining engineers.[11]

Ian R. Medlicott noted:

John Curr deserves recognition as one of the prominent mining engineers of his time. The introduction of rail roads, four-wheeled corves, conductors, tipplers, inclined planes, and flat ropes placed the Sheffield collieries among the most technologically advanced in the country. His improvements and innovations were widely adopted and made a major contribution to the expansion of the coal mining industry.[12]

Although a great deal has been written about Curr and his activities, much of it contradictory, there has as yet been no full-scale account. This book is an attempted remedy.

Two

FAMILY

Confusingly, three generations bore the name John Curr. In order to distinguish between them, each has been allocated a roman numeral. Thus, the Curr about whom this book is written becomes John Curr II; his father John Curr I, and his son John Curr III. A guide to some of the Curr family can be found on page 228.

Origins

The origins of both the Curr family and of John Curr II are obscure. What is known is that John Curr I was a practising Roman Catholic. Records of Roman Catholic families are often scarce before emancipation (full emancipation came in 1829 after the passing of the Catholic Relief Act). Both J. W. Fawcett,[13] before 1911, and E. C. Wilson, the Secretary of the Catholic Record Society, after 1931, attempted to find Curr's name in the baptismal registers of Co. Durham, but without success.

Neither the year of John Curr II's birth nor his birthplace has been determined. It is generally accepted that he was born in 1756. As to his birthplace, Kyo, Pontop Pike and Greenside (there is a Kyo near Greenside) in the parish of Ryton-on-Tyne[14] have all been suggested.

In 1858 Thomas Young Hall first identified Pontop Pike as the village where Curr was raised.[15] Hall is a credible source because his father was acquainted with the Buddle family. Pontop has a strong association with Roman Catholicism. A thesis by Leopold Gooch on Catholics in north-east England in the eighteenth century supports this:

> *In north-west Durham were Stanley, Lintz, Pontop and Hamsterley Halls, all Catholic houses within a short distance of each other and served by one, or perhaps two, priests. There were missions at Stella Hall, near Ryton, and Pontop Hall. A roll of 1705 lists 183 papists in the parish of Ryton. This return shows that the Catholics were mostly pitmen working in the Widdrington family collieries, and living with their families in Stella, Blaydon and Winlaton. By 1767 there were some 450 Catholics in the parish.*[16]

John Curr I was born in 1712 and died in 1777. Sometime around 1740 he married Elizabeth (1722–1801) whose maiden name is unknown. He was a coal viewer, almost certainly at Bushblades colliery.[17] If so, his religion

John Curr

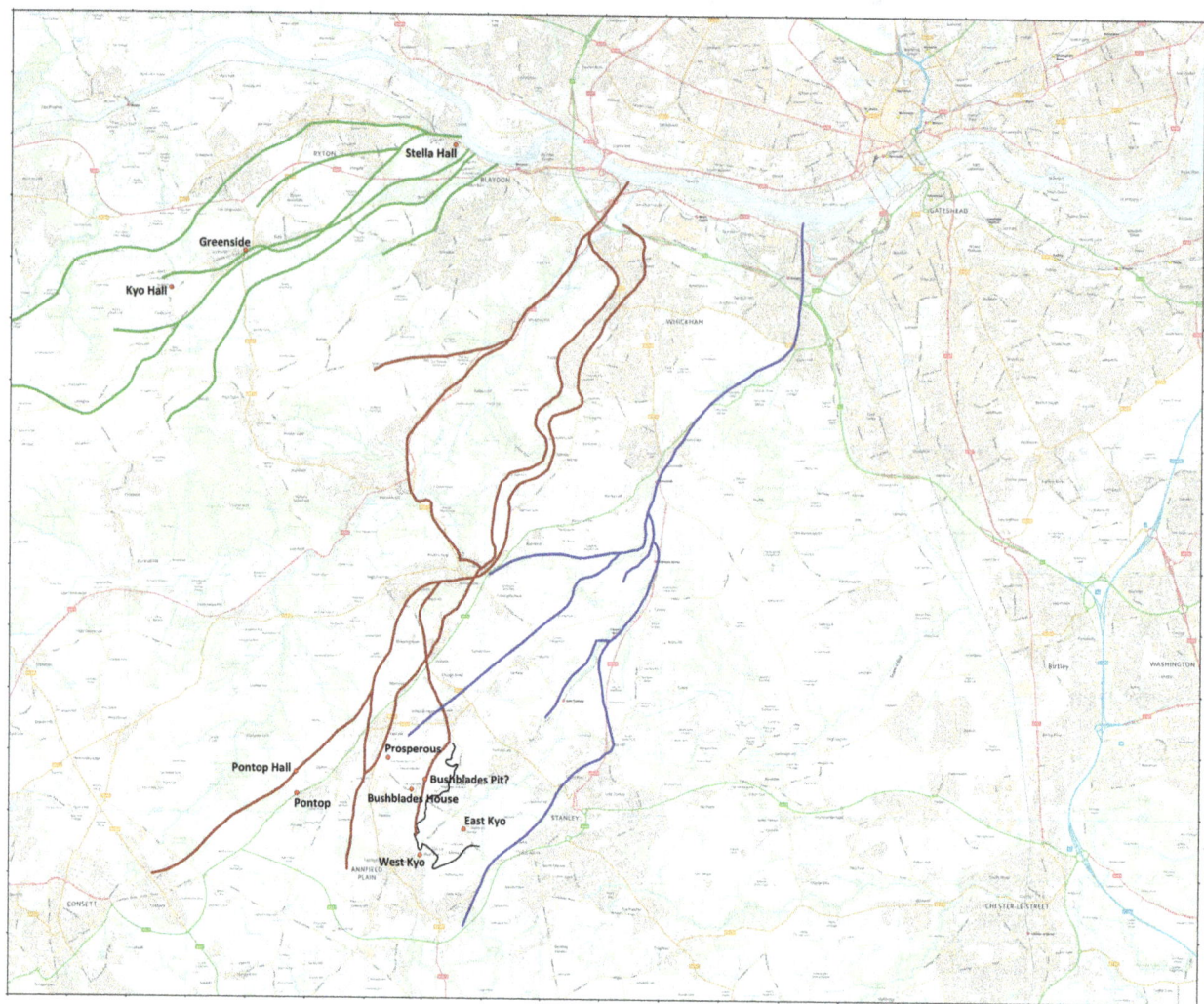

Figure 2.1. Relevant waggonways superimposed onto OS map of present-day County Durham. The short black line shows the surface outcrop of the High Main Coal, the modern name for the seam. It outcropped along the valley sides of Kyo burn from which Bushblades colliery worked it westwards under the high ground of Pontop Pike. Contains OS data© Crown copyright and database right (2022).
© John Hunter. All rights reserved.

may well have influenced his appointment because Bushblades, sunk in 1754, was owned by the Silvertops, a landed Roman Catholic family from Northumberland [see Figure 2.1]. There, John Curr [II] would have been exposed to coal mining operations from an early age.

John Curr I's burial inscription at All Saints Church, Lanchester,[18] gives his place of residence as Bushblades, which is situated near East Kyo. A photograph of the house, taken c. 1955, is shown in Figure 2.2.

John and Elizabeth had three sons: George (1749–1826), William (1753–1823) and John (c. 1756–1823). There may have been an earlier John, for there is a record for the burial of

John Curr, son of John Curr of Cadger Row. Buried 1752 at Holy Cross, Ryton.[19]

George, William and John were all employed by the Duke of Norfolk's Sheffield estate. Few records survive for George and William but both are mentioned in colliery accounts. George had a position with some responsibility and was to be found living in the Ponds area of Sheffield in 1790:

for sale by auction, by order of the assignees of Joseph Matthewman, a bankrupt. Lot 6: a large leasehold messuage, with outbuildings and

Figure 2.2. 'Bushblades, near East Kyo. Once the home of John Curr, an early genius of mining and an inventor of mining machinery.' Photo and caption reproduced by permission of Stanley Library and Durham County Record Office ref. DRO/5217.

gardens in the Ponds, leased to Messrs Birkes & Co, with another adjacent messuage, with offices and garden, now let to Mr George Curr (both held of the Norfolk family).[20]

The brothers' mother, Elizabeth, was perhaps living with one of her sons, for her sad end in 1801 is recorded:

A few days ago died, Mrs. Curr, of Sheffield. Though at an advanced age, her decease was unfortunately hastened by one of the most melancholy accidents, which, too frequently, cut off both the old and the young. In approaching the fire her clothes caught the flames, and she was burned so severely that she only survived for a few days.[21]

George was still living at the Ponds in 1810:

[A] violent thunder-storm … lightning struck the houses of Mr. Curr and Mr. Thompson [George, son-in-law of George Curr], in the Ponds, demolishing the windows, looking-glasses, picture frames, and cupboards in its course, and with a tremendous explosion rocking the buildings to their foundations. Though it passed through the bed-chambers where the families lay, providentially no person was hurt.[22]

Marriage

John married Hannah Wilson on 1 May 1781 at Sheffield Parish Church, by licence.[23] The witnesses were Hannah's brother Richard and John's brother George.

The Wilson family

Little is known about the Wilsons. Of Protestant persuasion, Hannah converted after marriage to Roman Catholicism.

Hannah's baptism is recorded in the Sheffield Parish Church register on 29 June 1759:

> *'Hannah, Daughter of Richard Wilson, labourer.'*

Hannah had an older brother, Richard, whose baptism is similarly recorded on 19 June 1751.

After his baptism the next record found for Richard is a 99-year lease from the Duke of Norfolk's estate, dated 1 January 1781, for a parcel of ground in Pond Street, Sheffield.[24] The lease lists his occupation as book-keeper. His burial entry reads:

> *Richard Wilson, Gentleman [of] Gell St., Sheffield 8th February [aged] 72 years. [Officiant] M. Preston.*

By the time of his death in 1823 Richard was a man of property. His will[25] bequeaths *'my share of houses at Pond Hill, my house and Garden in Broad Lane'* to his wife *'during her natural life'*. After her death all his wealth *'shall be equally divided amongst my Sister Curr [Hannah] and her Children save that my nephew Edward Curr to whom I gave Two hundred pounds when he went abroad to have that Sum'*. John Bernard Furniss [husband of the Curr's eldest daughter, Elizabeth] is named as *'sole Executor'*.

Something is known of Hannah's life after the death of her husband. Her grandson, Edward Micklethwaite Curr, wrote:

> *I knew her well in the summer of 1831 and winter of 1833. She was an old lady, benevolent, strict, and starch. She kept a good house in Sheffield. When I knew her she was blind in one eye, and so deaf that she could only hear with a trumpet. When there with the Miss Ellisons, (Mr. Ellison, I believe, succeeded my grandfather as steward to the Duke of Norfolk in 1831), we used to amuse ourselves with ringing all the bells in the house. The old lady, fancying she heard a noise, was unable to make out what it was. She was a very religious woman, and towards the end of her life removed from Sheffield to York to be near her daughter Harriet who was a nun at York Bar Convent. The poor old lady was always very kind to me, took care not to flatter my vanity, and died in about her 98th year.*[26]

A Mrs Curr is recorded as living at West Field Terrace around 1832.

The 1841 census (taken on 6th June) finds Hannah at West St., Sheffield, aged 80,[27] with one servant.

The 1851 census (taken on 30th March) finds her at No. 39 Blossom St, (parish of Holy Trinity, Micklegate) York, lodger, Wid[ow], 93, annuitant.

Hannah died at York and was buried at Bar Convent.[28] Her death was announced in the *York Herald* on 14th June 1851:

> *On Tuesday 10th Mrs. Hannah Curr, relict of the late John Curr Esq., of Belle Vue House, near Sheffield, in this county, in the 93rd year of her age. – R.I.P.*

The probate index lists:[29]

> Curr, Hannah, Aislaby Hall, Whitby, the Revd Andrew Macartney,[30] Egton Bridge. £790.

The death duty index lists:[31]

> Curr, Hannah formerly of Sheffield and, late of Aislaby Hall near Whitby. W[ill].Prog [Prerogative] T[estator], Cod[icil] £450.'

Aislaby Hall was advertised *TO BE LET* on 27th November 1852 in *The Farmer's Friend and Freeman's Journal* in York. Interestingly the description of the house notes:

> There is also a large Room adjoining, fitted up as a very NEAT CHAPEL – the family being Catholics – which would be a great acquisition to a Family of the same Creed, who would be preferred.

Death

The Currs left Sheffield for Paris in early 1820. On their return in August 1821, John wrote to John Buddle Jnr with news of the family:

> *Sheffield Aug 16th 1821*
> *After spending upwards of 18 Months in Paris I am returned with my Family to this place … After spending a few Weeks with my friends I am settled in lodgings in the Western suburbs of Sheffield with Mrs Curr, Harriet and Julia, and my daughter Mary Ann has changed her name to Beauvoisin and resides in Manchester. She was Married on Coronation Day, and Mrs Furniss and Harriet went with her to Manchester and spent a few Weeks. Mr Beauvoisin has been many years in Manchester and after spending a few more there, it is their intention to live in France. My son Joseph is just returned from a 5 weeks Tour in that Country. … I hear occasionally from Edward, who is now settled for some time in New South Wales … I am happy to say that Mrs Curr and myself continue to enjoy pretty good health …*[32]

His death was announced in the *Sheffield Mercury* for Saturday 1 February 1823:

> On Monday last, at Broom-spring, much respected by all who knew him, John Curr Esq. Aged 67.

Three

EDUCATION

How did it come about that a young man, of somewhat humble origin, and with little formal education, emerged as a fully fledged engineer at a young age? Norfolk estate accounts show that in 1775, when nineteen, Curr was commissioned to write reports on its Sheffield collieries and appointed 'Superintendent of the Coal Works' in 1781, aged twenty-five. In the next century Robert Stephenson, another native of north-east England, and admittedly with the advantage of being the son of George, was nineteen when appointed managing partner of the locomotive works, Robert Stephenson and Co., in 1823. This section explores how it was possible.

The coal viewer

Just what was a coal viewer? It was a term used to denote 'the chief man of every colliery'. He was required to prepare reports estimating output, costs and profitability of mines for the owners and had to overcome such problems as sinking shafts, pumping out water and hauling coal to the surface. The viewer had to be hard-headed, shrewd and practical. Citing the experiences of some well-known viewers, David Oldroyd explains:

> *John Barnes ... prepared ex-ante unit cost/profit figures for Fenham colliery in September 1717 ... There are many examples too in the notebooks of 19th century viewers, such as John Buddle (Jnr.). His father had calculated the cost of working Herraton Moor colliery in 1798 ... it is perhaps surprising that the underlying techniques were not set down in written form ... Works ... do exist, but these generally dealt with technical matters, or were arithmetically based ready reckoners, compiled by the viewers themselves ... John Curr's* The Coal Viewer, and engine builder's practical companion *(1797) combined both aspects.*[33]

Oldroyd also notes how knowledge was passed on:

> *A factor which emerges from the viewers' records is that ... they worked in association with each other, and learnt their trade as apprentices with established viewers. In this way the succession of knowledge based on practical experience was assured, perhaps reducing the need for textbooks ... Edward Smith (Jnr.) succeeded his father Edward Smith*

(Snr.), and in turn had five sons who were also brought up as colliery viewers. The basis of a viewer's training throughout the 18th century remained practical experience, acquired through formal apprenticeship, parental instruction or rising through the ranks in the pit. John Buddle (Jnr.), the doyen of colliery viewers, was himself said to have received only one year's formal schooling and to have been educated almost entirely by his father. ... [the viewers] were themselves subject to audit by their superiors, which was common practice at the time. For example, the accounts of John Curr were audited by Vincent Eyre, the land agent, who had overall responsibility for the mines.[34]

Christine E. Hiskey elaborated:

Whether a viewer was trained by apprenticeship, practical experience, or parental instruction, great emphasis was laid on two qualifications in particular: scientific and scholarly understanding, and practical 'breeding'. Numerous viewers already owed a great part of their fame to pit work – Thomas Barnes and Buddle were the most noted, but Edward Smith senior, William Brown of Throckley and, in Cumberland, Carlisle and James Spedding had already shown the way. Wider scientific interests were also becoming apparent – both Edward Smiths, like John Buddle senior, were 'celebrated mathematicians', and many viewers took a keen interest in geology. More tangible signs of formal education and qualification, however, were only just beginning to appear in the first half of the nineteenth century.

As the coal trade expanded and technical problems increased, the demand for good viewers was felt not only in the north-east but throughout Great Britain and abroad. The Newcastle coal-field had long been the source to which other areas looked for expert recruits or specialist consultation. William Dixon (1753–1824), for example, who by 1800 was a leading authority in the West of Scotland, had gone there from Tyneside; and John Curr, well-known as the Duke of Norfolk's viewer at Sheffield, was one of the remarkable group of viewers which originated in the Greenside area of north-west Durham.[35]

Michael W. Flinn concurs:

Any qualified viewer could be commissioned to make a report, but the reputation of the general viewers led to their attracting a disproportionate number of commissions. The prestige of the north-east viewers as the leading experts of the day meant that they received commissions from beyond their own coalfield. John Buddle senior reported in the 1780s on collieries in Yorkshire ...[36]

That this reputation and prestige were not shared by the layman is encapsulated in the following exchange, which took place in 1836 when John Buddle Jnr was giving evidence before a committee of the House of Commons. The cross-examination was by Mr Wilkinson:

You are not an engineer, I think, Mr. Buddle?
  – *I believe I am.*
Have you ever acted professionally as an engineer?
  – *I am so far an engineer that I was thought worthy of being appointed a member of the Institute of Engineers in London.*
Has that ever been your pursuit in life?
  – *It has.*
For what space of time?
  – *All my life.*
Do you mean to say there is no difference between the business of a coal viewer and an engineer?
  – *There is: one is a branch of the other.*

This attitude is reinforced in a letter written by 'A True Conservative' to Lord Londonderry in 1843, after Buddle Jnr's death:

There is much mystery and humbug thrown over the performance of the Craft call'd Viewing.[37]

The viewer was often nonconformist and, according to A. N. Wilson:

In England in the eighteenth century only those who subscribed to the 39 Articles of the Church of England could go to university. It is no coincidence that nearly all the great pioneers of science and technology in that time were nonconformists who were forbidden to go to university – men such as Joseph Priestley, a Unitarian minister who was the pioneer of modern chemistry. Josiah Wedgwood, Erasmus Darwin and Matthew Bolton would have learned nothing but a few limited bits of maths and Latin had they been to Oxbridge. As it was, their broad range of scientific curiosity and technological know-how transformed the world.[38]

The nonconformism was often Unitarianism:

The particular form of Protestantism remarkably commonplace among mid to late eighteenth-century British manufacturers and entrepreneurs was a strange hybrid of Christianity, ancient heresy and rationalism. This Unitarianism informed the lives of the Watts and the Wedgwoods just as it was 'strongly over represented' among Lancashire and Manchester cotton manufacturers or coal engineers like John Buddle.[39]

The Buddles

The relationship between the Currs and the Buddles demonstrates how knowledge, without formal schooling, was passed from one generation to another. John Buddle Snr (1743–1806) and his son John Buddle Jnr (1773–1843) loomed large in Curr's working life. The Buddles lived at West

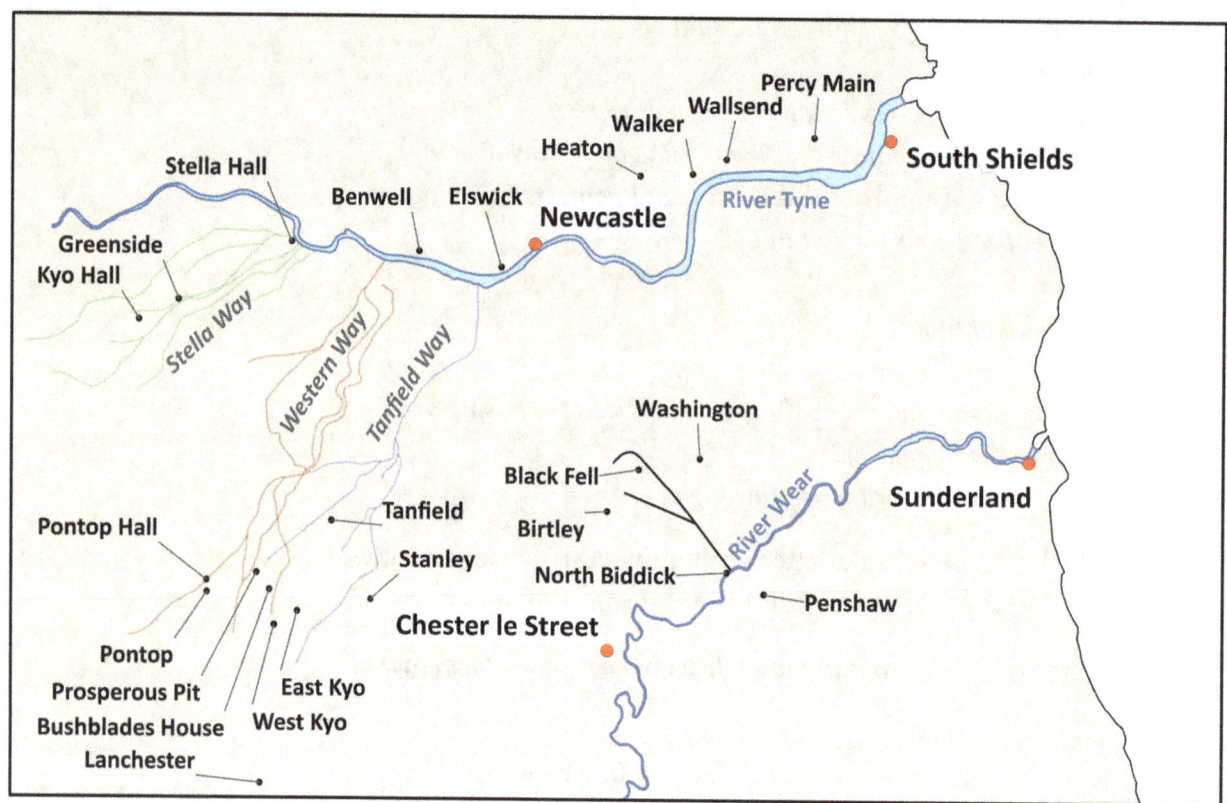

Figure 3.1. Tyneside and Durham region showing the places mentioned in the text. ©John Hunter. All rights reserved.

Kyo for some years, near to the Currs. The Roman Catholicism of the Currs was clearly no barrier to their close association with the staunch Unitarian Buddles. Buddle Snr was also to feature significantly during Curr's time at the Sheffield collieries.

John Buddle Senior

John Buddle Snr was born on 13 July 1743 at Chester-le-Street, Co. Durham, at the centre of the mining district of the River Wear [see Figure 3.1].[40] His father was George Buddles.[41] It is likely that the Buddles were reasonably well-to-do. According to Christine Hiskey:

> *The family preserved links with Chester-le-Street many years later: John Buddle senior was buried there in 1806 despite having become established at Wallsend, and so was his widow in 1827. At least until the 1830s, Buddle [junior] owned a tannery there which was leased out.*[42]

At the age of fifteen Buddle Snr was attending a local school, Mr. Robinson's at Biddick,[43] a village nearby. Thomas Robinson, as well as conducting a school, was a frequent contributor to such magazines as *The British Palladium* and *The Gentleman's Diary*, both as setter of, and responder to, mathematical questions. Buddle Snr followed in Robinson's footsteps. As well as answering a question in 1758, his replies to others are found in *The Gentleman's Diary* for 1760,[44] and also in 1760 he is to be found setting a question in *The Gentleman and Lady's Palladium*,[45] as well as answering some.[46] Buddle Snr also developed an interest in enigmas. In the same issue his name is given, among others, for solving one.[47] In the next issue, for 1761, under *New Queries*, he asks 'WHAT are Thunderbolts, and what is

Thunder?',[48] sets a rebus,[49] a mathematical question[50] and answers others.[51]

By 1762, aged nineteen, Buddle Snr was teaching mathematics in Chester-le-Street. The evidence for this is to be found in *Miscellaneous Correspondence*.[52] 'Mathematical Questions Answered' for May 1762 includes '*Question 376, answered by* Mr. James Watson, *at* Mr. John Buddle's *Mathematical School in* Chester-le Street' and, in June, '*Question 378, answered by* Mr. John Buddle, *Teacher of the Mathematics in* Chester-le-Street', although his name does not appear in schoolmasters' licences.[53] Richard Welford describes him as 'a schoolmaster of repute, a contributor to the Diaries'.[54]

R. L. Galloway claims that

At Bushblades, Mr. Buddle must still have kept on a school for a considerable time, if not, indeed during the whole of his residence there, inasmuch as it appears that the only schooling young Buddle got – who was born there in 1773 – consisted in a year's attendance at his father's school at an early age.[55]

He was evidently a man of wide interests:

The elder Mr. Buddle was a man of considerable attainments in mathematics; he was a correspondent of Hutton, Emerson [both local men[56]*], and other eminent men, and contributed many papers to the scientific publications of that period.*[57]

At the same time he was showing a keen interest in mining operations. Evidence of this is found in an article written by him for *The General Magazine* and printed in 1764, when he was twenty-one.[58] The article consists of a plate and description of a coal waggon and is addressed '*To the Authors of the General Magazine*' [see Figure 3.2]. Buddle signed his article from '*Loanings-House, near Chester-le-Street, May 1, 1764.*'

Buddle Snr may have removed in 1766 to West Kyo, a small village

Figure 3.2. A Representation of a Coal Waggon by John Buddle, from *The General Magazine of Arts and Sciences*, London (1764), p. 285.

nearby Bushblades colliery, in the vicinity of Stanley. In 1768 he married Ann Reay, daughter of a local farmer, in Chester-le-Street.[59] The entry in the parish register for the birth of their first child, also in 1768, records the couple as living at Kyo. Only after the birth of their fifth child, in 1777, is his occupation recorded as 'Coal viewer'.[60] All the baptismal entries record Kyo as the place of birth.

Buddle's name appears several times in the text of *The British Palladium* for 1771, answering the prize enigma, and further mathematical questions.[61]

A great lover of books,[62] in 1778 Buddle Snr published a new edition of *The Marquis of Worcester's Century of Inventions*, to which Buddle added *An Appendix; containing An Historical Account of the Fire-Engine for Raising Water; Which Invention originated from the above Work*.[63] This book, originally published in 1663, is remarkable for the addition of an account of the steam engine. According to Matthias Dunn in his book *History of the Viewers*, there were two common engines at work drawing water at Bushblades colliery in 1769, so Buddle Snr will have been familiar with them.[64]

Buddle Snr was perhaps involved in a reprint of a work by Thomas Bewick. According to Nigel Tattersfield:

> *Antiquary and bookseller John Bell of Newcastle had reason to believe Akenhead's reprint [1796] of an elusive original edition of 1655 smacked of sharp practice. According to him the late Mr. John Buddle of Bushblades, near Ryton, and of Wallsend ... having met with a copy of it came to my father's shop in Union Street one Saturday Morning, and proposed that my father and he should republish it ...*[65]

Another example of the breadth of Buddle Snr's interests is shown by an essay 'Observations on the Raising of Hedges: in a letter to the author, by John Buddle, Bushblades Colliery, February 5, 1785', which he contributed to a book on the subject written by Robert Callender.[66] These observations, Buddle wrote, were made over several years on the progress of raising hedges on Bulbeck and Lanchester Commons.

In 1788 Buddle Snr submitted to the *Gentleman's Magazine* an account of the extraordinary accident suffered by John Boys, a collier who survived falling down a pit shaft to *'a depth of 42 fathoms, or 84 yards'*.[67] The full account of this accident is provided in Appendix 1.

In 1789, of which more later, Buddle Snr published *An account of an improved method of drawing coals and extracting ores &c. from mines*.[68] In this is printed a copy of the Letters Patent and Specification for John Curr II's Patent 1660, published 12 August 1788, together with two plates signed 'J. Buddle Junr.', then aged sixteen.

An explanation for how Buddle Snr came by his knowledge of coal mining operations is provided by Richard Welford:

> *Living amongst men whose chief pursuits were the winning and working of coal, the elder Buddle became intimately acquainted with the business. His colliery friends, most of whom worked largely by 'rule*

of thumb', found him of great assistance in making their calculations, and his abilities as a mathematician being frequently laid under contribution for their benefit, he obtained a knowledge of colliery operations which was afterwards instrumental in raising him from the humble position of a village dominie [a schoolmaster] to the more exalted post of colliery manager.[69]

Jacob noted:

Eighteenth-century engineers who became overseers or viewers had not necessarily been trained in coal mining before they came into the business. As early as the 1740s the viewer John Watson, who had done an apprenticeship with his cousin, another viewer, still found that he needed to make a 'dictionary to explain the hard words & terms of art used in working coal mines.' In the next generation, also working throughout the Northumberland region, John Buddle had to make a dictionary for himself that literally translated coal-mining terminology. Much of it of medieval origin and used daily by workers at the site, words such as whim (an old word for gin), trouble, bob gin, kibble (a wooden bucket on a rope), corf (a coal-basket that would hold 10–30 pecks), and not least a ten (usually 440 bolls of eight pecks) needed to be explained to him. Such handbooks in printed form became a torrent in the generation after Buddle [no source given].[70]

When exactly Buddle Snr was first consulted by John Curr I at George Silvertop's Bushblades colliery is uncertain. However, a notebook on boring and sinking, dated 1765, bears his name.[71] In the collection of the North of England Institute of Mining and Mechanical Engineers (NEIMME) there are three notebooks, mainly compiled by Buddle Snr, which contain rough colliery memoranda, copy agreements, valuations and accounts.[72] The earliest of these proves that he was consulting at Bushblades in 1768. Nineteenth-century sources agree that Buddle Snr's first appointment as a colliery manager was at Greenside colliery, near Ryton, possibly in 1788.[73] It was only then, if Galloway is to be believed, that he ceased teaching.

In the meantime, Buddle Snr's reputation as viewer was growing[74] and, in 1773, he was commissioned by the trustees of the Duke of Norfolk's estate to report on its Sheffield collieries. Later, Buddle Snr was in the habit of taking his young son with him when he was carrying out his views and reports. It is feasible, therefore, that Buddle Snr took the young John Curr II, then aged seventeen, with him to Sheffield.[75] Two copies of the 1773 report exist, both unsigned.[76] The duplicate copy is written in a style that strongly suggests Curr II's signed report of 1779.[77] Buddle Snr was to submit further reports on the Sheffield collieries in 1787 and 1789. The tone of these reports shows that he provided support and encouragement to his protégé, John Curr II.

An example of Buddle Snr's interest in solving the problems facing coal mining is shown by Flinn:

In South Wales where the hilly country allowed the use of adits to lead water away from shaft bottoms, a device known as 'the balance' was quite widely used in the second half of the eighteenth century. This made use of the gravity of water-filled tubs as counter-weights to draw up loaded coal corves. As the water was released from the tubs at the shaft bottom it flowed away down an adit and the weight of the empty corf returned the empty water-tub to the surface.

An ingenious variant on the balance was devised by John Buddle senior in 1780. Recognizing that an adequate flow of water to activate a surface water-wheel for winding was rarely available in England, but that many collieries were plagued by feeders running into the shaft through strata well below the surface, he designed a balance in which the counterweight tub was filled from an underground feeder and the difference in heights to be travelled by the corf and the counterweight controlled by winding their ropes on a large and small drum respectively on the same axle. The filling and emptying of the tub was managed by valves controlled manually from the surface.[78]

In 1792 Buddle Snr was

> ... *selected by William Russell, Esq., of Brancepth, an excellent discriminator of talent, to superintend the difficult task of winning the celebrated colliery at Wallsend, a proof of the estimation in which he was held.*[79]

This growing estimation is confirmed by Hiskey:

> ... *in 1792 he was appointed viewer to the Bishop of Durham, and in 1800 to the Dean and Chapter of Durham, two of the largest coal lessors in the area; and he had a wide practice as a consultant viewer, including various Tyne collieries and the Wear collieries belonging to Sir Henry Vane-Tempest and to the trustees of John George Lambton, the latter of whom he represented at meetings of the Wear Coal Trade organisation.*[80]

At Wallsend, Buddle Snr, and his son, now nineteen, who worked as his assistant, were faced with a problem encountered by all the collieries in the Tyne basin, where, as R. L. Galloway explained:

> *The sinking of the pits was in most cases attended with much difficulty, on account of the watery strata passed through, from which large feeders of water were discharged into the shafts, necessitating the employment of tubbing to dam it out. In the case of Hebburn Colliery these feeders amounted to 3,000 gallons per minute. Up till this time the universal practice had been to construct the tubbing with timber ... the timber being applied either in the form of plank tubbing or of solid cribbing.*[81]

Buddle Snr, soon after his appointment as viewer at Wallsend in 1792, in damming back a quicksand in the A pit, was the first to substitute cast-iron for wood in the form of cylinders. The substitution of cast-iron for wood, however, posed a problem: because the tubbing was cast in circular bands as large as the shaft, their great weight made them difficult to handle. The solution proposed by the Buddles was to cast the bands in segments, with flanges turned inwards so that they could be bolted together. This system was first used by Buddle Snr in 1796–97 when sinking the Percy Main Pit.

The Buddles also applied themselves to the problem of ventilation. The old practice was to dig two shafts, one to let foul air out, the other to let clean air in. An iron basket of flaming coals was suspended in one which caused foul air to rise and fresh to be drawn down the other. This method was not sufficient for deeper mines. In trying to solve this problem the Buddles were probably first to apply a mechanical ventilator (of piston-pump type) at a British colliery. T. S. Ashton wrote of this:

> *Methods of ventilation were improved … and still more, in the nineties [1790s] when John Buddle brought into the Northumberland field his system of triple shafts and more elaborate methods of 'coursing'.*[82]

In contrast to Buddle Snr's progressive views, after being appointed viewer at Wallsend colliery, he destroyed those parts of the Roman Wall that were in his way. This is described by David Harrison in his book, *Along Hadrian's Wall*:

> *John Brand, in his* History of Newcastle, *said that John Buddle, senior (after whom, or his son, the modern street is named), had a house built in the south-west angle of the fort and that in 1783 two new wagon ways were made through the east part of the fort, which laid open the foundations and showed them running down to the river.*
>
> *'Stones with inscriptions were found, but the incurious masons built them up again in the new works of the colliery.' Mrs. Buddle told Lingard that in digging a cellar in their house they found a Roman well. John Buddle, Junior, … told Bruce that he had often noticed the wall running down from the fort to the river when bathing in the Tyne as a boy.*[83]

John Buddle Junior

Buddle Jnr's (1773–1843) achievements were such that he became known in North-East England as *'King of the Coal Trade'*. A paragraph published in the *Newcastle Journal* a month after his death possibly sheds some light on Buddle Snr's influence on Buddle Jnr's education:

> *Mr. Buddle was the only son of a colliery viewer of great eminence … The elder Mr. Buddle was a man of considerable literary and scientific attainments and he bestowed great care in educating his son in every branch of knowledge which could be advantageous to him in his intended profession. Mr Buddle therefore … was a well-educated gentleman from the beginning of his career.*[84]
>
> …

> *Buddle Jnr told Lord Londonderry, in connection with Lord Ashley's Bill on the employment of women and children in mines, that 'I myself was initiated into the mysteries of pit-work when not quite six years old' and, in cross-examination before a parliamentary committee, that he had been 'brought up a lad in a pit'.*[85]

As previously noted, he had frequently accompanied his father when he was carrying out his views and reports.

Conclusion

It is reasonable to assume that, given John Buddle Snr's geographical proximity to the Curr family, his occupation as schoolmaster and his consultation at Bushblades colliery, he took an interest in the education and nurturing of John Curr II and his brothers. It may be, too, that the boys' father paid fees to Buddle Snr for their education, allowing him to recognise John Curr II's potential at an early age.

Encouraged, no doubt, by Buddle Snr, John Curr II, aged fourteen, a year younger than his mentor, is first found answering mathematical questions in magazines. In his case he is acknowledged in the September 1770 issue of *The Town and Country Magazine* as one out of ten readers who answered all four questions.[86] His name is found in *The Gentleman's Diary* for 1771 [see Figure 3.3] in which, under Question 349, he receives a special mention:

> *The same answered by* Mr. John Curr; *who says* 'I presume there are superfluous Data in this Question, as the Difference of the Angles at the Base needed not to have been given;' *then we have the following ...*[87]

Young Curr answered many of the questions in *The Gentleman's Diary* for the following year, 1772,[88] when his name also appears in *The British Palladium* as having answered questions in *'the Last Year's Palladium'*.[89] Curr's name is not found in magazines for 1773, the year he accompanied Buddle Snr to Sheffield.

So, aged seventeen, armed with a sound knowledge of mathematics, and familiar with mining operations, the working of the common engines at Bushblades colliery and wooden waggonways in the area, Curr was ready to start making his contribution to the technology of mining coal.

38   Questions in 1771, answered.

(6.) Quest. 349, answered by Mr. *George Coughron*.

CONSTRUCTION. At any Point in the indefinite Line AB, make the Angle BDC=the Complement of half the given Difference of the Angles at the Base; with the given Radius describe the Circle HIK, to have its Center O in DC, and to touch AB, from O to AB; apply OE equal to the given Line drawn to the Middle of the Base; make EF =ED; from F to CD apply FG=2EH; and parallel to FG draw AC to touch HIK, meeting DC in C; and from C draw CB a Tangent to HIK: then will ABC be the Triangle required.

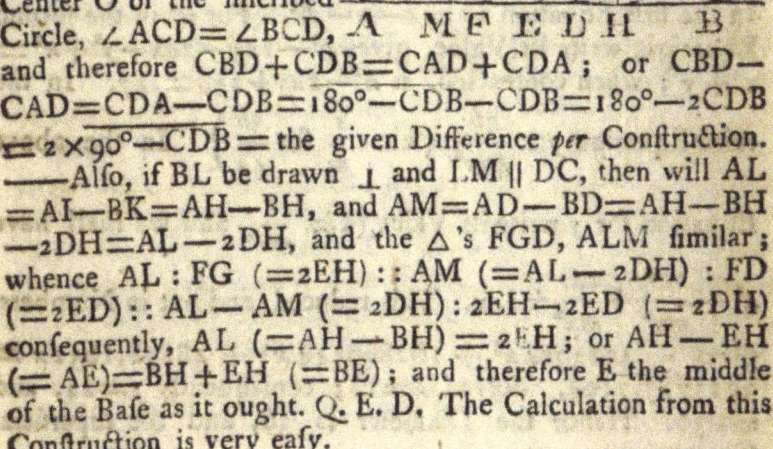

DEMONSTRATION. Since CD passes through the Center O of the inscribed Circle, $\angle ACD = \angle BCD$, and therefore CBD+CDB=CAD+CDA; or CBD—CAD=CDA—CDB=180°—CDB—CDB=180°—2CDB = 2×90°—CDB= the given Difference *per* Construction. ——Also, if BL be drawn ⊥ and LM ∥ DC, then will AL =AI—BK=AH—BH, and AM=AD—BD=AH—BH —2DH=AL—2DH, and the △'s FGD, ALM similar; whence AL : FG (=2EH) :: AM (=AL—2DH) : FD (=2ED) :: AL—AM (=2DH) : 2EH—2ED (=2DH) consequently, AL (=AH—BH) = 2EH; or AH—EH (=AE)=BH+EH (=BE); and therefore E the middle of the Base as it ought. Q. E. D. The Calculation from this Construction is very easy.

The same answered by Mr. *John Curr*; who says, "*I presume there are superfluous Data in this Question, as the Difference of the Angles at the Base needed not to have been given;*" then we have the following

CONSTRUCTION. With the given Radius of the inscribed Circle DF, and Line FE bisecting the Base, construct the right-angled Triangle DEF, continuing the Base ED both Ways, till AE=2ED, and DC=ED; draw AB and CB Tangents to the given in- 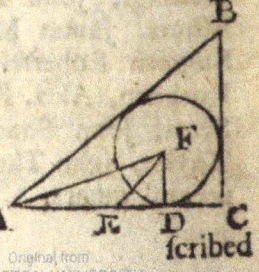 scribed

Four

CHARACTER

What was John Curr II like? Little is known. There is no known portrait, nor does any family correspondence survive [see Figure 4.1]. What is known is that religion, in the form of Roman Catholicism, informed the life of the family. The Currs were regular attendees at Mass and Curr was involved in plans to build a new chapel in Norfolk Row. Their children were educated at religious establishments: Ampleforth and the seminary at Ushaw College for two of the boys and Bar Convent in York for all five girls. One son became a priest and two daughters became nuns.

A grandson, Edward M. Curr, has this to say:

My grandfather, who was a well-educated man … was a man of considerable ability, self-reliant, original, and hard-headed. At least that is my impression from what I have heard my relatives say concerning him.[90]

I am very glad to say that my grandfather was a very exemplary and sterling Catholic. As an instance of his frame of mind concerning religion, I may relate that, when he was a young man and somewhat a beau, it was the fashion to wear a queue. This appendage required a hairdresser to arrange it before the wearer could appear in public. My

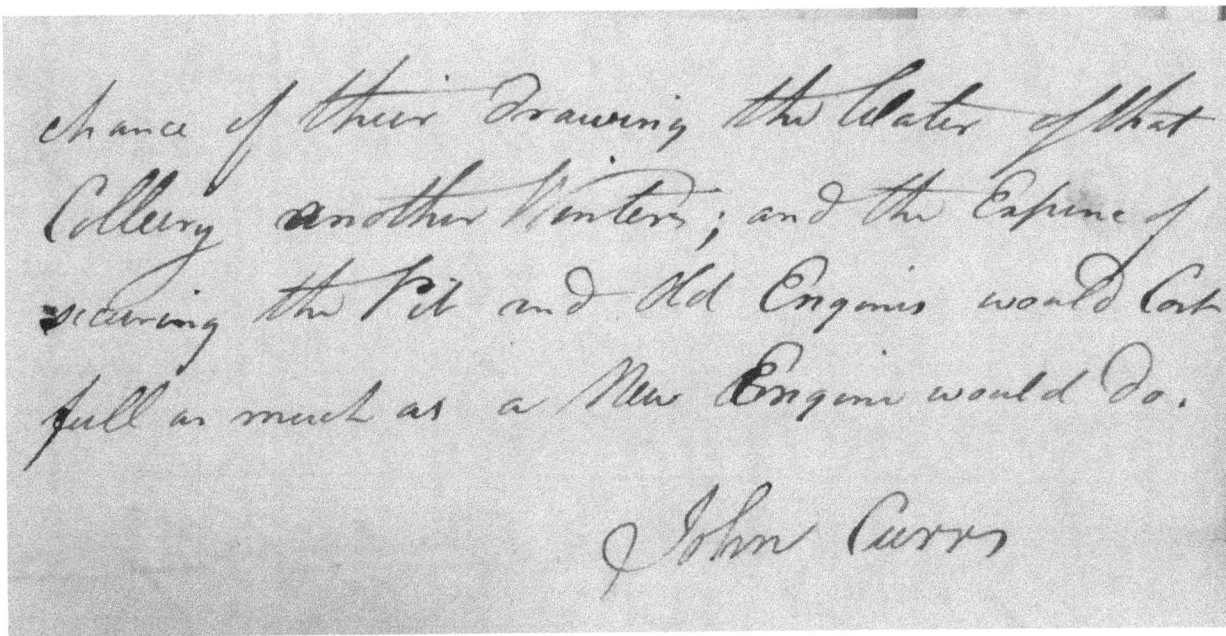

Figure 4.1. Example of Curr's hand and signature on a memo of 1795 (property of J. Hunter).

grandfather was of course a punctual attendant at Mass on Sundays. It is related that on one occasion the hairdresser arrived so late as to render his compliance with the Sunday obligation impossible. He reached the church long after the service had begun, and, on his return home said to his attendant 'I will never be late for Mass again on account of my queue, so take the scissors and cut it off'. His attendant did as he was required. Few now-a-days would estimate correctly the sacrifice of self which this act implied. It almost amounted to what a Hindu would call 'loss of caste'.[91]

His ability was publicly recognised: in 1796 he was granted honorary membership of the Literary and Philosophical Society of Newcastle-upon-Tyne, founded in 1793, alongside such luminaries as Sir Joseph Banks, and in 1797 made a freeman of the Company of Cutlers in Hallamshire.

Five

WEALTH

The probate value of Curr's estate was some £14,000, a considerable sum in those days (about £1.4 million today). His foundry supplied cast-iron goods not only to the Duke of Norfolk's Sheffield collieries[92] but to others elsewhere. How much he profited from his ropery is open to debate. He earned fees for his consultancy work and received rent from the streets of houses which he developed.[93]

Curr's will, which lacks the usual religious preamble, divided his assets into seven equal parts, Teresa having died before him.

An application made in 1873 for the appointment of new trustees in the administration of Curr's will refers to certain leaseholds in Sheffield [see Figure 5.1]:

> *The leaseholds consist of various properties in Sheffield Park, held under the Duke of Norfolk, for terms of 99 years from the date of the lease, and are now the only property subject to the trusts of the will.*[94]

Family lore has it that in the months prior to his death Curr lost £30,000:

> *Shortly before he died he entrusted a certain Pere Duchesne with thirty thousand pounds to invest for him in the French Funds. The poor priest,*

Figure 5.1. Colliers Row, Sheffield Park, back-to-back houses, 1940 (Sheffield City Archives: picturesheffield u00222).

with little knowledge of business, took it upon himself the responsibility of investing the whole sum in some mercantile bubble which he thought would be more profitable to my grandfather than the funds. The bubble, however, presently burst, and absolutely nothing was saved from the wreck. The poor priest returned to lay the sad news before my grandfather who, however, had just died. He went to the executors, and after stating what had occurred said 'Gentlemen, except the clothes in which I am dressed I have nothing to offer by way of restitution for the loss which I have incurred except this silk pocket handkerchief'. Years afterwards when I was a young man, this handkerchief was in possession of my mother who said to me 'Eddie, my boy, this handkerchief is all that your grandfather's family got for thirty thousand pounds – God willed it should be so.'[95]

A certain credence is lent to this story by the baptismal entry for Henry Beauvoisin, a grandson of Curr, on 29 March 1823, which records his godfather as Franciscus Nicholas Duchesne.

Six

BELLE VUE HOUSE, SHEFFIELD PARK

The map in Figure 6.1 shows Sheffield Park and Belle Vue House in 1823. The Currs lived for many years at Belle Vue House, Sheffield Park. Belle Vue, with its grounds, barns, stables and outhouses, was leased to John Curr II in 1803 for a term of ninety-nine years, at a yearly rent of £10 10s.[96] A further, counterpart, deed, also of 1803, granted Curr the lease of four fields adjacent to Belle Vue, together covering some twenty acres, at a yearly rent of £52 10s.[97]

When the Currs first moved to Belle Vue is not known, but it was prior to 1803. In his *Memoranda*, E. M. Curr claims of his father that 'He was born on 1st July 1798 at Belle-View House near Sheffield'.[98] This is supported by words in the deeds '… the Ancient Messuage, or Dwelling House now lately improved and Enlarged at the Expence of the said John Curr …'[99]

Figure 6.1. Plan of Sheffield in 1823: Belle Vue House and Curr's ropery highlighted in yellow. From www.sheffieldhistory.co.uk.

Figure 6.2. Clay property, showing the fields leased by Joseph Clay (Sheffield Archives ACM/SheD722).

and '... late in the Occupation Lessees of the said John Curr as Agent and Subtenant of the Sheffield Park colliery'.[100]

In the late eighteenth century, the area surrounding the house comprised enclosed fields which were let by the Duke of Norfolk's Hallamshire estate. The map in Figure 6.2 shows the fields leased by Joseph Clay, a notable Sheffield lead, steel and coal merchant in 1781. Plot no. 14 is Belle Vue House. Unfortunately, the fields on the map are not named. Curr's lease gives the names of those he leased as Cinder Hill, which included the garden; Three-cornered, which included the lane and the ponds; House; and Well. The first three likely equate to plots 15, 13 and 19, plot 18 being [Buck]well field.

The house was advertised for sale on 14th March 1820 in *The Iris, or, The Sheffield Advertiser*, which provides the following description:

> *The House, Hothouses, Green House, Stables, Coach-house, and other Out Offices, and part of the Pleasure Grounds, and gardens, and also Two small Tenements for labourers at the back, are held under the Duke of Norfolk for the remainder of a Term of Ninety-nine years, which commenced on the 29th of September, 1803, at a low annual rent.*
>
> *In improving and adorning this Property, Mr. Curr has laid out a considerable sum of money, and it may be fairly said he has succeeded*

in rendering it one of the most admired country residences in the picturesque neighbourhood of Sheffield.

Mr. Curr has also a Lease under the Duke of Norfolk for a term of 21 years, of which five were unexpired last Michaelmas, of 20a. 3r. 0p. of land adjoining the house …

In 1939 the house was occupied by Major A. J. Gainsford who explained that little was known of the early history of the house [Figure 6.3]:

It is early Georgian in style,' he said, 'and there is the date 1723 in the kitchen over the fireplace, though part of the house must have been built long before that date.[101]

Belle Vue house was celebrated in poetry in *Sheffield Park: A Descriptive Poem* by John Holland (1794–1872).[102] Holland was born in a cottage which *'stood at a distance of three fields'* from the house. Two stanzas from his poem follow:

XXXIII.
Yon cove of planting, drest in Spring's green hue,
Environs the neat mansion of Belle Vue;
Whence to discursive vision, at one glance,
Opens the outstretch'd scene, a gay expanse,
Where the loveliest landscape-charms in shade or light,
Harmoniously commingle and unite,
Till the far hills in far perspective rise,
And sweetly seem to slumber in the skies,
Where Evening leads her stars at twilight forth,
And the blue heavens are pillowed on the earth.
– Not more endear'd, though sung in sweeter rhymes,
Palermo's valleys or Ausonia's climes.

Figure 6.3. Belle Vue House, mid-20th century. Image © Sheffield City Archives, ref. y02170 (www.picturesheffield.com).

John Curr

XXXIV.
Years have with years, centuries with centuries roll'd,
And left unsung, unannall'd, and untold,
What scenes adorn'd, what incidents befell
The long-frequented precincts of Buckwell,
Where lovers oft around its daisied brink,
Recline, and gambol, and its waters drink;
– Love's ancient haunt! – The story now I sing,
Oft hath the crystal naiad of the spring
Pour'd on my ear, as o'er her urn, when young,
In fancy's pilgrimage, I musing hung, –
Pored, till the mind, as each vague theme it drew,
Half-wish'd, half-deem'd its own creations true.

The first edition notes '*Belle Vue – Beautiful Prospect – late the residence of Mr. Curr*'. The 1859 reprint notes '*The house named in the foregoing stanza is little altered, though its position has been improved by a diversion of the old road; it is now occupied by Mr. F.T. Mappin. Forty years ago it was the residence of John Curr …*'

It also notes Buckwell as: '*A well-known spring of water near Belle Vue, which supplies a considerable part of the neighbourhood during the summer-months, when a long drought has dried up the adjacent springs.*' Buck Well can be seen on the Clay property map of 1781 [see Figure 6.2].

In his 1877 *Memoranda*, Edward M. Curr claims that:

The eldest daughter [Elizabeth] married a Mr. Furniss who, after the death of my grandfather, resided at Belle-View.

It seems that this is correct, for an entry for 'J B Furniss Esquire, Belle Vue' is found in the 1822 edition of Baines [*History, Directory & Gazeteer of the County of York*, vol. I] and another for 'John Bernard Furniss. Merchant h. Belle Vue' in the 1833 edition of *White's Directory*.

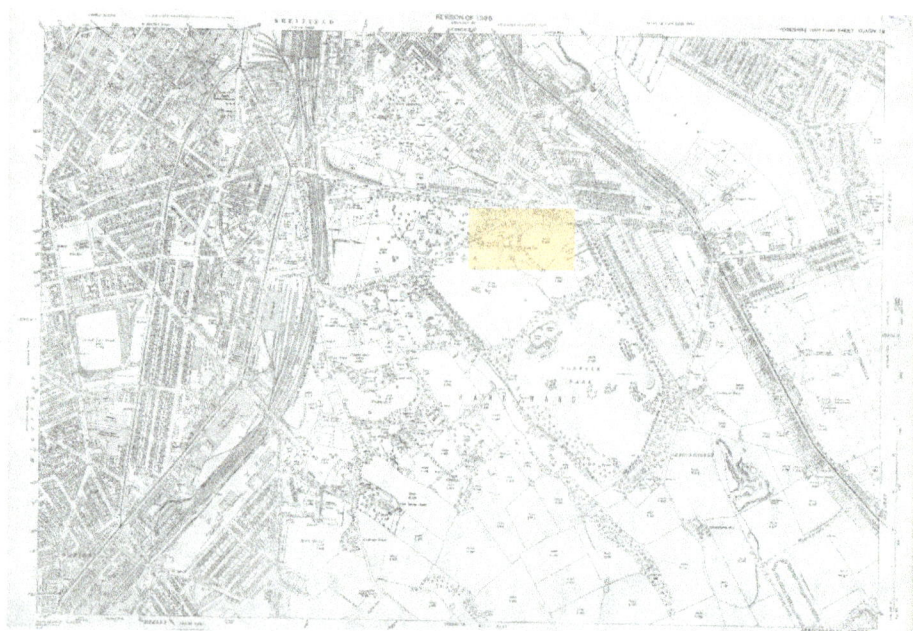

Figure 6.4. OS map 1937. Scale 1:2500. Belle Vue House highlighted in yellow.

Figure 6.5. Crater behind Belle Vue, after an air raid in December 1940 (Sheffield City Archives, ref. s01086).

The 1937 map opposite [see Figure 6.4] shows the progressive growth of Sheffield. In 1939 a plan was published which threatened the house with demolition. Its possible effect was described in the *Sheffield Daily Telegraph*:

Sheffield's plan to establish a great sports stadium on the [Belle] Vue Grounds, near Norfolk Park, Sheffield, published in today's Telegraph and Independent, *may mean the disappearance of one of the city's most distinguished houses – Belle Vue. Within five minutes' motor-car ride of the centre of the city, Belle Vue has retained its original rural atmosphere. Lying just off Granville Road, within the sound of busy traffic, [looking] out on to pleasant gardens and fields, and when the trees are in [leaf no] other house can be seen from its windows or grounds. Whether the actual house, which is now occupied by Major A. J. Gainsford, will be included in the stadium scheme is not known, but Major Gainsford told the* Telegraph and Independent *yesterday that a stadium would destroy the amenities even though the house remained. The fields around the house are owned by the Duke of Norfolk, but Major Gainsford owns the freehold of Belle Vue house and gardens.*[103]

Before being demolished the house was threatened with a near miss by a bomb in 1940 [Figure 6.5].

Figure 6.6. Opencast mining close to Belle Vue House in 1957. From: K. Haddock, *British Opencast Coal, a photographic history, 1942-1985*, Ipswich: Old Pond Publishing, 2015 (figure 2.38).

In 1957 the house was threatened again by nearby encroaching opencast mining [Figure 6.6].

Belle Vue is thought to have been demolished by 1966. In 2008/9 it was the site of the Norfolk Park Hostel and is now the site of the South Yorkshire Probation Trust.

Seven

WORSHIP

E. M. Curr notes in his *Memoranda*: *'My grandfather was a very exemplary and sterling Catholic'*.

Faith informed the lives of John Curr and his family; as noted, three of his eight children were to take Holy Orders. The Curr family's involvement in the Roman Catholic chapels and church of Norfolk Row lasted for three generations. The role of the Catholic priest in Sheffield at the time, Richard Rimmer, loomed large. He christened John and Hannah's children in the small room behind the Lord's House [see Figures 7.1 and 7.2] in Fargate where he first ministered.[104] He married Edward and Elizabeth:

> *The marriage ceremony was performed at Belle-View House on the 30th June 1819 by the Reverend Richard Rimmer, a catholic priest, and was repeated the day after in the Protestant church in Sheffield by the Reverend Matthew Preston. This was necessary, as, at that time, a marriage by a catholic priest in England had no force in law.*[105]

Richard Rimmer, who was born in 1754, took charge of the Sheffield mission in 1787 where he remained until his death in 1828. His arrival was on the eve of better times: the passing of the second Catholic Relief Act in 1791.[106] Canon William Odom, *'whose own religious views scarcely leant towards Rome'*,[107] described him thus:

> *Mr Rimmer is said to have been a very remarkable man, universally esteemed in the town by all classes. He was noted for his great kindness and constant charity; it is said that he devoted all he had to charitable causes.*[108]

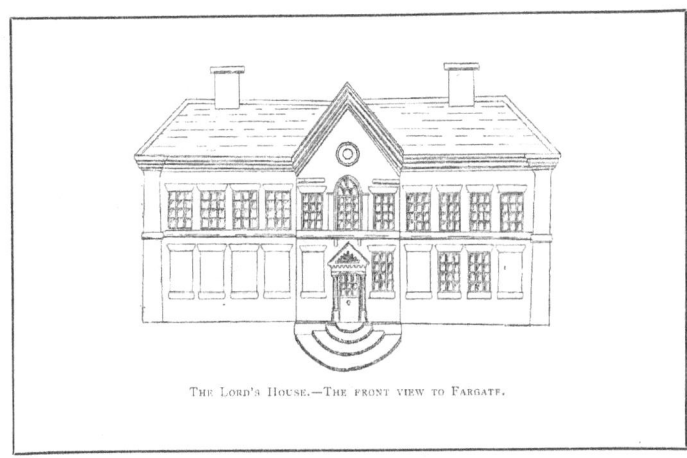

Figure 7.1 (left). Lord's House, drawn by Alfred Gatty. From Wikipedia, original from: Alfred Gatty, *Sheffield Past and Present* (1873), p. 94.

Figure 7.2. The Lord's House – The front view to Fargate. From: Hadfield 1889, opposite p. 40; with the accompanying text: 'The illustration given is taken from a porcelain model of the Lord's House …'.

In 1889 *A History of S. Marie's Mission and Church, Norfolk Row, Sheffield*, by Charles Hadfield, was published. Thanks to Hadfield much is known about where the Currs worshipped in those days.

The Old Chapel

There cannot be living more than a few persons who remember the 'Old Chapel' in the Lord's House; and the particulars given are enhanced in value, being the statements made to the author by the venerable Canon Frith, ..., now in his 81st year, ..., and by Mr James Brown, ... At the advanced age of 84 his recollections of Mr Rimmer and the old chapel ... are very clear and accurate. The accounts given of the old chapel by both are identical, and are borne out by Fairbank's plan [see Figure 7.3][109] *... alluded to, as to the entrance, which, it is stated, was close by the entrance to the yard behind Mr Rennie's shop, or, as the plan indicates, exactly opposite the Catholic Book Repository. On the reproduction of this plan are marked the probable site of the 'Old Chapel' and its entrance, the sites of the new chapel erected by Mr Rimmer, and the present church.*

It would appear that the old chapel was about 50 feet by 28 feet within, and with the gallery may have given accommodation to about 600 persons. It was entered at once from Norfolk Row through a lobby or porch, by a flight of steps, which gave access to the ground floor. At the Fargate end, opposite the altar, was a gallery, reached by two flights of stairs, in which was the Duke of Norfolk's pew, approached from the mansion through an ante-room used as a vestry, with a window looking into the chapel and gallery. In this vestry Mr Rimmer heard confessions, until, in later years, a new vestry was built over an adjoining yard on the chapel floor level.

... Music has always held a high place in the services of the Catholic Church. And it was not neglected at Sheffield, though the action of the penal laws had put a stop to musical services for the time, ... The first notice of music in connection with Catholic services in Sheffield was in the 'Old Chapel'. It was necessary to avoid disturbance, and to keep the services quiet and retired. Indeed, it is said that, at the period we are considering, an old man named Eyre, a Shrewsbury Hospital pensioner, stood at the door in Norfolk Row, and was very particular whom he admitted, his duty being to scrutinise every applicant. One Christmas Day Mr Rimmer wanted to have some singing, and the service was held after midnight. It had been decided to sing the Portuguese hymn 'Adeste Fideles', as published by Novello, with the first verse arranged as a treble solo, which was sung by Mrs Valentine ... The singing was at first without accompaniment ... Afterwards Mrs Morton ... accompanied the singing on a piano, until a small organ was purchased ...[110]

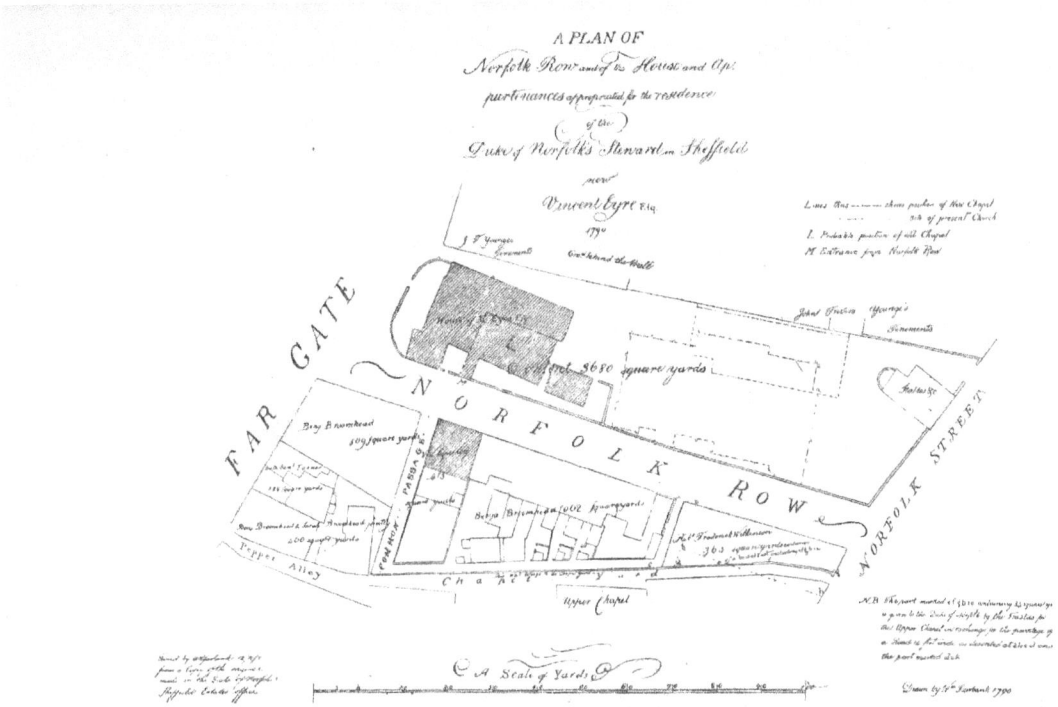

Figure 7.3. Norfolk Row in 1790, drawn by Wm Fairbank. From: Hadfield 1889, opposite p. 38. Deciphered from the bottom left: Traced 12/4/8[9] from a copy of the original made in the Duke of Norfolk's Sheffield Estates Office.

The New Chapel

It became clear that a growing congregation meant that the capacity of the building behind the Lord's House which was used as the chapel was inadequate. So, around 1813, Rimmer embarked on an ambitious plan to build a new chapel. The first move was to purchase the land on which to build it. The Lord's House, together with its grounds,[111] 3,680 superficial square yards, was purchased from the Duke of Norfolk's estate in May 1814. Unusually, the duke, presumably sympathetic to the plan for a church, agreed to sell the freehold. The purchase was made, for the sum of £3,045, by a syndicate of three local Catholics: John Smilter, John Curr and his son-in-law, John Bernard Furniss, who bought the property (in equal shares) as a speculation.

Complex financial and legal transactions ensued. The syndicate borrowed £2,800 from Messrs Walkers, Eyre and Stanley, bankers. The Walkers were not Roman Catholic, and the loan would have been a commercial transaction. In 1816, possibly as a result of the downturn in business in Sheffield caused by the Napoleonic wars, Curr personally took on the bulk of the loan, repaying Messrs Walkers, Eyre and Stanley £2,000.

It was not until 18th January 1837 that ownership of the property was conveyed to trustees. Among these were a son and grandson of Curr: the Rev. Joseph Curr and the Rev. John Furniss. The very elaborate financial proceedings are detailed in Appendix 2.

Hadfield continues:

The old chapel in the Lord's House was not pulled down until the new chapel had been completed, some two years later, in 1816. Meanwhile

John Curr

Figure 7.4. The New Chapel, around 1830. From: Hadfield 1889, opposite p. 32.

Figure 7.5. South-east view of St Marie's Church, Norfolk Row. From: Hadfield 1889, opposite p. 72.

Mr Rimmer started the good work with energy, and raised a sum of money amounting, it is stated, to nearly £3,000, partly by subscription from his congregation, and the Duke of Norfolk, who contributed, it is said, £300. The new chapel, as will be seen from the illustration taken from a contemporary oil painting [see Figure 7.4], was a neat and substantial building, and considered a handsome one at the time; and at its completion, it has been stated, was all paid for.

Internally it had a spacious gallery at the west end above the entrance. Here were placed the organ and choir, and the mention of some of the names of old parishioners who had sittings [these were reserved pews] in bygone days may not be uninteresting:- Shuttleworth, of Cannon Hall; Smilter, of Richmond; Curr, of Bellevue;[112] *Furniss,*

Gainsford, ... Michael Ellison. It was opened with some ceremony, and dedicated to Our Lady, on 1st May, 1816.[113]

Henry Blackwell described the cemetery:

In 1816 ... the Catholics ... erected ... the present place of worship ... Annexed is a small enclosed cemetery, neatly palisaded from the street, and planted with trees and flowers, which give it that pleasing air so uncommon in the burial-grounds of large towns. It contains a neat monument to the memory of John Curr, Esq., late of Belle Vue House in Sheffield Park; a man distinguished as an engineer and especially deserving of remembrance as the inventor and original patentee of flat ropes alike calculated for coalpits and ships' cables.[114]

St Marie's Church

In 1845 a new church was proposed and on Lady Day (25th March) 1847, its cornerstone was laid. St Marie's was to be larger than the New Chapel. A contemporary view of St Marie's is reproduced in Hadfield's book [Figure 7.5]. This required disturbing some of the graves, in particular those of John Curr and his daughter, Elizabeth, and her husband, John Furniss. John Curr's tomb, no. 11 on the plan in Figure 7.6 (the nearest to the area marked 'old chapel'), was underneath, or very close to, where the present-day altar stands.[115]

Hadfield described the disinterment:

Very early on the morning of the 23rd of March, at four a.m., this labour of love was commenced, ... and the remains of the faithful departed were reverently and tenderly, one by one, disinterred, the

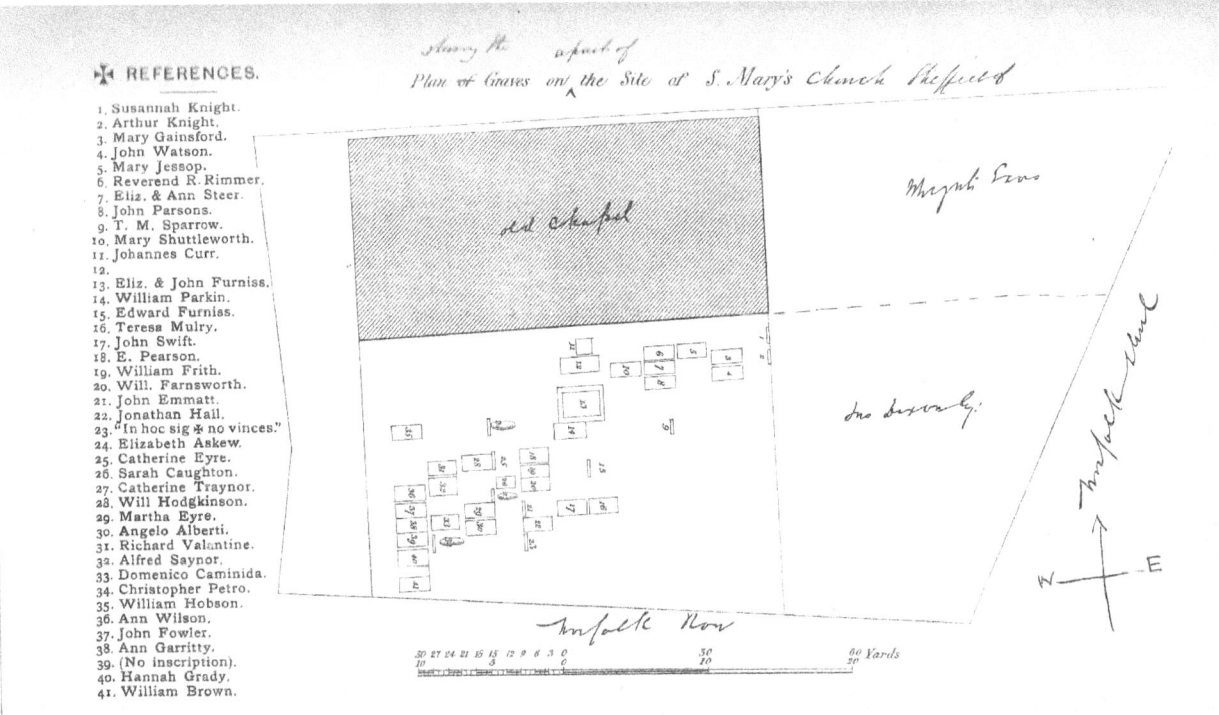

Figure 7.6. Plan of graves next to the New Chapel (marked Old Chapel on the diagram), dated sometime between 1845 and 1847. The Old Chapel (in the Lord's House) was replaced by the New Chapel which was, in turn replaced by St Marie's. From: Hadfield 1889, opposite p. 68.

prayers of the Holy Church recited the while, and consigned to their last resting place, beneath the floor of the new church; …[116]

According to R. A. Mott, John Curr's tombstone [actually ledger stone] occupies the place of honour in a porch at the corner of Norfolk Row and Norfolk Street, in what is now known as St Marie's Cathedral, Sheffield.[117] At present it is thought to be covered by carpet.

The Latin inscription on Curr's ledger stone [see Figure 7.7] is reproduced in Hadfield's book. The Rev. Geoffrey White has translated the inscription as follows:

Figure 7.7. Tombstone Inscription: John Curr. From: Hadfield 1889, p. 179.

Appendix.

INSCRIPTIONS
FROM TOMBSTONES NOW IN S. MARIE'S CHURCHYARD,
which were formerly in the Chapel Burial Ground.

✠
In Memory of
CHRISTOPHER PETRO,* who departed this life December 10th, 1823,
— Aged 85 years. —

D. O. M.

Hic jacet
Beatam spem expectans,
JOHANNES CURR,
Qui obiit die Januarii XXVII.
A.D. MDCCCXXIII., Ætatis ejus LXVII.
exiguum
Magnæ pictatis
Monumentum
Dilecti mariti
Optimique parentis
Memoriæ
P.C.
Amantissimi
Conjux ac liberi.
Beati mortui qui in Domino moriuntur Etiam dicit Spiritus, ut requiescant a laboribus suis: et opera eorum sequuntur eos, Apoc.

* His it is said was the first interment in the Burial Ground. It is a remarkably well lettered slab, and bears the name of the Mason—John Emmat.

Here lies
in expectation of a blessed hope
JOHN CURR
who died on the 27th January
in the year of our Lord 1823, and of his age the 67th.
To the memory of
a beloved husband
and the best of fathers
(this) modest expression
of much piety
has been set up
(by his) most affectionate wife and children.

Blessed are they who die in the Lord
(so says the Spirit)
for they rest from their labours;
and their deeds shall follow them.
Apocalypse.

A memorial in St Marie's Mortuary Chapel is reproduced by Hadfield [Figure 7.8].

Figure 7.8. Mortuary Chapel: list of clergy. From: Hadfield 1889, p. 189.

In the Mortuary Chapel is a memorial of the clergy (not a complete list) who have served the Mission. It was erected by the Rector about three years ago:—

✠ "Of your charity pray for the souls of the clergy who have laboured in this parish, whose names and anniversaries are undernoted, and on whose souls may Jesu have mercy. May the Angels lead you into Paradise."

Rev. Thomas Lynch, January 4th, 1868.
,, Matthew Kavanagh, January 18th, 1863.
,, Andrew Macartney, 27th January, 1884.
,, C. J. Pratt, 27th February, 1849.
,, Daniel Harrold, 19th February, 1879.
,, Bruno Rigby, March 18th, 1872.
,, Richard Rimmer, 12th May, 1828.
,, Thomas Loughran, May 6th, 1875.
,, Joseph Curr, 30th June, 1847.
,, James Atkins, 10th July, 1877.
,, R. Bimson, August 9th, 1830.
Right Rev. Bishop Sharples, August 11th, 1850.
Right Rev. Robert Tate, 25th August, 1876.

The guidebook describes the Mortuary Chapel as having 'names of deceased parishioners on the tiled floor'. Looking carefully between the pews, inscriptions may be found, in particular:

To the Memory of
ELIZABETH MARY,
The Wife of John Furniss, who died August the 28th, 1823,
Aged 42 years.
Also of
JOHN BERNARD FURNISS,
Who departed this life the 7th day of September, 1834, Aged 60 years.
R. I. P.

A memorial to past priests, in the form of tiling, is found on the wall of the Mortuary/All Souls Chapel [Figure 7.9]. The name of Rev. Joseph Curr is found under June.

Figure 7.9. Memorial to past priests in the Mortuary/All Souls Chapel (Rev. Joseph Curr in June). From: Bob Rae; © the Cathedral Church of St Marie, Sheffield.

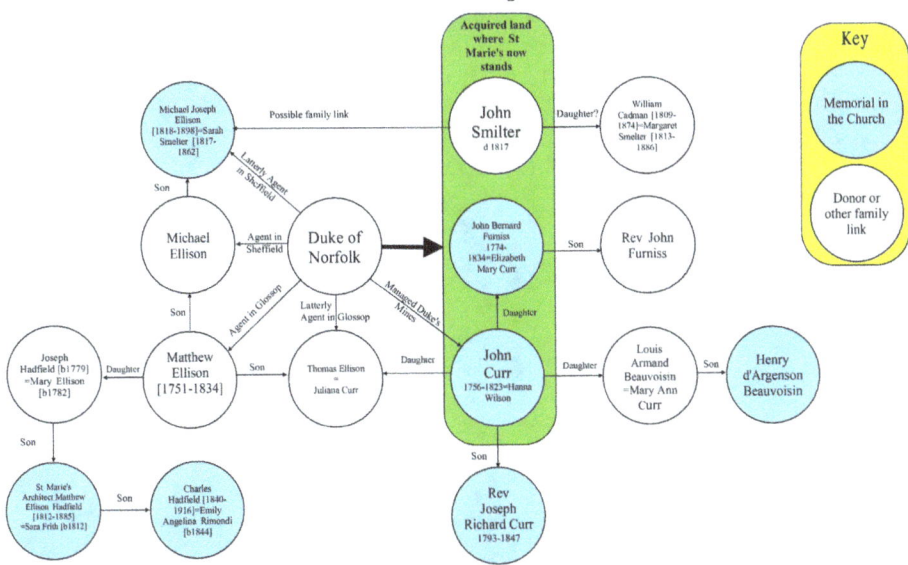

Figure 7.10. Curr family connections with St Marie's. From: Schematic by Bob Rae, © the Cathedral Church of St Marie, Sheffield.

The interrelationships between the Roman Catholic families of South Yorkshire and Derbyshire are complex. The schematic in Figure 7.10 provides some indication of this, as well as Curr family connections with St Marie's.

Eight

EARLY CAREER

Chronology

1756 Born, Co. Durham.
1761 Henry Howard appointed steward of Duke of Norfolk's Hallamshire estate.
1773 Accompanied John Buddle Snr to Sheffield.
1774 Sheffield waggonway constructed.
1775 First mentioned in Sheffield colliery accounts.
1777 Succession of 10th Duke of Norfolk.
      Charles Howard (later 11th duke) becomes Earl of Surrey.
1779 Vincent Eyre III appointed steward.
      Curr submits his first report.
1781 Appointed 'Superintendent of the Coal Works'.
1786 Earl of Surrey succeeds to dukedom.
      Shafts sunk at Attercliffe Common Colliery.
1787 Railroad installed underground at Sheffield Park Colliery.
      Steam pumping engine installed at Attercliffe Common.
1788 First patent, no. 1660, published (guides in shafts &c.).
1790 Consulted by Sir Thomas Gascoigne.
1792 Patent no. 1924 published (method of working ropes and pulleys in mining operations).
      Curr's iron foundry established at Sheffield Park.
1793 Consulted by George Overton re Merthyr Tydfil to Abercynon tramroad.
      Consulted by William Reynolds re Coalbrookdale transport system.
1797 Publication of *The Coal Viewer*.
1798 Patent no. 2270 published (manufacture of flat ropes).
1801 Death of Vincent Eyre III.
      Dismissed by the Duke of Norfolk.

A guide to the various working relationships can be found on page 229.

The Waggonway of 1774

Early days at Sheffield, I

It may never be established when Curr first visited Sheffield Park collieries. Although the first reference to Curr in estate accounts is in December 1777,[118] it may be that he accompanied John Buddle Snr on the latter's visit to the colliery in October 1773.

When preparing his paper on early pumping engines, John Hunter examined the archival documents held at Sheffield City Library.[119] He notes that following Buddle Snr's visit in 1773, a plan of Sheffield Park Colliery was prepared, accompanied by a written report which, although unsigned, bears Buddle's name [Figures 8.1–8.3].[120] The style of penmanship is the same on both documents, indicating that Buddle was probably the author. There is a duplicate of the 1773 report, also unsigned, but written in a different hand which, Hunter suggests, strongly resembles Curr's. Its style is similar to that of Curr's report of 1779. Neither report mentions Curr by name. In his reports Buddle mentions Curr only once, in his report of 1789.[121]

Could the young Curr have accompanied Buddle in 1773? Buddle's claim for '*Expence of a Journey to Sheffield to do Business for his Grace the Duke of Norfolk*'[122] shows that he travelled by post-chaise from Durham to Sheffield. The post-chaise was a carriage for two or four passengers, with a postilion, so it is possible that Curr accompanied Buddle (the journey took two days at a cost of £5 7s 11d).[123] Hunter notes that it was Buddle who recommended

Figure 8.1. Plan accompanying John Buddle Senior's original 1773 written report. From: Sheffield Archives ACM/S/215.

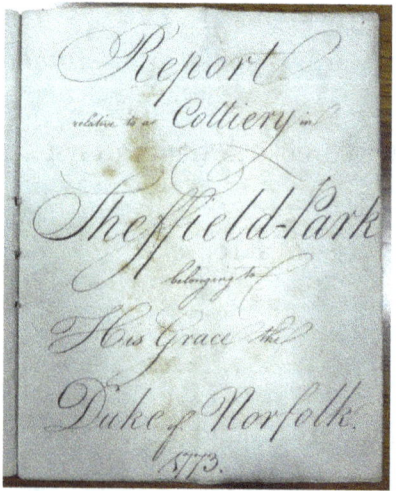
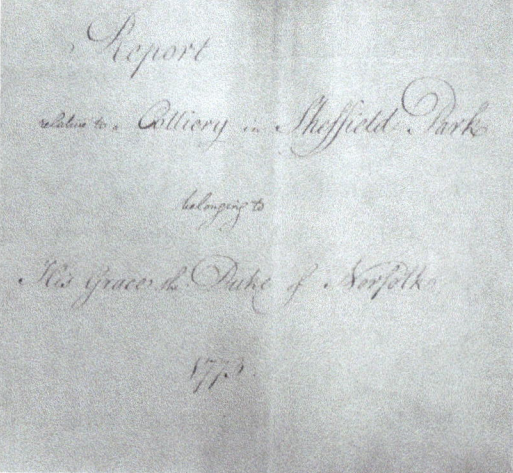

Figure 8.2. Title pages of original & copy 1773 report.

The Colliery the Object of the present Enquiry, belongs to his Grace the Duke of Norfolk and lies under a Tract of about 270 Acres in Sheffield Park, as near as can be estimated from the present Direction of the Water level continuing through the Seam of Coal, an ancient Hollow or Waste on the North-West Side, and the Contiguity of the vendable Coal with the Out-burst on the South-West Side thereof.

By Virtue of the abovementioned Water level, this Colliery is drained, and its Workings extended; the Rise of the Seam of Coal being between the South and West, in Consequence of which the said Level ranges in a Direction between the South and East, and discharges the Feeder of the Colliery into the River Sheaf about a Mile and Half North from the present Workings, and close to the South Side of the Town of Sheffield.

The Colliery the Object of the present Enquiry, belongs to His Grace the Duke of Norfolk, and lies under a Tract of about 270 Acres in Sheffield Park as near as can be estimated from the present Direction of the Water Level continuing thro' the Seam of Coal, an ancient Hollow or Waste on the North West side, and the Contiguity of the vendable Coal with the Outburst on the South West Side thereof.

By Virtue of the abovementioned Water Level, this Colliery is drained & its Workings extended, the Rise of the Seam of Coal being between the South and West in Consequence of which the said Level ranges in a Direction between the South and East; and discharges the Feeder of the Colliery into the River Sheaf about a Mile and an half North from the present Workings, and close to the South side of the Town of Sheffield.

This Colliery supplies all the Cutlers, and about one half of the Houses in Sheffield with Coals, the other part of the Houses being supplied from a Colliery about two Miles East from the Town belonging to

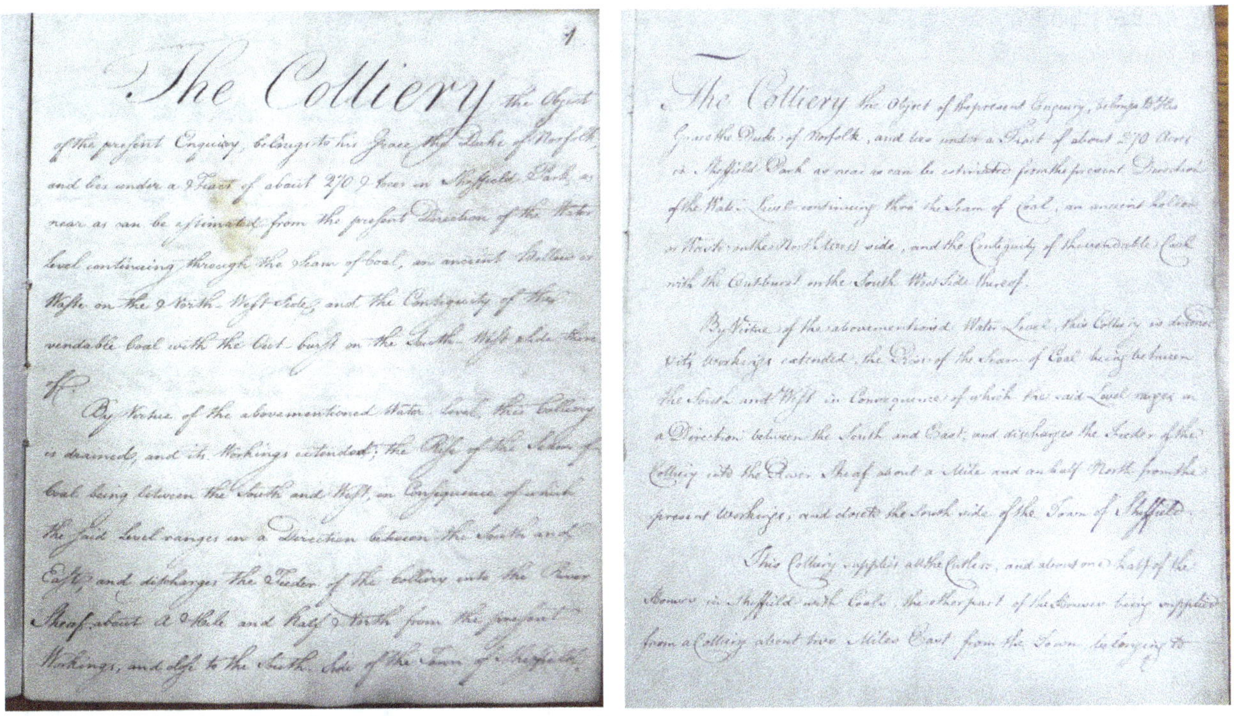

a waggonway and it is feasible that Curr returned to Sheffield to assist in its construction. Certainly, in the first report to bear his name,[124] Curr says that, in 1774, '*a plan was produced for conveying the coals to the Town of Sheffield by the Newcastle method which was agreed upon and effected before the Christmas by the undertaking at the expence of £3,250*'.

Figure 8.3. Page 1 of original & copy of 1773 hand-written report. From: Sheffield Archives ACM/S/215.

Background

Shallow coal mining in the Durham coalfield goes back to the 1600s. A network of wooden waggonways was built linking the mines with the rivers Tyne and Wear. Young Curr will have been familiar with those that

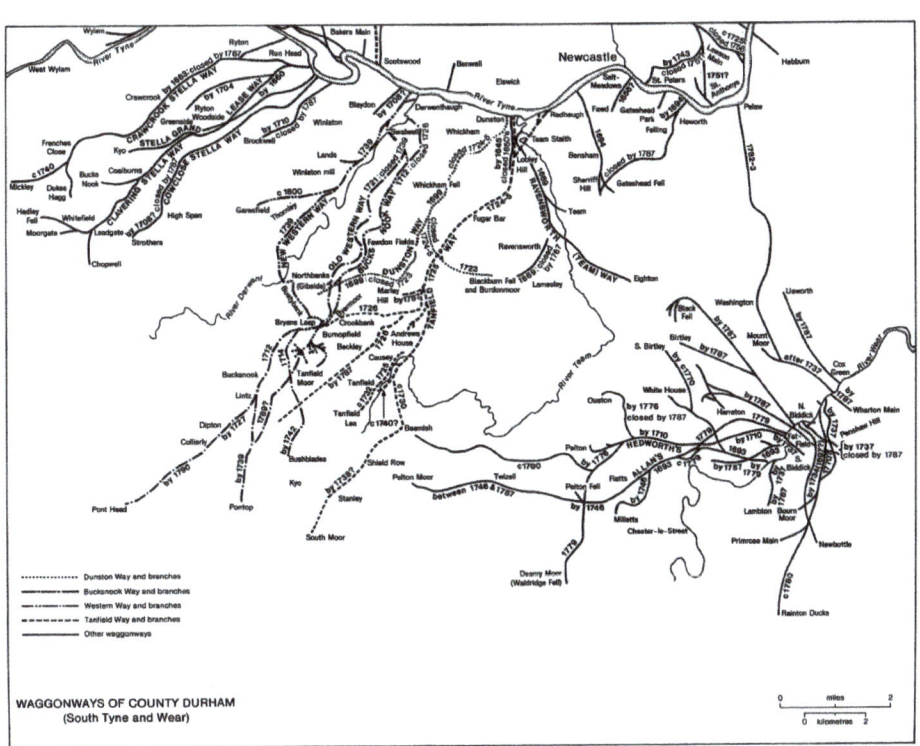

Figure 8.4. 'An historical atlas of County Durham', map produced by Michael Lewis for the Durham County Local History Society, 1992 (which holds the copyright).

Figure 8.5. Wooden waggonway remains, cover image, *International Mining & Minerals*, May 1998, vol. 1, no. 5.

reached the Tanfield–Bushblades–Pontop area, where the Currs lived. These converged on the Tyne and those around Houghton-le-Spring on the Wear [see plan in Figure 8.4].

Fortunately, in 1998 remains were uncovered (and afterwards reburied) at a former Lambton pit. The cover of the *International Mining and Minerals* journal in Figure 8.5 shows what they looked like.

Extraordinary events

A waggonway had been built at Sheffield Park Colliery in 1722, but it was disused before the 1770s. The Norfolk estate offered Messrs Townsend and Furniss a lease on the colliery on condition that the existing method of transporting the coal, by carts, was improved. John Buddle's 1773 report recommended:

> *The site of the Colliery in Question naturally points out a Waggon Way to be laid to a Coal Stage near the Town, there to deliver the Coals for the Regular supply of a great part thereof.*[125]

Construction of the waggonway, which was built of oak with beech rails, was completed in November 1774. Sales of coals at the pit were then forbidden. From the point of view of the lessees it was a success: the cost of carriage from pit to town was reduced from 2s 8d to 1s 2d per ton. For poorer parts of the community the result was very different. They had been accustomed to buying their coal at the pits and carrying it home themselves: the increase in price was beyond their reach. The result of their anger was described in *Universal Magazine*:

> *December 2.*
> *Extract of a Letter from Sheffield, Nov. 28.*
> *A few days since we had a very riotous meeting of the populace of this town, occasioned by an imposition which some persons belonging to the Duke of Norfolk had the cruelty to put in practice, with respect to the price of coals got in the Duke's estate, adjacent to the town. These merciless wretches formed a scheme the beginning of last summer to bring all the coals got in the Duke's land about the town into a yard at the outside of the town, by means of large carriages on low wheels, which run on a road made of timber, in imitation of one at Newcastle. This road, together with their carriages, being ready for use, they have begun to bring coals to the intended place, where they exact from the buyers almost double the price they were sold for before at the pits; and in order to compel the people to comply with their exorbitant demands, they have totally put a stop to the usual delivery of coals at the pits, and also refuse to sell less quantities than a horse load, which is a great hardship upon the poorer sort of people, unable perhaps to spare more than one penny at a time to buy coals with. These were alterations so disagreeable to the populace, that they assembled in a prodigious number, and destroyed several of their carriages, totally pulled down a watch-house and compting-house in their new coal-yard, and set fire to all the timber machinery erected for discharging their loading, brought one carriage in triumph through the town, afterwards kindled a fire in, and sent it flaming into the river.*[126]

The waggonway, which ran behind Belle Vue house, slightly higher above the hillside, went from Sheffield Park Colliery to the Sheffield coal stage, a distance of about 1¾ miles.[127] The location of the coal yard is today in the space between the railway cutting and the Park Square roundabout – at the base of the Park Hill flats [see map in Figure 8.6].

Curr family lore describes what happened. Curr's son, also John, gave his version in his book, published in 1847:

> *I have before me a book printed in 1797 ... He [John Curr, his father] was the inventor of the cast iron railway, – then called a railroad; in*

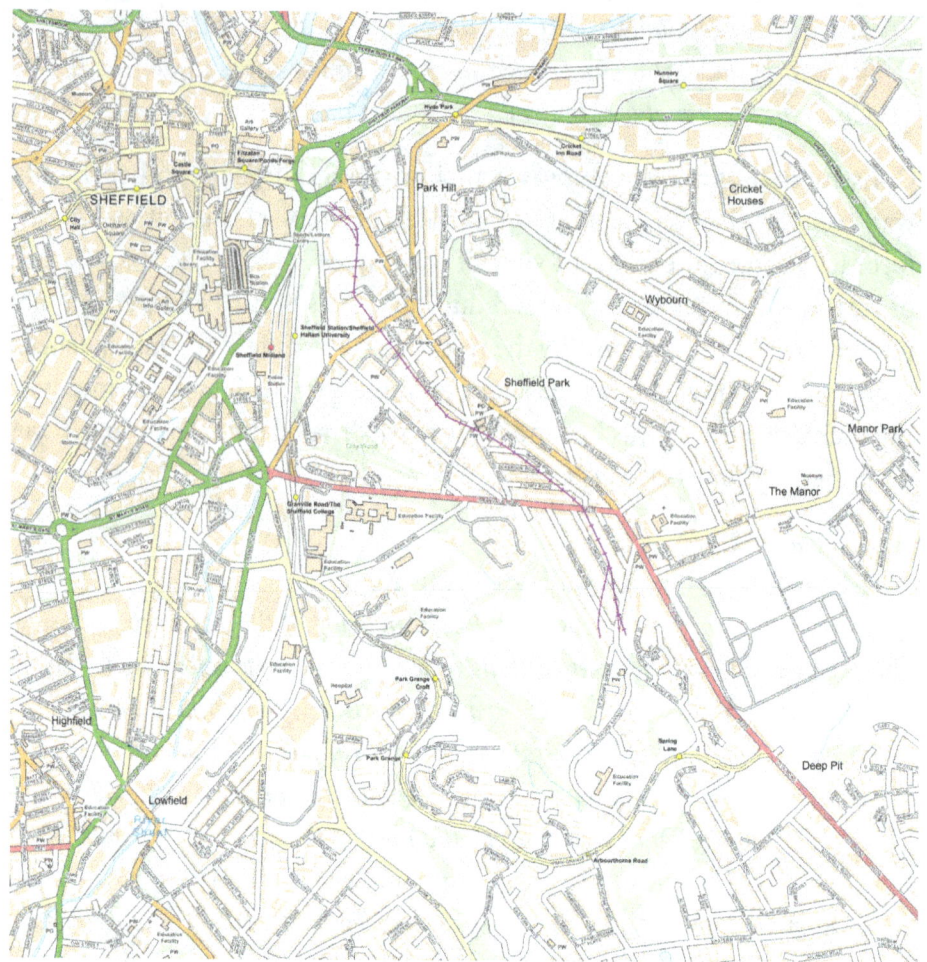

Figure 8.6. Curr's 1774 waggonway overlaid on an OS map of present-day Sheffield. Several of Fairbank's maps show parts of the route. John Hunter combined these. Contains OS data© Crown copyright and database right (2017). © John Hunter. All rights reserved.

the book are plans of every description of rail, with particulars of their length on passing curves, given to the one hundredth part of an inch. He therein foretells what in time it would become; but note, – the plan was opposed; it meddled with the interests of a certain body, – a riot ensued, – the railway was torn up, – the coal staith burnt, and the inventor, my father, reduced to the necessity of concealment in a wood, for three days and nights, to escape the fury of the populace.

At the same time he was persecuted, if I may use the expression, in an opposite quarter, – his father differed with him in opinion as to the result of the invention, and disowned him, and the letter to that effect was retained in his family for 30 years or more.[128]

By 1877 family lore had both changed and expanded. Curr's grandson, E. M. Curr, tells the tale this way:

At that time, coal was brought from the mouth of the pit for a certain distance in vehicles which ran on wooden rails, and it occurred to my grandfather that a great saving would result from the substitution of iron rails which up to that time had been unknown. Pondering on this subject, he wrote to his father telling him that it was his determination to substitute the wooden tramway with iron rails.

To this the old Northumbrian replied that what his son contemplated was an innovation which in no wise met with his approval, ending his short letter with the words 'If you put down iron rails, Jack, I curse you, and here is my hand on it'. On the opposite side of the page my great grandfather laid his stalwart hand, around the edges of which he drew his pen. His father's anger was due to the fact that he owned forests and supplied timber to the mines.

In view of this denunciation my grandfather abstained from taking further steps in respect of his iron rails until some time afterwards, when, mentioning the matter to a priest who was dining with him, he learnt that, notwithstanding what his father might think, it was his duty to act as seemed to him best for the interests of the Duke of Norfolk. He then laid down the rails which created such feeling amongst the colliery population that they threatened to take his life, so he hid himself in a wood for three days until the ferment had subsided.[129]

In 1926, in the Tasmanian newspaper, the *Advocate*, H. R. Haslock wrote:

And on the other side of the page appeared the outline of a hand with thumb and fingers out-spread, showing where the old man, in his indignation at the thought his son was under the impression that he could teach his forbears and improve on methods which had been in vogue for generations, had dumped his stalwart left hand, and, having taken his pen in his right hand, had traced the wide-spread fingers lying on the open sheet, thus setting his seal to a curse which he, doubtless, firmly believed would operate, should his son dare to disobey what was a virtual command.[130]

Doubt has been cast on the veracity of the tale, but, as previously noted, it is feasible that the young Curr was involved in construction of the waggonway. Whichever, it makes for a great story, oft repeated, not only by his son and grandson, but also by the likes of Samuel Smiles[131] and C. L. Mateaux.[132]

In his report of 1779,[133] Curr noted that repair *'was effected before the Christmas by the undertaking'*; that is less than a month after the riot. M. J. T. Lewis notes that:

The story that Curr rebuilt the track with less combustible materials – iron rails and iron (later stone) sleepers[134] *– is quite apocryphal, for the colliery accounts show that the waggonway was still using wooden rails in 1787, and probably later.*[135] *It was closed before the end of the century.*[136]

The Norfolk Estate at Sheffield

Edward, 9th Duke of Norfolk, who was the last of the senior (Worksop) branch of the Howards, died on 20 September 1777 without issue. Under the 1767 settlement, he was succeeded by Charles Howard (1720–1786), the 6th duke's great-nephew from the (second) Greystoke branch.[137] The 10th duke was a cultured man, but he lived as a semi-recluse in Surrey.

John Curr

Vincent Eyre II, the duke's steward for the Hallamshire estate, died in 1761 and the 9th duke appointed his cousin, the Hon. Henry Howard of Glossop (1713–1787), in his place. Henry Howard belonged to the third (cadet) branch of the House of Howard.[138] As steward, he lived at the Lord's House in Fargate, where the lords of the manor had lived until 1701 when the 7th duke died. The 10th duke was not dependent on its revenues, because he enjoyed an income from the Greystoke estate. His son, also Charles (1746–1815), became earl of Surrey, receiving the profits of the Sheffield estate, which remained under trustees until he succeeded to the dukedom in 1786.[139] The trustees were the Earl of Strafford, of Wentworth Castle near Barnsley, and the Earl of Scarbrough, both of whom were owners of collieries.

Nathaniel Eyre, the younger brother of Vincent Eyre II, was Henry Howard's steward at his Glossop estate, and Nathaniel's son, Vincent Eyre III, replaced Henry Howard as Hallamshire steward in 1779.

Vincent Eyre III, 1744–1801, was born in Glossop, Derbyshire on 20 September 1744. He began working life as a lawyer and, after admission as a barrister, joined the chambers of George Wilmot, attorney for the 10th Duke of Norfolk. As steward he was responsible for the duke's estates in Sheffield and Worksop and was to become the duke's partner, providing some of the capital, in the development of the Sheffield collieries. A busy man, with fingers in many pies, in 1791 he became a founding partner in Walkers, Eyre and Stanley, a bank with branches in Sheffield and Rotherham.

The map in Figure 8.7 shows the collieries of the Rockingham–Fitzwilliam and Norfolk estates at this period.

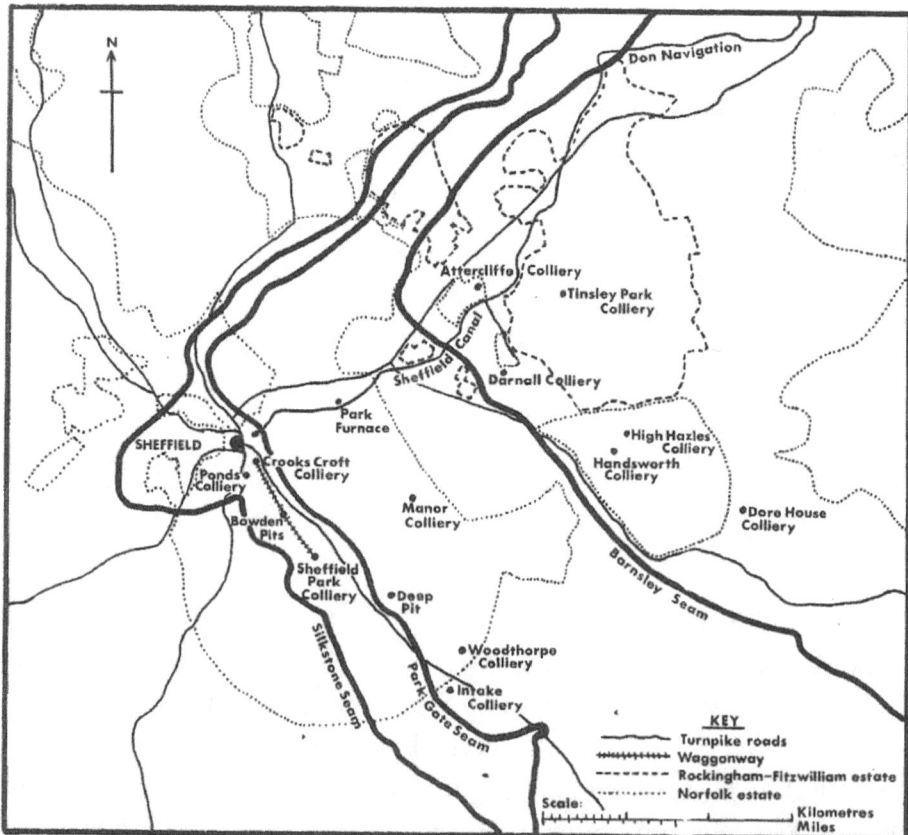

Figure 8.7. Map showing Norfolk estate collieries 1750-1850. From: *Yorkshire Archaeological Journal*, 1987, vol. 59, p. 104.

Early Days at Sheffield, II

As the Duke's agent Howard had commissioned Buddle in 1773 to make his report and, presumably, authorised Curr's participation in the construction of the waggonway of 1774. It seems that Howard was aware of Curr's precocity, for mention of his employment in 1775, aged 19, is found in the estate accounts for 20 December 1777:

> *Received of Mr John Curr the Remainder and in full for the overend both of the Wood & Manor Pits due Xmas 1775 and 1776 – £134 1s. 6d.*[140]

The Estate accounts show further payment to Curr during this period:

> 2 December 1778, £4 5s 2d.[141]
> 3 December 1778, *'By John Curr for viewing and examining several collieries £7 4s. 0d.'*[142]

On the death of the 9th duke in September 1777, and because Townsend and Furniss's original lease of 1765 was nearing the end of its term, Howard commissioned, on behalf of the trustees, a series of reports into the state of the collieries at Sheffield. Two were drawn up. The first is signed by Curr and the second, although unsigned, is in his hand:

> *Report on Sheffield pits in the tenure of Messrs Townsend and Furniss, dated 27 August 1779.*[143]
> *State of Sheffield Colliery from the improvements in 1774 to this time, dated (25th) March 1780.*[144]

Henry Howard kept the young Curr busy. Curr prepared two further reports at that time: on Mr Swallow's colliery at Parkin Wood, Ecclesfield (in 1779),[145] and another, unsigned, but in his hand, dated 6 November 1780.[146]

By 1779, the year of Curr's report of 27 August, the lessees of Sheffield Park Colliery, Townsend and Furniss, were struggling; they had been in financial difficulty since 1774. That year they had been granted a new lease on Sheffield Park and Manor Collieries at £1,000 and £50 per annum (p.a.), respectively, with royalties payable on coal production above a certain level. The result was that the lessees were paying £460 p.a. more than in 1765. Further, in 1774, the year of construction of the waggonway, they had borrowed £5,200.[147] There were *'No profits from small coal'*: the proportion of 'small coal' relative to 'hard coal', which fetched a higher price, was increasing.[148] In 1774 the ratio of hard to small had been three to five; by 1779 it was one to three. The result was that the colliery was £60 in deficit at Christmas 1778, which deficit was forecast to rise to some £250 in 1779. The lessees continued reluctantly in 1779, but it was clear that the cost of extraction exceeded the sum achieved for the sale of coals when interest on their capital invested was taken into account. As the workings moved deeper under the hill of Sheffield Park, by 1784 they were twice the depth of a decade earlier, and farther from the town, the capital costs could only increase.[149]

Curr's report stated:

> *The Colliery at this period of time was wrought at a moderate expence by reason of the shallowness of the pits, the quantity of hard in consequence produced, and the short distance to the Town and notwithstanding the Rent being somewhat advanced to his Grace, left a little profit to the undertakers but not sufficient encouragement after sinking such a sum.*[150]

On Lady Day (25th March) 1781 Townsend and Furniss surrendered their lease. The mines thus reverted to the direct control of the trustees. In that same year Curr was appointed '*Superintendent of the coal works*', aged 25. The appointment of Curr, a Roman Catholic, to this position was consistent with the practice of granting leases for Sheffield Park Colliery to Catholics such as John Bowden, George Townsend and Mark Furniss. Another aspect to this practice is that Curr's elder brother, George (1749–1826), was in the estate's employ before his brother John's appointment:

> *… he was joined in Sheffield by an older brother called George. His name is not mentioned very often in archived estate documents, but he is noted in other sources, such as church registers, and he is usually qualified with the term 'engineer'.*[151] *He was paid a salary for engine attendance by John at Attercliffe Colliery, and he was noted by Charles Hatchett as accompanying John and himself on a tour of Pond's Colliery, in Sheffield, in June 1796. In early 1780 George Curr was named agent for the sale of a major share in a colliery, with fire engines … [see Figure 8.8]. In the 1790s, the Attercliffe Colliery accounts also show a salary being paid to one William Curr, in addition to John and George.*[152]

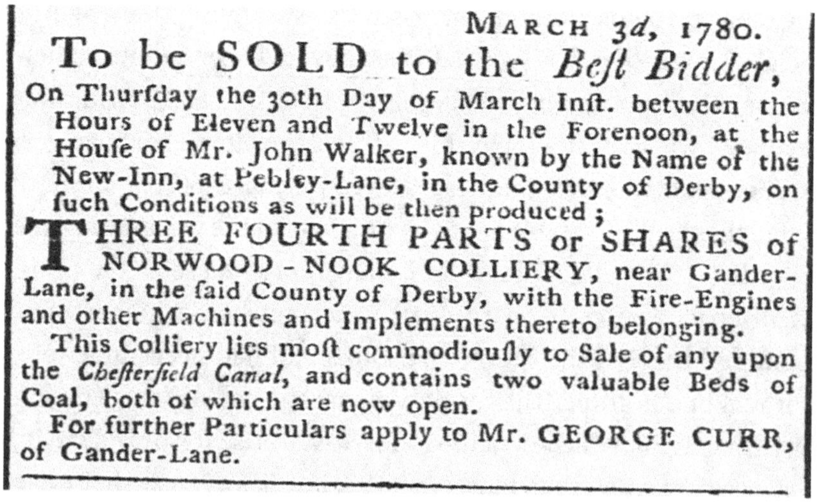

Figure 8.8. Auction notice, *Derby Mercury*, 3 March 1780.

> MARCH 3d, 1780.
> To be SOLD to the *Best Bidder*,
> On Thursday the 30th Day of March Inst. between the Hours of Eleven and Twelve in the Forenoon, at the House of Mr. John Walker, known by the Name of the New-Inn, at Pebley-Lane, in the County of Derby, on such Conditions as will be then produced;
> THREE FOURTH PARTS or SHARES of NORWOOD-NOOK COLLIERY, near Gander-Lane, in the said County of Derby, with the Fire-Engines and other Machines and Implements thereto belonging.
> This Colliery lies most commodiously to Sale of any upon the *Chesterfield Canal*, and contains two valuable Beds of Coal, both of which are now open.
> For further Particulars apply to Mr. GEORGE CURR, of Gander-Lane.

Medlicott provides further information about George Curr's employment by the Norfolk Estate:

> *In 1791 George Curr was paid £40 for labour and assistance at the new engine and in 1793 received £10 10s 0d for one year's attendance at the engine of Sheffield Park Colliery and £63 as agent for the colliery.*[153]

Thus was the stage set for Curr to put into practice his ingenuity and inventiveness. Curr's openness to improvements made by others in the field and the introduction of his innovations into the Duke of Norfolk's Sheffield collieries over the next 20 years led to advances in mining technique which were adopted, not only in Great Britain and Ireland, but on the continent of Europe. For instance, Charles E. Lee notes:

> *… the improved English railway, as developed principally in our coalfields, was referred to as an englischer Kohlenweg*[154] *when it was introduced into the Ruhr district at the end of the eighteenth century.*[155]

Supporting this view, in his 2018 review of Anthony Burton's *Railway Empire*, Vitali Vitaliev writes '*… we cannot ignore the fact that they* [the British] *came up with the very concept of transport on rails … as well as that of "the metal plateway", precursor of the railtrack, first introduced by the mining engineer John Curr.*'[156]

After Townsend and Furniss

Why did direct control of the collieries revert to the trustees? In view of the difficulties besetting the collieries – geological faults, depth of workings, proportion of small-to-hard coal, problems of drainage, competition and, at that time, a slump in the price of coal – it was unlikely that new lessees were to be found. In light of this, the importance of coal supply to local industry and the requirement for substantial capital investment, the trustees chose to assume control. This decision marked a crucial point in the development of the collieries. The trustees and, after succeeding to the title in 1786, the 11th duke, displayed a willingness to invest.

Due to the substantial capital investment required[157] the 11th duke entered into partnership with his agent, Vincent Eyre III.

> *The agreement*[158] *granted Eyre the right to mine coal for 21 years in Handsworth and Sheffield, and to be paid one quarter of the clear profits per annum after the deduction of any capital expended. However, later evidence states that the costs of working the collieries and their profits were to be shared equally between the partners, with Eyre receiving 10% per annum on the 'Capital' or 'gross sum' of any money he invested.*[159]

Superintendent of the Norfolk Estate collieries
Responsibilities
Curr was allowed considerable latitude in his running of the collieries. While his management abilities are open to question, of necessity he had to be something of a jack-of-all-trades, although nobody could question his ability as an engineer.

He was accountable to the steward of the Norfolk's Hallamshire estate, at first Henry Howard, until 1779, when Vincent Eyre III was appointed. The steward carried overall responsibility for the collieries, reporting until 1786 to the trustees and afterwards to the 11th duke. Curr enjoyed a good relationship with Eyre and, at a remove, with the duke, as evidenced by Curr being granted permission to establish an iron foundry in his own name in Sheffield Park in 1792, and, later, a ropery. After his dismissal Curr continued to supply the duke's collieries with goods from both foundry and ropery. He dedicated his book *The Coal Viewer* to Charles Howard, Duke of Norfolk.

Curr's salary obliged him to keep an assistant clerk and, it appears, he delegated some of the responsibility for the management of the collieries: in 1802, the year after his dismissal, the agent for Attercliffe was John Jeffcock; for Manor, Martin Elliott; for Ponds, Robert Bower; for Hesley Wood, John Locke Snr; and, for Handsworth, John Locke Jnr. Additionally, each agent had an agent underground.[160]

Remuneration
Curr's salary, which was to remain unchanged until his dismissal, amounted to £190 per annum. This was made up of a fixed rate per colliery: £100 for Sheffield Park and Manor collieries, £70 for Attercliffe and £20 for Hesley Wood. He was paid a further £25 when called upon to view and measure the coal and ironstone mines under lease.[161]

Development of the railway
Historical overview

1556   Publication of Agricola's *De Re Metallica* [see Figure 8.10].

1700   Onwards, establishment of wooden-rail waggonways in North-East coalfield.

1734   First-known illustration of flanged-wheel vehicle on rails (Prior Park waggonway, Bath).

1749   First recorded use of iron-edged boards (side piece nailed to horizontal board to form L-shape at Ladywash mine, Derbyshire).

1767   First recorded covering of existing oak rails with flat cast-iron plates (Coalbrookdale).

1776   First railway at Froghall (laid with flat-iron plates).

1783   Second (replacement) railway at Froghall.

	Wooden waggonway recorded underground at Sheffield Park Colliery.
1787	Probable introduction of Curr's flanged cast-iron rails at Sheffield Park Colliery.
	Probable introduction of Curr's re-design of the corf.
1788	Curr's railroad first recorded above ground (at Wingerworth).
1793	Railroad installed at Coalbrookdale, replacing iron-edge rails.
	Benjamin Outram and Wm Jessop begin to spread Curr's tramroads throughout much of England and South Wales. This continues for some 30 years.
	First tramroad from Merthyr Tydfil (Dowlais Ironworks) to Abercynon.
1795	Railroad installed at Little Eaton, Derbyshire.
1797	Publication of Curr's *The Coal Viewer*, which contains detailed instructions for the casting of rail plates &c.
1803	Third (replacement) railway at Froghall, using Curr's rails.
	Surrey Iron Railway (the world's first public railway) opens, using Curr's rails.
1804	Trial of Trevithick's locomotive breaks many of the cast-iron plate rails on the Merthyr tramroad.
1821	Darlington & Stockton Railway opts for wrought-iron edge rails.

Railway technology advanced in a series of steps, each coming on the back of a previous innovation. Curr contributed a small, but crucial, step to this advance with his 'invention' of the flanged cast-iron rail and the introduction of the 'rail road', as he termed it, underground (the rail road probably emerged above ground in 1788). As with his innovations in steam engine technology (the external boiler) and the design of the corf, little was original. Never one afraid to blow his own trumpet, in his book *The Coal Viewer*, he states that '*the making and use of railroads and corves were the first of my inventions*', a claim he was to repeat in his letter to the Duke of Norfolk after dismissal. Significantly, Curr himself appears not to have considered his 'inventions' original: he was to take out ten patents but neither his rail plate nor his corf was among these. These 'inventions', though, demonstrate the depth of Curr's knowledge, his powers of observation and his ability to adapt and improve on what had gone before.

The history of the development of the railway is problematic. As Charles E. Lee explained:

> *The subject of railways … has resulted in a vast output of published material … Much of it includes sheer conjecture as statements of fact, and even more repeats erroneous particulars gleaned from secondary sources.*[162]

... Initially it is desirable to define what is meant by a railway in my remarks. I submit that the one common characteristic of the many forms of track which may be regarded as a railway is that the combination of track and vehicle is self-steering.[163]

Lee follows with some discussion of early rail roads and illustrative examples [see Figures 8.9–8.11]:

Not until the publication in 1556 of Agricola's classic De Re Metallica, *however, do we have a clear description and illustration of the self-steering principle. ... It seems unlikely that, if the flanged wheel were in use by the middle of the sixteenth century, it would have escaped Agricola's comprehensive survey. ... The first illustration showing beyond all doubt a flanged-wheel vehicle on rails is that of 1734 in Dr. John Theophilus Desagulier's* Course in Experimental Philosophy, *depicting the Prior Park wagonway at Bath. The equally clear drawings of a flanged-wheel coal wagon given by Gabriel Jars [in* Voyages Metallurgiques*] are of 1765.*[164]

Figure 8.9. Mining truck with four iron wheels running on rails, from Sebastian Münster's *Cosmographiae Universalis, Libri VI*, 1550 (Note: names of equipment in Latin, but names of workmen in German). From: Lee 1960–61, p. 5, fig. 2.

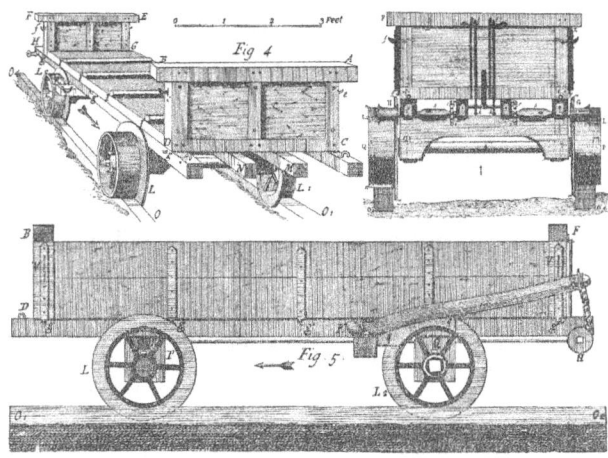

Flanged, albeit wooden, rails, iron strips laid onto wooden rails and the railway had been in use before Curr introduced his rail road at Sheffield.

Jim Rieuwerts has written about earlier use of the first in Derbyshire:

In August 1749 they were 'planking the Cartgate' at Ladywash Mine, Eyam (SA/OD/1152). An improvement was the plateway in which a side piece was nailed to the horizontal board to form an L shape and thereby prevent the tubs from slipping off. They are said to have been introduced about 1756 by the noted engineer John Smeaton but 'edged boards' were in use at Milnes and Middleton Engine Mine by late 1748 (SA/Bag 379) and both Ladywash Mine and Stoke Sough by late 1754 (SA/Bag 389). Edged planks are also recorded at Milnes and Middleton Engine Mine for a short period in 1752 (SA/Bag 379).[165]

Brian Bracegirdle has written about the second:

On curves and gradients strips of wrought iron were fixed on top of the wooden rails, a practice said to have been introduced in England 'at least as early as 1716', but we have no contemporary evidence for the data. In a few cases it appears that the whole surface of the rail was covered. The covering of existing oak rails with flat cast-iron plates, 5 ft. long, 4 ins. broad, and 1 in. thick, began at Coalbrookdale in 1767 but the practice was not widely copied.[166] *A few years later an entirely different use of cast iron for rails was introduced by John Curr.*[167]

M. J. T. Lewis has written of an early railway in Staffordshire:

The longest and most celebrated railway built by a canal company in our period was that of the Trent and Mersey ... In 1776 they obtained an Act for a branch line canal to Froghall, with a three-and-half-mile 'Rail-Way' thence to the quarries. ... it proved unsatisfactory. ... So in 1783 a new Act was obtained. The second line was better, but not much, and was replaced by, a third, an iron plateway, built by John Rennie in 1803.[168]

The first two lines were laid with flat iron plates. Curr wrote (*The Coal Viewer*, p. 13):

Figure 8.10 (left). Mining truck illustrated by Agricola in *De Re Metallica* (1556). (A) *Capsae quadranguli ferrei* (iron-bound rectangular truck); (B) *Eius bacilla ferrea* (its iron straps); (C) *Axiculus ferreus* (iron axle); (D) *Orbiculi lignei* (wooden wheels); (E) *Parvi clavi ferrei* (small iron keys); (F) *Magnus clavus ferreus obtusus* (large blunt iron pin); and (G) *Capsa eadem inversa* (same truck inverted). From Lee 1960–61, p. 6, fig. 3.

Figure 8.11. Flanged-wheel vehicle of 1734 for the Prior Park Wagonway at Bath. Reproduced from Plates 21 and 22 of *A Course in Experimental Philosophy* by Dr. John Theophilus Desaguliers. From Lee 1960–61, p. 8, figs 4 & 5.

> *These roads, which are upon the plan of what is called Newcastle waggon roads, are laid in a firm manner upon wood, (after having been at a great expence of stoneing about ten or twelve inches thick for a foundation;) upon this wood is laid cast iron an inch and a half thick, a part of which weighs in every single yard forward one hundred and forty-one pounds, and other models weigh only eighty-one pounds.*

His book advocates the adoption of his own cast-iron plates, with trains of waggons, and, by the time the line built by Rennie in 1803 was proposed, use of these had become widespread.

On the subject of wooden barroways, so widespread in Co. Durham, Lewis has this to add:

> *The wooden barroways were not satisfactory, especially since the thin iron tram wheels quickly reduced the guiding flanges to matchwood. A Tyneside viewer, telling a Parliamentary Commission of the old wooden rails, said, 'the trams were often getting off these boards, and then it was very difficult to get them on again'.*[169]

Curr's legacy

Curr's ingenuity

In addressing the shortcomings of the Sheffield collieries on his appointment as Superintendent, Curr first applied his ingenuity and inventiveness to improving the method of hauling, or hurrying, coal from coalface to surface. As collieries had become deeper, the cost of sinking shafts rose and, in consequence, fewer were sunk, and the distance coal had to be hurried grew longer.

At this stage Curr's solution was threefold: first, the introduction of the cast-iron railroad; second, his design of a four-wheeled waggon, or corf, to replace the hazel kibbles or corves heretofore employed to transport coal underground; and, third, vertical guides in the shaft to prevent wear and tear caused by the corves bumping into each other and the sides.

The terminology can be confusing. For instance, the terms rail road, railway, plateway, tramway and tramroad have been used interchangeably. 'Railroad' was the term Curr himself used exclusively for his system employing cast-iron flanged tram plates. He also used the terms 'waggon ways' and 'Newcastle roads' to describe wooden railways which accommodated flanged-wheel vehicles. Curr's rail road, with the flange on the rail, must be distinguished from the edge-rail of *'fish-bellied'* form, which had its flange on the waggons.

Benefits and unjust accusations arising from Curr's innovations

In 1840 Parliament established the Royal Commission of Inquiry into Children's Employment in Mines and Manufactories. Its first report,[170] together with two appendices, was published in 1842. The part which led to criticism of Curr's innovations was the Children's Employment Commission's *Appendix to the First Report of the Commissioners, Mines, Part I: Reports and Evidence from Sub-Commissioners*. The Commission interviewed 4,018 witnesses and a line from the evidence given by one of these – *'It was when the iron railways came in that they were putting away the horses and brought boys in to draw'*[171] – may have led some historians and others to conclude that an unfortunate result of Curr's 'invention' was that it encouraged the employment of boys and girls to haul or push the loaded waggons.[172] This was reinforced by the publication, in the same year, of a commentary, a furious reaction sparked by the government report, written and privately published under a very similar title,[173] which included copious extracts from the evidence. It also included four horrific engravings showing children hauling or pushing baskets, sledges and corves in narrow passages underground. Together they inspired a justifiable outcry.

The accusation that Curr's 'invention' of the cast-iron rail led to the exploitation of child labour in mines is not justified. There is little evidence, other than that of the witness quoted above, to show that his railroads were responsible.

That boys were employed in mines is not disputed. M. W. Flinn explained: *'The economics of substituting horses for human traction underground were complicated. Boys' wages were low, while horses were expensive in terms of both capital and running costs.'*[174]

The introduction of Curr's improved corves undoubtedly led, in some instances, to the abuse of child labour. Lewis puts it thus: *'Curr's corves were not an unmixed blessing. Sometimes they brought back hand-tramming where horses had pulled sleds and rolleys before, and this led to a wider use of boy labour.'*[175]

Ashton and Sykes note that although *'the wheeled corves could be moved by young children, and though at most places horses were retained to draw a train of corves along the main gates, in some Scottish pits boys were substituted for horses.'*[176]

Matthias Dunn, a government inspector of mines, commented on the report of the Children's Employment Commission in his 1844 book on the coal trade:

> *Height of Roads – The committee justly observe, that no coal mine can be comfortably worked 'whose main roads are less than 5 to 6 feet, and the side roads less than 2 ½ feet'; and they cite a good deal of evidence to shew, that many collieries are carried on below these standards, consequently requiring the employment of very young operatives.*
>
> *In this respect, it is notorious that in proportion as the collieries are conducted by skilful persons, so are the roads made commodious for the application of small ponies or putters; and in the Newcastle collieries, be the height of the seam what it may, artificial height is always provided up to the standard of 3 1/8 feet, and the horse roads from 5 to 6 feet. …*

> *With respect to the sort of carriage by which the coals are conveyed along these low passages, in many districts they use sledges, which are both oppressive to the workmen and wasteful to the proprietor.*[177]

This goes to show that such exploitation, when it took place, was almost certainly limited to pits working thin seams where the passages, especially the side ones, were of low height. It is unlikely that the profitability of such pits would have allowed for investment in cast-iron rails.

At the duke's Sheffield collieries boys were employed before the installation of rail roads as hurriers and trap boys underground and as jinny boys supervising the horse-drawn whin gins above ground. The first payments for the boys who led the horse-drawn corves were recorded in the accounts in January 1784.[178] Elsewhere, for instance in the north-east coalfield, putters usually worked in pairs, a younger boy (the 'foal') assisting an older one (the 'heedsman').[179]

The introduction of Curr's rail roads marked a great improvement in the *'slavish labour'* [in R. L. Galloway's words] of the putters who conveyed the coal from the coal face to shaft bottom.

Lewis wrote:

The miseries of those days and their solution are best brought out in The Pitman's Pay, *a long dialect poem written by Thomas Wilson,*[180] *a Tyneside collier, and first published in 1826–30. Two veterans are yarning in a pub about old times:*

> *When Dicky's corf was fill'd wi' sic,*
> *He let his low*, and stuck't agyend*-*
> *Ax'd Deddyte lay down his pick,*
> *And help him te the heedwis* end.*
>
> . . .
>
> *The wark now placed, and pit hung on,*
> *The Heedsmen, whether duin or nut,*
> *Mun iv'ry man and mother's son*
> *Lay doon the pick and start te put.*
>
> *Now then the bitter strife begins.*
> *All pullin', hawlin', pushin', drivin',*
> *'Mang blood and dirt and broken shins,*
> *The waik uns wi' the strang uns strivin',*
>
> *Aw mind a tram byeth waik and slaw,*
> *Just streen'd te rags te keep her gannin'.*
> *Frae hingin'-on* till howdy-maw,**
> *Ye hardly knew if gawn or stannin',*
>
> *Just pinch'd te death, they're tarn and snarly,*
> *A' yammerin' on frae morn till neet-*

> *Jack off the way, blackgairdin' Charley,*
> *For at the corf nut lyin' reet.*
>
> *The bits o' lads are badly us'd -*
> *The heedsman often run them blind -*
> *They're kick'd and cuff'd and bet and bruis'd*
> *And sometimes drop for want o' wind.*
> . . .
> *Sic, then, was the poor putter's fate,*
> *Wi' now and then a stannin' fray,*
> *Frae yokens,* cawd pies,* stowen bait,**
> *Or cowp'd* corves i' the barrow-way,*
> . . .
> *But heavy puttin's now forgotten,*
> *Sic as we had I' former days,*
> *Ower holey thill and dyels a' splettin':-*
> *Trams now a' run on metal ways.*
>
> *God bliss the man wi' peace and plenty,*
> *That furst invented metal plates!*
> *Draw out his years te five times twenty,*
> *Then slide him through the heevenly gates.*
>
> *For if the human frame te spare*
> *Frae toil and pain ayont* conceivin',*
> *Ha'e ought te de wi' getting there,*
> *Aw think he mung gan strite te heeven.*
>
> **Low = light; agyend = again; heedwis = headway's; hingin'on = start of work; Howdy-maw = last corf of day; yokens = trams meeting head-on; cawd pies = broken axles; stowen bait = stolen food; cowp'd = overturned; ayont = beyond.*

Lewis continues:

> *The man who, one hopes, came to enjoy this bliss in the hereafter was John Curr... His scheme for underground railways was to replace the cumbrous Tyneside system of carrying a single basket corf on a tram or sledge inby, and transferring it to a horse-drawn outby sledge or rolley* ....[181]

Curr's book

Much of what follows is to be found in Curr's book, *The Coal Viewer and Engine Builder's Practical Companion*, which he published in 1797. One aspect of its significance is explained by Flinn:

> *Harris has pointed out that there were no published works in English of importance in this field* [British mining technology] *between 'J.C.'s' The Compleat Collier of 1708 and John Curr's The Coal Viewer of 1797.*[182]

The book is dedicated to Curr's employer, the Duke of Norfolk, and is the earliest known work to describe cast-iron railway track. It is also the first practical handbook on the construction of steam engines and boilers. John Farey, in his *Treatise on the Steam Engine*, wrote:

> *It is a useful guide for those who require to construct such engines, being a complete manual for their instruction.*[183]

The text is accompanied by five plates, as well as 26 pages of tables. Charles E. Lee considers that:

> *The tabular material is extraordinarily detailed; few modern writers would regard it as necessary to carry to the fourth place the quantity of water in pumps.*[184]

The year in which Curr began to write his book is not easy to establish. From a paper written by John Liffen we learn:

> *In 1997 the Science Museum Library, London, acquired a manuscript copy of* The Coal Viewer *from the dealer William Duck.*[185] *It is not a first draft, with crossings-out and additions. Rather it is a hand-written fair copy, bound within covers. It comprises a title page and 143 pages of text, together with five hand-drawn and hand-coloured sheets of drawings. These are designated 'plates', numbered Plate II to Plate VI, suggesting that Plate I is missing.*
>
> *... The text is fairly close to the printed text but is sometimes paraphrased, with the occasional paragraph not present in the book as printed. The 'introductory preface' of the book is not printed in the manuscript.*
>
> *Most significantly the date on the title page is 1795.*[186]

The fact that this is a fair copy leads to the conclusion that there was at least one earlier draft.

In 1931 C. F. Dendy Marshall tells us that:

> *I have recently acquired a copy dated 1796 which is evidently the date of the first issue.*[187]

But, in 1938 he wrote:

> *The author's copy is, however, dated 1796 and was probably a proof. The 1797 copies, of which there are not many, have a more elaborate title page.*[188]

Thirty-five copies are known to have survived but, fortunately, the reprint published by Frank Cass & Co in 1970 is easily available.[189] It contains a new and informative introduction by Charles. E. Lee, one of the foremost experts on the early history of railways.

Several years prior to the book's publication Curr published a prospectus. This is known to have been printed in at least two newspapers: *The Sun*, a London newspaper, on Friday 8th November 1793, and later in the *Leeds*

Intelligencer, on 2nd December 1793. The full prospectus (see Appendix 3) describes the proposed contents in great detail, which are very much as in the printed book.

Curr advertised his book extensively, and at prices up to £2 12s. 6d., it was not cheap; for instance in the *Star and Evening Advertiser*, London, on Tuesday 2nd May 1797 [see Figure 8.12], the *Manchester Mercury and Harrop's General Advertiser* on 6th June 1797 and in the *Newcastle Courant* on 6th January 1798.

A contemporary review was positive:

> *This work contains practical remarks on the conveyance of coals under ground, with descriptions of machines and rail roads, contrived by the author for those situations: proportions of the different materials used in constructing fire engines; tables of their several powers and expences; tables of the quantity and weight of coal in a statute acre; and general estimates of opening collieries: - with descriptive plates. The whole is detailed with great minuteness; and, being professedly the result of much experience in conducting works in the neighbourhood of Sheffield, the publication is likely to prove of considerable utility to those who are concerned in similar undertakings.*[190]

It may be assumed that Curr's intention in publishing his book was twofold: to promote his professional services; and to sell rails and engine components manufactured at his foundry.

In R. L. Galloway's opinion:

> *While the use of iron in railway construction had been going on for some considerable time, and was gradually extending as this material became more abundant and cheaper, it cannot be doubted that a considerable impulse was given to the movement by the publication of Mr. Curr's book in 1797.*[191]

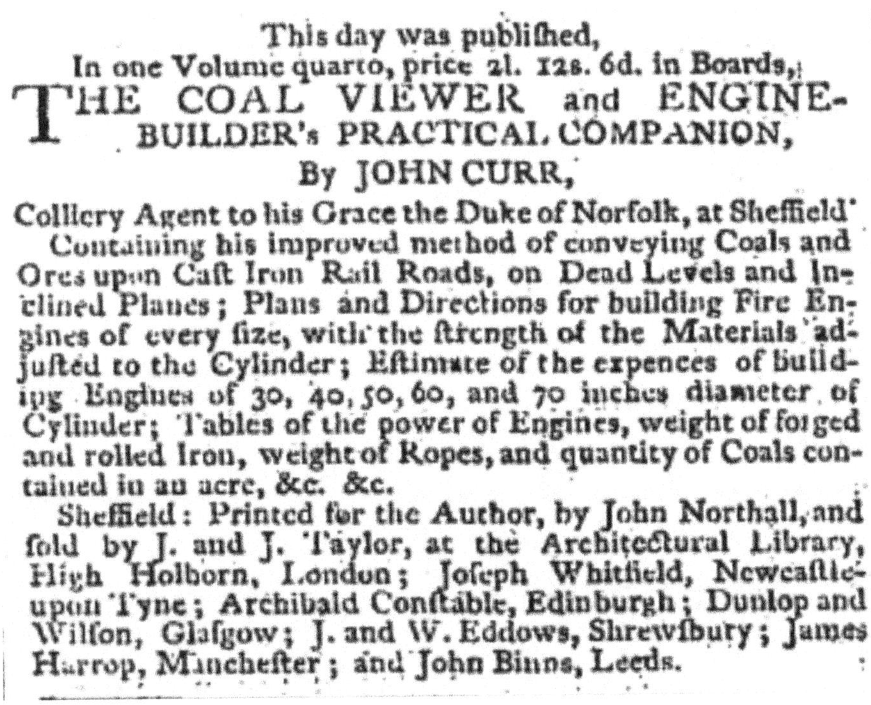

This day was publiſhed,
In one Volume quarto, price 2l. 12s. 6d. in Boards,
THE COAL VIEWER and ENGINE-BUILDER's PRACTICAL COMPANION,
By JOHN CURR,
Colliery Agent to his Grace the Duke of Norfolk, at Sheffield.
Containing his improved method of conveying Coals and Ores upon Caſt Iron Rail Roads, on Dead Levels and Inclined Planes; Plans and Directions for building Fire Engines of every ſize, with the ſtrength of the Materials adjuſted to the Cylinder; Eſtimate of the expences of building Engines of 30, 40, 50, 60, and 70 inches diameter of Cylinder; Tables of the power of Engines, weight of forged and rolled Iron, weight of Ropes, and quantity of Coals contained in an acre, &c. &c.
Sheffield: Printed for the Author, by John Northall, and ſold by J. and J. Taylor, at the Architectural Library, High Holborn, London; Joſeph Whitfield, Newcaſtle-upon-Tyne; Archibald Conſtable, Edinburgh; Dunlop and Wilſon, Glaſgow; J. and W. Eddows, Shrewſbury; James Harrop, Mancheſter; and John Binns, Leeds.

Figure 8.12. Publication notice for *The Coal Viewer*, in the *Star & Evening Advertiser*, 2nd May 1797.

Curr's railroad, I

The rails

Was Curr the inventor of the flanged cast-iron rail? Perhaps not, as noted earlier.

The chief builder of rail roads was Outram & Co. It is from Outram that the false derivatives of 'tramways', used interchangeably with 'rail road', and 'tram' have arisen. Samuel Smiles, for instance, in his *Life of George Stephenson* (1857) concluded that *'tram'* was derived from Outram.[192] Although he withdrew this statement from later editions, the damage was done and the theory is still popular.

Benjamin Outram (1784–1805) himself was in no doubt as to their inventor. In *'an interesting and little-known manuscript, undated, but probably about 1803 or 1804, preserved in the Outram family*: Minutes to be Observed in the Construction of Railways,' Outram wrote:

> *In Mines and other works underground where very small Carriages only can be used, very light rails are used, forming what are called Tram Roads, on a system introduced by Mr. Curr and these sort of light Railways have been much used above ground in Shropshire, and other Counties where Coals and other Minerals are gotten.*[193]

Of the railroad, R. A. Mott wrote:

> *It is certain that William Jessop and, perhaps particularly Benjamin Outram, both competent civil engineers, were pioneers in applying Curr's railroads from 1793. It is unfortunate that Curr did not patent his system and unlikely that he derived much financial reward from his invention. We are only concerned to honour him to whom honour*

Figure 8.13. Rails. From: The Coal Viewer, plate 2.

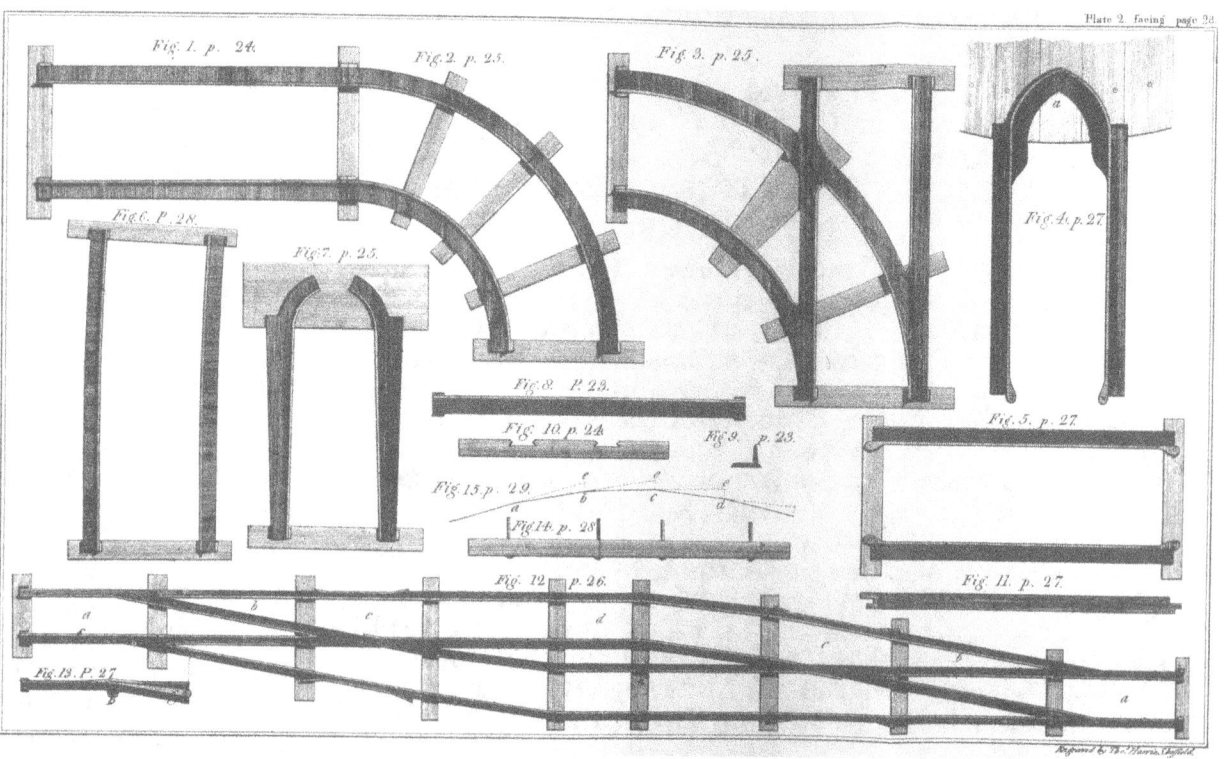

is due, and John Curr has never been given due acknowledgement.[194]

The term used by Curr for his flanged rails was *'rail plates'*. These came to be known commonly as *'plates'*, hence a *'platelayer'* is, in modern parlance, the man who lays rails.

Curr's *The Coal Viewer* gives seven pages of technical detail for the construction of his plate rails and their associated sleepers, together with a plate showing the various types [see Figure 8.13].[195] In the book he claims that in the iron rail he has:

> *hit upon a mode superior to anything hitherto practised, as the result of seven years experience informs me. (p. 9)*

In *The Coal Viewer* Curr describes his rail plates in great detail. It begins as follows:

THE COMMON PLATE.

The plate of general use shewn plate 2, fig 8, (which suits both sides of the road) is 6 feet long, 3 inches broad on the trod [tread of a wheel], and ½ an inch thick. The margin [wheel flange] stands 2 inches high above the plate, and is ½-inch thick where it joins upon it, but is tapered to the top (which is rounded) to ⅜ of an inch thick, for the convenience of moulding. There must be counter sunk nail holes within 1 inch of each end, and the lugs for fixing the plate in the sleeper may be 1 ¾-of an inch long, and measure when put on 4 ⅝-inches broad over the bottom. ... The weight of this plate is 47 to 50 lb. [23 ½ to 25 lbs per yard].

This rail plate is well adapted to the corves heretofore described, and hurrying *or* putting *by horses; and when greater burdens are necessary to be taken in each corf, the plates may be strengthened by casting them 4 or 4 ½-feet long, and the margins may be raised ½ an inch higher in the middle, and tapered down to 2 inches at the ends; and if very great burdens are required, the metal may be made in general ⅛ of an inch thicker; and on the contrary, if the corves are lighter than those herein described, the plates may be made ⅛ of an inch thinner.*

THE SLEEPER.

Shewn fig. 10, on an enlarged scale, is 3 feet, 4 inches long for the wide corf heretofore described, and 3 feet 2 ½ inches for the straiter one; should be sawn out of Oak, 4 ½ or 5 inches broad, by 2 ½ inches thick, and the plate must be sunk down 1 inch deep into the sleeper, and the road must be laid down 22 ½ inches wide to suit the narrow corf, and 24 inches wide to suit the wide one, affording about ⅝ or ¾-inch play in the corf wheels, which will be found quite enough in the straight or easy bending roads, but for very quick turns the play requires to be 1 ½ inches.[196]

Other plates were made, for example, to go round bends, or to allow corves to pass and Curr's description of these was summarised by Bland:

The narrow ends suit straight track Fig. 1, and the broad ends 1 inch extra to join the turn plates, as Fig. 2, when the tread is 4 inches broad and forms the quadrant and must be laid 1 ½ inches 'straiter' than the straight road.

The margin of the inside part has a radius of 3 feet 2 inches and the outside plate has radius of 4 feet 11 inches suitable for the narrow corf with 1 ½ inches extra width for the wider corf.

Fig. 3. For turning into 'benks or boards.' Here the treads on the straight portion are 3 ½ inches broad and the circular part 4 inches broad forming also a quadrant and for turning into 'benks,' where more room can be given, the breadth may remain the same, but the margin of the inside plate may be 4 feet radius and must form a quadrant also. Fig. 7 is useful for taking the corves from the road. These may be about 4 feet 6 inches or 4 feet 9 inches – the pointer ends are 5 ½ inches broad to admit of being laid 2 inches closer together of that end when the corf passes with less friction.

Fig. 12 – 'Pass bye.'

Plates should be 4 inches broad as they must be laid 'straiter than the common roads.'

The plans show a turning to right without switch points. A full and detailed list of lengths and sizes is given for the construction of this 'pass bye', in fact his book generally shows very carefully thought out instructions to both corf makers, 'sawers,' carpenters, blacksmiths and founders.[197]

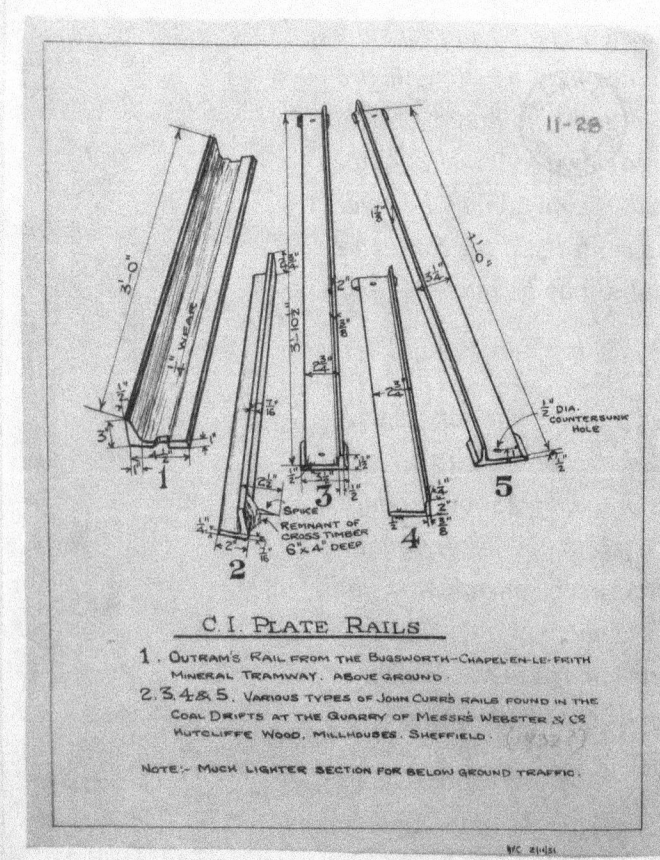
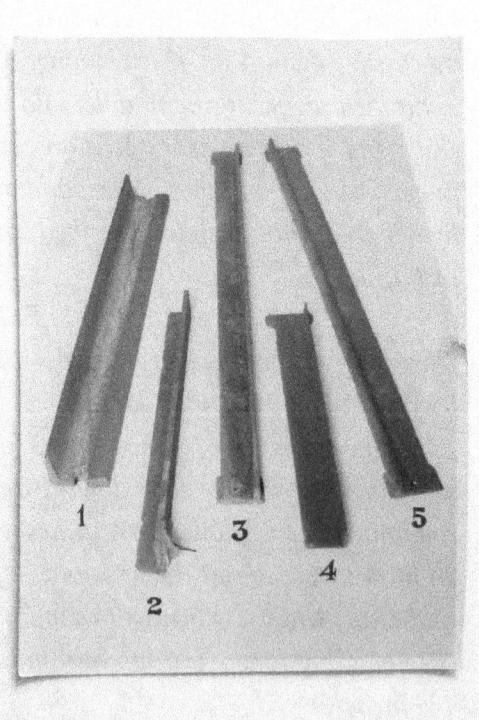

Figure 8.14. Examples of Curr's cast iron rails, thought by John Hunter likely to be from the 1820s. From: Sheffield Archive (Bland collection) SA/BC11-28.

Bland states 'Curr's rails', but 'Curr type' is probably a better description [see Figure 8.14].

M. J. T. Lewis provides another:

His [Curr's] track was light, since the weight of the coal was spread over many wheels. It was simply a conversion of the wooden flanged barroway into iron, not, as has been said, an addition of the flange to existing flat plates for flanged wheels. The rails were L-shaped and usually 6 ft. long, weighed 47 to 50 lb. each, and were nailed to wooden sleepers at a gauge of around 2 ft. The principle was, he pointed out, quite suitable for surface use as well.[198]

The corves

Before Curr's time coal was conveyed underground in a *'corf'*, or basket, on a *'tram'*, a wheelbarrow or sledge, to the shaft, where it was wound to the surface. At Sheffield Park Colliery, before Curr introduced his new system, coal was conveyed in *'kibbles'*, or tubs, on *'sledge trams'*.

R. L. Galloway provides the background:

Though small four-wheeled carriages, running on wooden railways, had been in use in coal mines in Shropshire, as we have seen, long before their introduction by Mr. Curr into the Sheffield district, we only hear of them being employed there in cases where the mines were entered by horizontal drifts. To Mr. Curr appears to belong the credit of being the first to use these carriages in connection with vertical pits.[199]

Curr replaced the hazel wickerwork kibbles [bucket drawing coal up a shaft], or baskets, that held up to 14 cwt. of coal with a four-wheeled corf constructed of solid wooden planks attached to an iron frame and with a capacity of 5 ½ to 6 cwt. of coal or 12.2 cubic feet. The wheels were from 10 to 13 inches in diameter (one was 13 ¼ in. diameter at the edge with 8 spokes, a 11 ½ in. rim, ¾ in. thick and weighed 14 ½ lbs). The corves were fitted with loops and *'coupling-chains'* were used to attach them together. For hauling up the shaft they were hooked onto the rope by the *'corf-bow'*. Up to then, hard hauling, or dragging, of kibbles, sometimes on flat-bottomed wooden sledges, with some exceptions,[200] had been usual practice. A further benefit was that the necessity of reloading at pit bottom was no longer required. Although the plate rails were flanged, the corf's wheels were flat so that it could run with or without rails above ground. Curr claimed that where the ground fell ½ inch in the yard, up to 24 corves could be hauled.[201] This compares with an average of two or three that could be hauled at any one time in the Newcastle area.[202]

The introduction of Curr's corf brought its own problems:

Being high, they were difficult to fill at the hewer's working place. At the junction between the branch railways and the horse roads they required to be lifted from the small trams on to the horse rolley by

John Curr

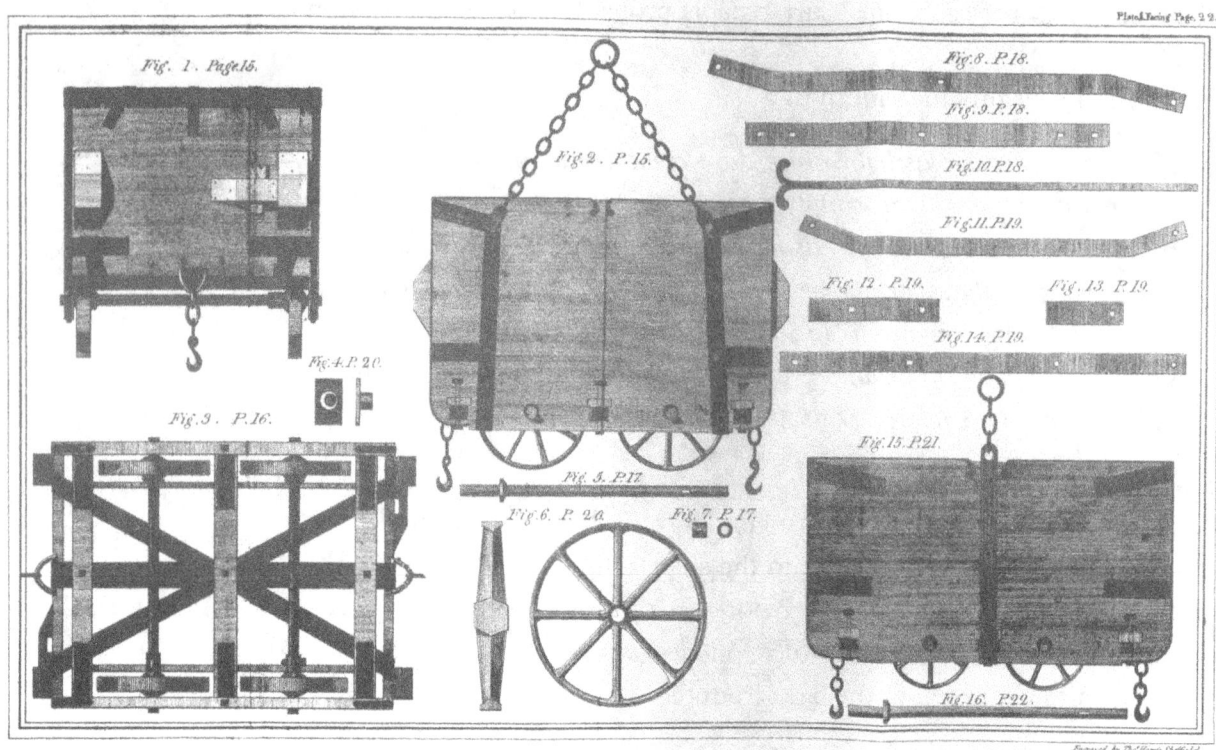

Figure 8.15. Corves. From: *The Coal Viewer*, plate 1.

means of a crane. They were unwieldy to handle, both at the top and the bottom of the shaft; they required endless repairs ...[203]

Also, the increased speed of winding after the installation of steam engines led to a greater risk of damage to both corves and the sides of the shaft. Curr was alert to such problems. He continued to work on the design of the corf, as evinced by a letter to John Buddle Jnr in 1800,[204] and the problem of damage led to his invention of shaft conductors, which he patented in 1788.

The Coal Viewer has eight pages of technical detail for the construction of his corves, together with a plate showing its components [see Figure 8.15].[205] Medlicott summarised Curr's lengthy description:

He [Curr] developed two sizes of corf, one for the 'long way' working, where the roads ran along the bank face (coal face), were narrow and the coals small, that weighed about three hundredweight (cwt) and held approximately five and a half to six cwt of coal. The outside dimensions were 40 inches in length, 30 inches wide, and 21 ½ inches high (30 inches high on its wheels), and cost an estimated £2 16s 6d each to produce. The corf for 'short work' where the coal was extracted in large pieces and the seam not very thick, was 42 inches in length, 31 ½ inches wide and 19 inches high (36 inches high on its wheels), weighed two and three quarter cwt and cost £2 13s 6d.[206]

Hatchett, in his diary describing his visit to Sheffield Park Colliery in 1796, noted:

... the coal is drawn up in square wooden vessels bound and cased with iron called Corves. Each of those used at Sheffield

hold between 500 and 600 Cwt. [sic MS – 5 or 6 cwt.] These corves have at the bottom four small Iron Wheels by which they run in the Cast Iron rail roads both above and underground.[207]

Curr's railroad, II
Underground

Ian R. Medlicott notes that wooden waggonways had been constructed underground at Sheffield Park Colliery in 1783. Evidence for this is to be found in colliery account books: payments for boys to lead horse-drawn corves are recorded in January 1784;[208] expenses for road levelling and sleepers, 19th–26th February 1785;[209] and an account for December 1785 refers to laying 596 yards of beech rail at twopence per yard.

It is hard to be precise about the year of introduction of Curr's railroads underground. Curr is responsible for much of the confusion; in the *The Coal Viewer*, published in 1797, but most likely written in 1794, he states:

The making and use of railroads and corves were the first of my inventions, and were introduced at the Sheffield Colliery about twenty-one years ago.

and, in his report of 1801:

The First of My Inventions was the Rail Roads which happened 2 or 3 years prior to the Duke of Norfolk's being concerned in Collieries.[210]

These two sentences have led almost all historians of railways to date his 'invention' to 1773, with notable exceptions. Galloway wrote:

Mr. Curr has indeed been said to have introduced these rails as early as 1776, but this statement, though frequently repeated, appears to rest on a misquotation of his words...[211]

M. J. T. Lewis concurs:

... there is no doubt that the introduction of Curr's famous iron angled rails is to be dated to 1787. ... His six reports on the Sheffield collieries from 1773 to 1785 make no mention whatever of iron rails;[212] every report from 1787 on emphasises their novelty and their usefulness. The only surviving account book of this period, from 1781 to March 1787, gives no hint of Curr's rails and corves until almost the very end. In February and March 1787 we find '4 Sole Trees ⅛' – which must be for corves and trams, since surface waggon soles cost at least four times as much – and 'to Moddle Wheels ... 5sh', meaning patterns.[213] The clinching evidence is to be found in John Buddle's report, dated 7 April 1787.[214]

Ashton and Sykes add:

It has been questioned whether Curr actually made use of cast-iron rails below ground before 1790, but a report prepared by John Buddle

Snr. puts the matter beyond doubt; for a comparison of the costs of 'the new scheme of hurrying the coals' with those of hurrying with horses includes 'Expenses of Cast Iron Plates and Barrow-Way'.[215]

Buddle's report was commissioned by Vincent Eyre III, the Duke's agent, in 1787. Notably, he was tasked as follows:

You are required to give your opinion on the Scheme of Hurrying the Coals which Mr. Curr has lately introduced in one of the Pits, and compare the Mode of Hurrying by the Wheel Corf with that of Hurrying with Horses.[216]

In his report Buddle records that rail roads were in operation at only one pit at Sheffield Park Colliery; at Attercliffe *'Mr Curr's new Scheme of Hurrying'* was not yet applied. Again, it's hard to be precise, but the introduction of cast-iron plates at Sheffield Park had only just begun. According to Medlicott, colliery account books first record payments for these shortly before April 1787.[217] Buddle was wholeheartedly in favour and recommended their instant adoption. He calculated that the new method would save £312 10s 6d per annum. Buddle's calculations in the extract from his April 1787 report to the Duke of Norfolk are shown in Figure 8.16.

Figure 8.16. Extract from Buddle Snr's 1787 report. From: Sheffield Archives ACM/S/223.

	£ s d
For one Year by Old Mode with Horse	
Keeping horse, 52 weeks @ 15—0d	39—0—0
Driving by a boy 52 weeks @ 5—0d	13—0—0
Wear & Tear, extra cleaning @ 1—0d	2—12—0
Hanging on 1612 waggons the average work of one horse when only 2 are kept @ 2¼d	15—2—3
	69—14—3 = 10½ per waggon
Charge of the new mode	
Hurrying and hanging them on p.wag	4
Odd expense of C.I. plates & Barrow way	1
Banking them	1¼
	6¼
Saving per waggon by the new mode	3¾
The average years sale from the Colliery of 14,000 waggon loads if the whole quantity was hurried by Mr. Curr's new Method Say on 8,000 waggons @ 3¾ (2 horses employed)	125—0—0
Say on 6,000 waggons with advanced work, 3 horses reqd @ 7½	187—10—0
Yearly saving	312—10—0

The average production in 1787 was 14,000 waggon loads instead of 13,000 in 1780; the waggon loads were those of the Newcastle road or surface wagonway.

Buddle's recommendation was soon acted upon. Dr Lewis and W. N. Slatcher noted:

> [Curr's] account with Binks, Booth & Hartop, the Sheffield Park founders showed orders for corf wheels and 'road plates' beginning on 20 November 1787 and increasing greatly in May 1788.[218]

Medlicott gives the following information:

> By 28 August 1789 Messrs Binks, Booth & Hartop of the Park Iron Works, Sheffield, had supplied rails with a total weight of 92 tons 1 cwt 3 qtr 4 lbs at a cost of £1,084 19s 9d. For Sheffield Park Colliery there is an entry dated 16th–23rd February 1788, 'To Messrs Walker & Co. for Rails 63. 11. 7'.[219]

Lewis describes these events as follows:[220]

> *The first use of Curr's iron rails in Sheffield Park Colliery was evidently an experiment limited to one pit, and the corves were moved singly by hand, not in trains by horse, as he later advised. Attercliffe Colliery was not immediately equipped with iron rails throughout, for John Buddle and John Stephenson,*[221] *taking another view there in April 1789, estimated*
>
>> Iron plates wanted to complete the Heading and Gait-Ways of Attercliffe Common … 2600 yards at 5/6 p.[222]
>
> *But, fortified by Buddle's approval Curr went ahead and completely fitted out Park Colliery with the new rails. His account with Binks, Booth and Hartop, the founders, shows heavy orders for corf wheels and 'road plates' beginning on 20 November 1787.*[223] *Curr is also said to have been supplied with rails by Joseph Outram, a Ripley ironfounder and Benjamin Outram's father; …*[224]

Curr's railroad, III

Ashton and Sykes are clear about the benefits arising from the introduction of Curr's railroads underground:

> *[Curr] and his predecessors in the development of the railway certainly deserved well of their fellows, for they did more than any philanthropist of the day to lighten the lot of the most heavily pressed grades of underground labour, the youthful putters and drivers. Not only was the individual relieved, but the proportion of workers engaged in this onerous branch of mining was substantially reduced. At the beginning of the eighteenth century far more labour was involved in moving, than in getting, the coal. At Bo'ness, in 1681, there were 37 bearers to 13 hewers, and at Dunmore, in 1769, 74 bearers to 28 hewers.*[225] *Even in Northumberland and Durham where the crude system of bearing had long been given up, there were at Charlaw, in 1769, 10 barrowmen*

to 10 hewers; and at Stanley Kinghill Colliery the coal hewn by 70 pitmen required the service of 50 putters and 27 drivers to move it to the pit bottom.[226] *But, as the direct result of the improvement in underground carriage, by the early years of the nineteenth century the hewers almost always outnumbered the drawers of coal: at Heaton Colliery, in 1806, there were 143 hewers to 84 putters; at Middleton (Yorks), in 1808, 90 hewers to 60 putters; at Washington, in 1813, 67 hewers to 40 putters; and at Gatherick, in 1823, 12 hewers to 6 putters.*[227]

Such was the immediate result of the work of John Curr.[228]

Lewis concurs:

However, there was no doubt about the value of his rails. On Tyneside in particular the outby rolleys remained, and Curr's rails simply replaced the previous wooden ones inby, though still carrying the old type of tram with separate corf. It was here that the real benefit was felt, as The Pitman's Pay *so graphically describes. The labour force was also much reduced. Whereas with wooden rails there had been about as many putters as hewers, the number of putters could now be cut by a third or even a half.*[229]

R. L. Galloway has the last word:

Mr. Hall[230] *also effected a great improvement in the underground rails, by substituting light malleable-iron-edge-rails and flange-wheels for the cast-iron tramplates and sharp wheels previously employed between the horse roads and the working faces. Thus he completed the improvements which Mr. Curr had begun and established the system of drawing and conveying coal in carriages now universally practised.*[231]

A growing reputation
Gascoigne collieries

Word of Curr's innovations was by now spreading and his services began to be called upon as a consultant. Sir Thomas Gascoigne (1745–1810), a Roman Catholic and Yorkshire landowner, was a friend and political ally of the 11th Duke of Norfolk.[232] Like the duke, Gascoigne employed a Catholic steward, James Catton, and, like the duke, had become Anglican in order to stand for election to Parliament. Curr was to send his daughters to be educated at Bar Convent, York, where a school for girls had been founded in 1686 by an earlier Sir Thomas.

The Gascoigne family was notable for taking upon itself the exploitation of their assets, of which coal mining was the principal concern, after agriculture. Sir Thomas was a member of the Society for the Encouragement of Arts, Manufactures and Commerce, and had sought the advice of the Co. Durham inventor, George Dixon (1731–1785), in 1774 regarding the possibility of building a canal. He had also consulted John Smeaton

(1724–1792) in 1780 about the reconstruction of an engine at Sturton Colliery.

In December 1790 Gascoigne invited Curr to advise upon the opening of new workings at Garforth Colliery. In all likelihood, Curr's services had been recommended to Gascoigne by the Duke of Norfolk. Curr had previously advised a local colliery owner, Thomas Fenton, and Gascoigne sought Fenton's 'candid opinion' as he had 'experienced the advantage of Curr's designs'.

For Garforth, Curr proposed the installation of a water-returning winding engine and rail roads. If this be done, according to Curr:

'the neat saving into Sir Thomas Gascoigne's own pocket' ... would amount to upwards of '£800 per annum ... after paying the premium [of £400 annually] to the patentees'.[233]

Gascoigne was almost convinced for, in January, he wrote to George Townsend:

My Garforth colliery is nearly exhausted as to its present working and I am under the necessity of opening new works, the expense of which will entail a considerable sum, particularly if I employ the machinery and patent obtained by Mr Curr coal agent for the Duke of Norfolk; he is strenuous for it, and well he may, but I am not quite so ready to adopt as formally ... the expense in one way [of working] *amounts to near £3,000 in Curr's, to upwards of £4,000. The difference at the outset is considerable, but if in the process the outgoings are diminished the colliery ultimately will repay itself.*[234]

Alexander Lock questions whether much advantage was gained by Gascoigne from Curr's advice:

Due to a lack of evidence it is unclear what, if any, savings were made by John Curr's improvements to the collieries. [...] [A sole surviving account book for the 1790s suggests] *that no real savings were made by the collieries nor that their vend was increased as a result of John Curr's improvements.*[235] *The same was true of Curr's activities in Sheffield.*
[...]
  Having said this, however, Curr's inability to reduce Garforth's mining costs substantially should be attributed more to the near-exhausted state of the colliery than to any inherent flaw in his project. In 1791, when John Curr was employed, Garforth colliery was believed to be 'nearly exhausted' and John Curr's 'schemes' successfully extended the working life of the colliery at least until 1810. As such, Curr's schemes were effective. They enabled a near-exhausted mine to be exploited for a further nineteen years and, although costs did increase due to the growing difficulty in extracting coal, the colliery still managed to maintain a substantial net profit and vend. Indeed

> *it is likely that without Curr's 'schemes', mining costs at Garforth in the 1790s would have been considerably higher and the mine's closure somewhat sooner.*[236]

It seems, though, that Gascoigne was satisfied with Curr's advice for, in 1801, and after Curr's dismissal by the Duke of Norfolk, he again sought it,[237] this time on the opening of new works at Parlington. Lock further states:

> *Curr thought that Gascoigne would make similar savings with the installation of two steam-winding engines at the new colliery recently opened at Parlington. However, whilst he recommended a water-returning engine for Garforth in 1791, by 1801*[238] *Curr felt that the developments made in steam winding engines made any 'further introduction of the waterwheel' ill-advised. Though the initial costs, of £630, to install steam-winding engines at Parlington were considerably higher than the initial costs of installing '3 large coal gins' for horses, at £230, the steam-winding engines would make considerable annual savings in running costs ... According to Curr, the 'annual saving drawing with engines' rather than horses would amount to £367. 6s. and this, when combined with all his other hurrying and winding technologies, would afford Parlington Colliery a considerable 'total saving' of at least £567 6s. a year.'*[239]

Coalbrookdale

Word of Curr's abilities continued to spread. In 1793, a busy year for Curr, he was to advise John and George Overton about construction of a tramroad from Merthyr Tydfil to Abercynon on the Glamorgan canal and was also invited by the ironmaster William Reynolds (1758–1803) to Coalbrookdale. It will be seen that Curr's work at Coalbrookdale was to result in a wide dissemination of his plate rails.

The Brierley Hill terminus of the Shropshire Canal near Coalbrookdale used a system of tunnels and shafts for the movement of materials, which is described by Lewis:

> *There was also a two-and-three-quarter-mile branch, opened in 1792, from near the Windmill plane to Brierley Hill just above Coalbrookdale itself, and linked to Horsehay by a railway specially built by William Reynolds from the nearest point of the canal. At the Dale end, the traffic was taken from the terminus to the works in an unusual way. Each boat carried four iron crates of 2-ton capacity. But instead of being let down an incline the crates were craned down (and up) vertical shafts. At the bottom they were received on to railway waggons standing in a tunnel ('footroad' as it was called) 120 ft. below, whence they were driven into the open air and to the Dale works. The shafts were modelled on similar ones on the Bridgewater Canal ... After two years the shafts were replaced by a railway incline (the*

'Old Wind') for ordinary waggons, which may have been the first of the new-style plateways in the area. Finally, in 1800, a railway was laid along the tow-path from the incline head to the wharf where the Horsehay link railway went off.[240]

The Coalbrookdale Company brought the system into operation on 24th January 1793. Early in May these operations came to a halt and Curr's advice was sought. He provided this in his 25th May letter to Richard Dearman (shown in Appendix 4).

Barrie Trinder explains:

Curr outlined his proposals in a letter on 25 May. His most important suggestion was the installation of platforms or cages in the shafts, in which rails would be laid to carry two trucks side by side. When containers arrived in boats at the terminus, they were to be lifted off on to railway trucks which would then be pushed along rails into the cages. When the cages reached the tunnels at the bottom of the shafts the trucks with their containers would be pushed out into a siding where they waited until a train of six waggons was ready to be drawn out of the tunnels by a horse. Meanwhile the men at the shaft bottom were to push into the cage waggons and containers from another siding. Curr's letter suggests that work was already in progress to alter the system for he claimed that his suggestions would 'not necessitate one brick to move in the Tunnel or any other openings made'.[241] *It seems likely that Curr's alterations were carried into effect, for, apart from the provision of the cages, they necessitated only the re-siting of the cranes at the shaft tops and some changes in the layout of the railways in the tunnels. A drawing in the William Reynolds Sketch Book shows the system according to John Curr's suggestions.*[242]

The terminus re-commenced operations on 11 July 1793 and was heavily used.[243]

As further noted by Barrie Trinder:

The Coalbrookdale partnership's railways were transformed by the substitution of the L-shaped plate rails developed by John Curr of Sheffield which replaced the earlier pattern of iron edge rails. The Curr type ... was reckoned to be cheaper to lay and to operate than the earlier forms, and economies were anticipated when it was introduced.[244]

Here Curr was to work with William Jessop I (1745–1814)[245] and Benjamin Outram (1764–1805). Jessop had, from the age of 16, been a pupil of John Smeaton (1724–1792), while Outram had, aged 25, been appointed assistant engineer to Jessop. A year later the partnership of Outram & Co. (with Jessop, John Wright, and the latter's brother-in-law, Francis Beresford) was formed. Jessop and Outram were primarily canal builders but used Curr's tramroads for links with their canals. In 1792 Outram & Co. built a coke blast furnace at Butterley, near Ripley on the Cromford Canal. Outram &

Co. were to become known as builders of tramroads all over the country.[246]
Lewis noted:

> *It was on the canal feeders that the all-iron rail first appeared, sweeping across the coalfields in a wave of new railways and converted wooden ones. ... In the Midlands, and especially in Nottinghamshire and Derbyshire, it so happened that two of the canal engineers whose services were much in demand were partners in Butterley Ironworks: William Jessop and Benjamin Outram, both of them staunch advocates of railways. It is not surprising that, wherever they went, they left iron railways behind them. It is no accident, too, that the appearance of the all-iron rail in the 1790s and its rapid spread over the first years of the nineteenth century coincided with the Revolutionary and Napoleonic Wars. The price of wood soared, and with the intensification of industry as a result of the wars, the cost of iron fell.*[247] *All things being equal iron proved itself far superior in technical and economic terms, for though the initial cost was considerably higher than for wood, the subsequent maintenance and replacement bills were vastly smaller. And so the wooden track gave way to the iron rail and the stone sleeper.*[248]

Spread of the railroad

In *The Coal Viewer* (p. 6), Curr claimed that his rails

> *... have been generally imitated, and made use of in most collieries for the last three years, especially in the southern parts of the kingdom.*

The history of the spread of Curr's railroad is hard to establish. The first known record of its emergence above ground, in print, is given by John Farey:

> *I have heard it said that the earliest use of these flanched rails above ground (for they were first introduced in the underground Gates of Mines, it is said), was on the south of Wingerworth Furnace; leading to the Ironstone Pits, of Joseph Butler, about the year 1788.*[249]

Lewis concurs:

> [John Curr's plate-rail] *came to the surface in 1788 at two places: Ketley Canal incline* [in Shropshire] *and* [Joseph Butler's] *Wingerworth Furnace near Chesterfield. After a brief pause, it went forth in the mid-1790s to sweep the country, with Benjamin Outram of Butterley Iron-works – his father Joseph had cast underground plates for Curr at his Ripley foundry*[250] *– as its main protagonist. Outram, more than any other man, determined the shape of industrial railways for a full thirty years. He was responsible, by direct advice or by example, for countless plateways throughout the Midlands. When he was consulted about the Monmouthshire Canal railways in 1799, he recommended a wholesale change-over from edge to plate rail,*[251] *and practically every*

railway-owner in South Wales followed his advice. Even after his death in 1805, his disciples carried on the work, building new railways by the score and adapting old ones, until by the 1820s, south of County Durham, the edge rail was very much the exception, the plate rail the rule. ... The Coalbrookdale empire had fallen before the conqueror at some indeterminate date, probably around 1800.[252]

Lewis adds:

Underground, the Middleton Colliery at Leeds was quick off the mark. In 1790 it

> 'Paid John Curr a Gratuity for his Trouble &c respecting the Hurrying & Drawing the Coals agreeable to his Patent £10.0.0' and next year paid 'J Curr for sundry Castings had from Sheffield £17.6.7.'[253]

Henceforth cast-iron plates were a regular feature here. The introduction of Curr's rail to Shropshire mines and surface railways has been dated to 1793, when he was called in by the Coalbrookdale Co. to advise on the operation of its railway and shafts at the end of the Shropshire Canal. Although his report makes no mention of his plates, this date may well be correct for underground, where wheeled corves were in use by 1794 and Dale-made Curr rails by 1796; but on the surface the conversion from edge to plate, with the possible exception of the 'Old Wind', probably occurred somewhat later, maybe even after 1800.[254] *We first hear of Curr's rails in 1796 in South Wales, about 1800 in Cumberland and Scotland, and only in 1802 on Tyneside,*[255] *where they must have been used some years before – again, the lack of iron foundries probably delayed their widespread adoption.*[256]

R. A. Mott identified eleven out of the twenty rail roads which Curr, in his letter to the Duke of Norfolk of 23rd October 1801,[257] said had been undertaken. Mott's list of these eleven follows:[258]

The decision of the Stockton & Darlington Railway Company, in July 1821, to adopt wrought-iron fish-belly edge rails for the railway virtually marked the end of angled tramroads for major railway undertakings using locomotives, but tramroads proved to be satisfactory for horse-drawn waggons for some time to come. The conversion of many tramroads to edge rails and the confusion between 'waggonways', 'tramroads', and 'railways' makes it difficult to give an accurate list of the 20 tramroads which John Curr said had been undertaken by October 1801. Eleven of those can be identified with reasonable certainty and the remainder listed as possibles or probables as follows.

> *Nos. 1 to 4 were underground 'rail plates' as Curr called them, the first at Sheffield colliery in 1778. In 1801 Curr said that his corve-winding method was in use at Sheffield, Attercliffe Manor, and*

Hesley Wood collieries, which implies that underground plate-rails were also in use. Since, in 1796, Hatchett described corve-winding at both Sheffield and Attercliffe collieries, it is possible that plate-rails were then in use both above and below ground in all four collieries.

No. 5, the Wingerworth tram road (or boxes on trams) from the blast furnaces to Woodthorpe ironstone pit, was built in about 1788, possibly in 1786.

No. 6, the Lings and Ankerbold tram road, connected Lings colliery (where coke was made) to Wingerworth furnaces, through Ankerbold, with another link to the Chesterfield Canal. Baxter dated this 1788.

No. 7, the 'roads' (as Curr called them) advised to the Coalbrookdale group to enable 'crates' wound up a pit shaft to be transferred to a canal. His report, dated 25 May 1793, from Sheffield, was probably made in connection with the development of the Shropshire Canal in that year. The Coalbrookdale group, subsequently, replaced the cast-iron coverings on its own railways by Curr's system.[259]

No. 8 was the tram road of Outram & Co. from the Crich limestone quarry to the Cromford Canal, dated 1793 by Baxter.[260]

No. 9 was the Little Eaton 'gang way' from the Derby Canal at Little

Figure 8.17. Cast-iron rail from the Surrey Iron Railway at Merstham. Photo: Courtesy Annamarie Critchard.

Eaton to Marehay (6 miles) with branches, authorised 1793, opened 1795. The corves from this gang way, after passing along the canal, entered Derby market place. The plates were cut at Wingerworth.[261]

No. 10 was the Merthyr tramroad (or dramroad as it was sometimes called) from Merthyr Tydfil to Abercynon on the Glamorgan Canal, jointly owned by the Dowlas, Penydarren, and Plymouth ironworks companies (1793). An illustration of it is reproduced in Fig. 1 [see Figure 8.18]; the gauge at this period was 4 ft. A curious lack of appreciation of Curr's connection is shown by the statement 'George Overton was the engineer for the line and the plates were laid by an engineer named Curl.'[262]

No. 11 was the Surrey Iron Railway from Wandsworth to Croydon authorised 21 May 1801; William Jessop was appointed engineer 4 June 1801 and the line was opened 26 July 1803. It was the first public railway and had a gauge of about 4 ft. 2 in.[263] *It had plates 4 or 5 in. wide, with an angle curved downwards at one edge and, at the other, an angle curved upwards.*[264]

A preserved rail from the Surrey Iron Railway is shown in Figure 8.17.

Mott added suggestions for the remainder, as follows:

To these well-authenticated examples of the use of Curr's cast-iron angled rails (tram-roads as Outram and Jessop called them), there are other probables and possibles which have to be sought for in Lancaster, Stafford, Warwick, and the 'neighbourhood of Newcastle upon Tyne.' For the latter the 'iron rails used underground at Walbottle colliery' in 1794[265] *appears to be the first, but the others cannot be identified, nor can the others implied by Curr to have been laid in Wales.*

The Stafford tramroad could be Consall Plateway (7 miles to near Stoke-on-Trent.)[266] *The Lancaster tramroad might be that at Wigan (Balcarres Railway) surveyed in 1788.*[267] *The Warwick tramroad might be that from Griff colliery to the Coventry Canal branch at Arbury, opened in 1793.*[268] *The Dale Abbey tramroad, from Dale Abbey furnace to the Cromford Canal, authorised in 1792*[269] *and connected to the Nutbrook Canal in 1798,*[270] *is a probable. So too is the line from Bugsworth to Peak Forest limestone quarry, engineered by Outram and opened in 1796.*[271] *Aikin, writing in 1795, said*[272] *'There are several railways for the conveyance of coal to the canal from Barnby and others from Barnby-bridge. It (the Barnsley Canal) is no cutting.' Since this was engineered by Jessop, it seems likely that the 'rail way' from the Barnsley Canal to Barnby furnace, said by Baxter to be opened in 1800, was a tramroad.*

The ones mentioned bring the total to 18 which, with the uncertainty attaching to others near Newcastle upon Tyne and in Wales, is reasonably satisfactory. For most of these after No. 8 Jessop or Outram, or his assistant, John Hodgkinson, were connected as

Figure 8.18. Track of the Merthyr Tramroad, 4 ft. gauge (1793) (Figure 1 in the quoted text). From Mott 1969-60, p. 14.

engineers or, for example for Merthyr, supplied the plate rails.

Of the tramroads mentioned, the Merthyr and the Surrey Iron Railway were the most important and had a great influence on tramroads built in the next decade before the influence of the locomotive became predominant. The Merthyr tramroad had stone blocks and not cross-sleepers, this making it easier for a horse to move freely. The deepening of the upward angle of the plate rail at Merthyr, to give increased strength, is evident in Fig. 1 [see Figure 8.18]. This was supplemented in the Surrey Iron Railway by an extra angle downwards at the other edge of the plate. The plates of the Surrey Iron Railway[273] *3ft. 2 in. long, have a calculated weight of 80 lb. (one angle) and 108 lbs. (two angles), much heavier than the plates Curr described for his 18 or 24 in. gauge underground lines but appropriate for the heavier loads to be carried.'*

Mr. J. K. Major said he wanted to add to Dr. Mott's list of twenty. The earliest underground piece of plateway that he had come across in the Lake District Mineral Mines was 1798 – a typical L-section plate 2 ½ in. x 3 ¾ in., probable gauge 20 or 24 in. It was definitely authenticated as 1798 by Jonathan Ockley writing in the 1830's. … [Curr] did not mention Cumberland in his list.[274]

Curr's railroad, Conclusion

Curr's system, with a flange on a cast-iron rail, revolutionised the transport of coal, particularly underground, and spread quickly throughout most of the United Kingdom. It was to remain in use for thirty years or so, until the coming of the locomotive.

His system has its detractors. As early as 1915, W. W. Tomlinson wrote '[Curr's] plate-rail was not a necessary intermediate form.'[275] Barrie Trinder is of the opinion that 'its use in Shropshire … was probably the reason why the stream of innovations in railway technology for which the area was famous for two centuries came to a stop.'[276] Lewis is particularly damning:

The plate rail proved a cul-de-sac in railway development. In spite of all the enthusiastic support it received from expert and amateur, it had serious defects. Its cross-section, with most of the metal thinly disposed in a flat plate and little strengthened by the vertical flange, was deplorably ill-equipped to carry heavy rolling weights. The wide running surface with its retaining flange might almost have been designed to catch and to hold the dirt kicked up by the horses' hooves, to the great increase of the friction on the wheels. The iron edge rail, in contrast, being a vertical beam, was stronger by comparison, and its comparatively narrow and elevated running surface harboured little dirt. Nonetheless, the plate rail did mark an improvement on plain wood, and in most cases one horse could pull a train of several waggons at a time.[277]

In 1804 Richard Trevithick's locomotive, built at Pen-y-Darren, made several runs on the Merthyr tramroad after its famous initial run on 21st February. The weight of the engine caused many of the cast-iron plate rails to break and the tramroad reverted to horse power after the test runs. Put simply, cast-iron was not a satisfactory material for rails.

A good example of this was in the north-east coalfield: although the cast-iron rail made its first appearance in 1797 it did not sweep the board as in other areas, and here they remained faithful to the edge rail and the flanged wheel. So it was in this area that the next great advance in railway technology took place. Nevertheless, for the board of the Stockton and Darlington Railway the choice between the use of much cheaper, cast-iron over wrought (malleable) iron was to take from 1818 to 1821 to resolve. Even then, despite a resolution that the whole of the line should be laid with malleable iron, it was not strictly carried out.

Curr's mining innovations

Winding

How was coal raised from the pits when Curr first arrived at Sheffield Park colliery in the early 1770s? Matthias Dunn provides a description:

The coals were drawn from the mines by machines, called 'gins'; the earliest known construction ... being called a 'cog and rung gin', the horse-wheel being vertical and toothed. In fact, a spur wheel and pinion turned an horizontal shaft lying over the pit, to which the ropes were attached. ...

The 'whin-gin' was an improvement upon the complex combination of the cog and rung, and has universally superseded it.[278]

As collieries increased in depth the limitations of the horse-gin became apparent. Steam engines had been used for drainage for some years and had been adapted for use in winding by the 1790s.[279] Dunn commented:

Although, however, the coal might be said to be drawn by steam, it was not by direct application, but by the intermediate agency of a water-wheel, supplied by water pumped into a cistern by means of the common steam-engine ...[280]

The final part of Curr's trinity of innovations was his invention of conductors. In *The Coal Viewer* (p. 9) Curr states that he had:

> *introduced machines for drawing coals at two of his Grace the Duke of Norfolk's collieries, near Sheffield ...*

In 1787 Vincent Eyre III asked John Buddle Snr to give his opinion *'re Mr. Curr's Scheme of drawing 2 corves abreast up a shaft 8½ or 9 ft. diameter by means of steadying conductors'*. Buddle was enthusiastic; in his report of 7 April he replied that he had *'not the least doubt of Mr. Curr being able to carry into execution the wished for Effect, not only drawing of two corves abreast but all the other designs he has projected'*.[281]

In the collieries of the North of England, prior to Curr's invention, putters brought single baskets from coal face to horse station. There, they were, by means of a crane, loaded onto a rolley, which carried two or three baskets, which conveyed them to the bottom of the shaft. Here they were attached to the rope, by means of a corf-bow, and swung up the shaft to the surface. For many years, damage caused by the baskets striking each other, or buffeting against the side of the shaft had been a problem. Such damage, until the introduction of the steam-engine for winding, had been minor because winding by horse-gin was slow. After the introduction of the steam engine the greatly increased speed of winding, coupled with an increase in the depth of shafts, became a major problem.

Curr devised a system whereby two corves abreast were raised with the aid of conductors, or guide rods, so that the danger of each pair colliding or buffeting the sides of the shaft was much reduced. He describes them in *The Coal Viewer* (p. 36) thus:

> *nothing more than two or three upright rods of deal 4 inches by 3, braged [sic] upon opposite sides of the pit, forming mortices or channels, by which the corves are conducted, being suspended upon cross-bars with rollers at their ends, which run within the mortices.*

The conductors, or guide rods, consisted of two pairs of deal rails, measuring four by three inches, fixed on opposite sides of the shaft. The corves were suspended below crossbars which were attached to the rope, and which ran with the aid of rollers in a channel formed by the conductors. The shaft widened mid-way, at the point known as 'meetings', and the speed of winding slowed to allow for by-pass without the corves clashing. On arriving at the surface the corf was raised a little bit above the mouth of the pit in order to allow a wooden platform to be introduced beneath it, onto which the loaded corf was lowered. It was next detached from the rope, an empty corf replacing it. The empty corf was then raised and the platform withdrawn so that winding could recommence. Curr invented a system of emptying the corves by tipplers, whereby two empty corves were pushed forward ready to go down the shaft, while at the same time the full corves were moved towards the tippler for mechanical unloading. The corves were unloaded automatically at the coal staith by a tippler.[282]

A contemporary account of Curr's invention in operation in 1794 was given by William 'Strata' Smith:

We found the small steam-engines much better applied to raising coal in Yorkshire than in Somersetshire, where not more than one (I believe), badly constructed was then in use.

Flat ropes were in use, and at Hisley Wood or Whitelane Colliery, on Earl Fitzwilliam's estate, I stood on one end of a cross-bar to which the corves were suspended, and Mr. Perkins on the other, and we were very smoothly let down a little more than mid-depth of the pit to see Mr. Curr's, so called, sliding rods; when, on being stationary and directed to look up, we saw the ascending corves over our head without knowing that they had passed us at the mid-depth enlargement of the pit.

On reversing the motion of the engine we were soon again on the surface, and simply and easily as this seemed to be effected, by grooves down from top to bottom of the pit, for opposite ends of the cross-bars to move in with an enlarged place mid-way for their passing, my learned friend could not understand the mechanism, though I was occupied nearly the whole of a long stage in explaining it to him.[283]

Another was given by Hatchett. On his visit to Sheffield Park Colliery in 1796 he wrote:

By an ingenious machine worked by water and invented by Mr. Curr two of these corves are drawn up and two let down at the same time, by the same machinery a Platform slides over the mouth of the shaft and the two full corves are pushed forward by two empty ones which are next to go down, and the full corves are then rolled by men to the edge of a Platform where by a discharging machine also invented by Mr. Curr they are emptied into carts etc. Each corf is guided by friction rollers during the descent and ascent in the shaft, and as these meet about the middle of the shaft this divides there somewhat in this manner. Mr. Curr and myself in one corf with Mr. Curr's brother and Mr. Savaresi in the other thus descended the shaft which was 25 yards deep in about ½ a minute whilst two corves went up loaded with coal.[284]

The system had its disadvantages. Hemingfield Colliery, in South Yorkshire, which operated from 1849 to 1920, used Curr's patented shaft conductors. The problems were threefold. Where the shaft widened to allow cage by-pass it was liable to become unstable, necessitating repair. Weakness in the shaft wall was also created by the use of numerous *'buntons'* (horizontal wooden beams) which were set into the wall in order to hold the conductors in place. Commonly, in Curr's time, the shaft served a dual purpose: winding and upcast ventilation. Ventilation was achieved by a coal furnace placed at the shaft bottom to create a column of rising hot air which, in turn, drew fresh air downcast. The third disadvantage was that the hot air in passing over the dense web of wooden buntons posed a risk of fire. When Hemingfield was used as a pumping station, from the 1930s to the 1940s, the shaft was re-lined with concrete.[285]

An elaborate description of Curr's system was presented on 30 August 1887, as part of the President's inaugural address to the Midland Institute of Mining, Civil & Mechanical Engineers:

GUIDES IN SHAFTS

The first attempt to introduce guides into shafts was by Mr. John Curr, 'Colliery Agent to his Grace the Duke of Norfolk at Sheffield Park and Attercliffe Common Collieries in the county of York'. This would be about the year 1787. He obtained for this invention His Majesty's Royal Letters Patent. Mr. Curr says that by his invention 'a greater quantity of coal can be drawn in a given time and from equal depths than has hitherto been done, and the dashing and wear of the corves, and waste at their meetings, are entirely prevented'. His plan was to place in the upper and lower parts of the shaft one set of conductors as far as necessary, and in the vicinity of the meetings the passing of the corves was effected by a long switch which opened when the ascending corf passed and closed when it descended, in a similar way in which wagons pass each other on an inclined plane at the surface when laid with three rails above and below meetings. To diminish the friction of the guides, rollers were fixed to what he called triangle balls, that is, the bar which supports the corf by two chains. A platform was drawn on an inclined plane over the shaft, on which the corf was landed, while the empty corf was placed on another platform and lowered onto the

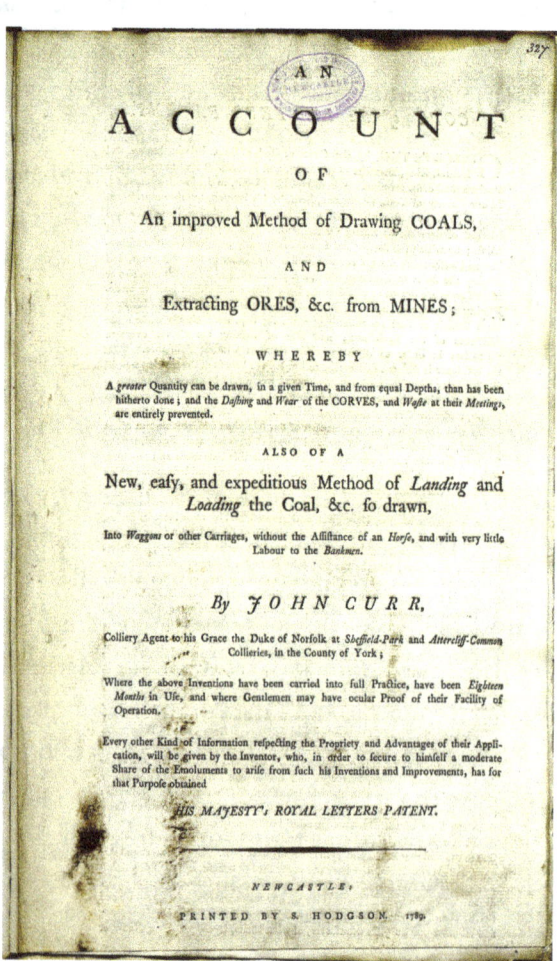

Figure 8.19. Title page of 1789 pamphlet describing patent no. 1660. From: TCR NEIMME Bell/3/327.

platform from which the loaded corf had been withdrawn. The original engraved details of Curr's patent are signed 'J.Buddle, Jnr.' This plan was adopted in many pits in Yorkshire up to 1860, with various modifications. It would therefore appear that previous to 1787 coals had been drawn to the surface with the ropes and corves free in the shaft. ... Indeed in the North of England and perhaps elsewhere, shafts not fitted with guides were so used for many years after the above date; indeed, till about 1833, in the North guides were not used, the ropes and basket corves hanging free in the shaft. Oftentimes the ropes twisted together and work was stopped until the ropes swung free. The men descended and ascended by placing a leg in a loop made on the chain by attaching the hook to a link of the chain above them. But now cages are attached to the winding rope and men walk into them. The cages on reaching the surface are rested on 'keps', and these are withdrawn when the cage is lowered. Subsequently to Curr's plan, iron rods were used instead of wood, screwed together like boring rods ...[286]

The Buddles, father and son, were quick to grasp the advantages. In 1789, less than a year after the granting of Curr's patent no. 1660, John Buddle Snr published a pamphlet, in Newcastle, *An account of an improved method of drawing coals.*[287] The title-page of the pamphlet is shown in Figure 8.19. It also contains two plates, both of which are signed 'J. Buddle Junr.', who was then aged 16. These, too, are shown in Figures 8.20 and 8.21.

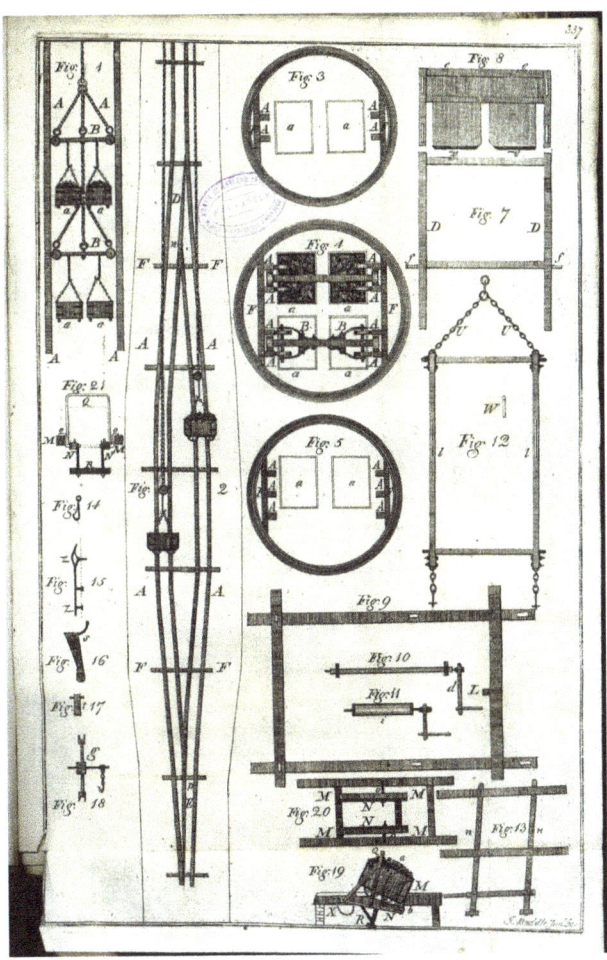

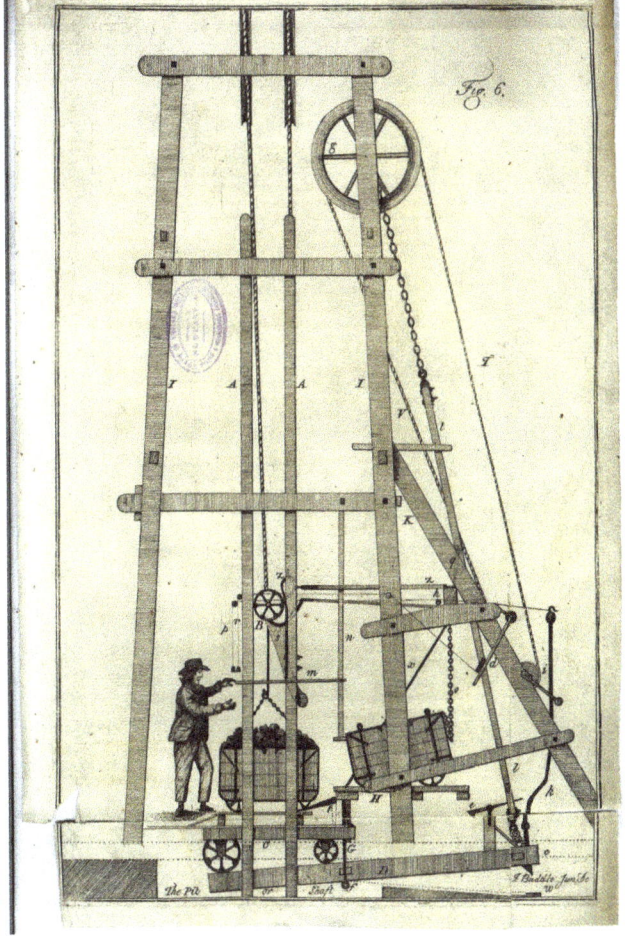

Figure 8.20 (left). Plate from 1789 pamphlet showing Figures 1–5 & 7–21 of patent 1660. From: TCR NEIMME Bell/3/337.

Figure 8.21. Plate from 1789 pamphlet showing Figure 6 of patent 1660. From: TCR NEIMME Bell/3/333.

A letter, dated 29 December 1800, written by John Curr II to Buddle Jnr at Wallsend Colliery reveals that the Buddles planned to install conductors there. The relevant part of the letter is as follows:

> *With respect to the Conductors you are about to try at Walls end, before we parted every thing was concluded upon, I think, excepting the Hanging on at the Bottom of the Pit, and having now had an opportunity of turning it in my mind, have in closed you a scetch [sic] which I have a high opinion of with respect to success, and doubt not but it will meet with your approbation. We had at first proposed the empty Corf lighting down perpendicular on the Bottom and the full Corf standing its own Breadth out of the perpendicular, which in its first ascent would cause a Dash agst. the Pit side, but what I now propose is that the empty Corf and full Corf shall each of them stand half their Breadth out of the Perpend[icula]r, by which means the dashing will be in a great measure done away, especially if the Hanger on lay his hand upon the Corf when it is ascending. The empty Corf will be thrown into its place when lighting down by the full Corf and the Hanger on will stand alternately at a and b to do his business and will never have to reach far into the Shaft. It strikes me that upon this plan the Hanger on will have much less work than he has at present, as the empty Corf will only be to move its own Breadth, and he will*

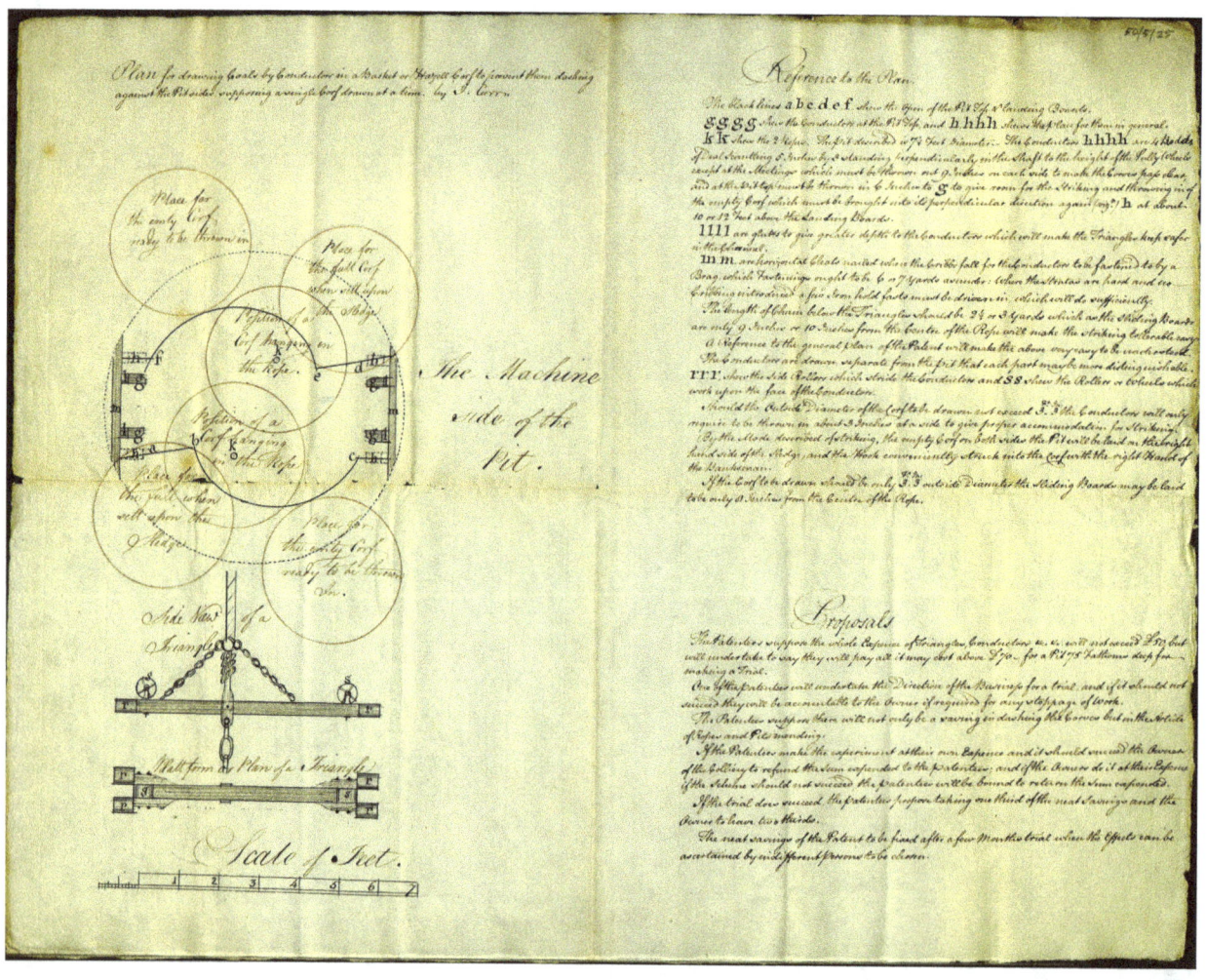

Figure 8.22. Plan, partial sketches and estimate for Curr's device (hand-drawn original from NEIMME collection). From: TCR NEIMME Bud/137/5/25.

have only one end of the Chain to move on one side which is now to be lifted about a good deal. The horse going through one side of the Center of the Shaft (about 2 or 2 ½ feet more into the Center than he goes at present) will on the Case of Conductors I hope be attended with no material inconvenience, but if any doubts are left a small chain 4 or 5 yards long to drag by would put him out of danger, and he will only have to move the Corves every other time.[288]

The 'scetch' mentioned no longer accompanies the letter, but the drawing shown in Figure 8.22 may give some idea.

In his *History*, Galloway elaborates on the description above of how men and boys descended and ascended the shaft:

Under the corf system, the arrangements for lowering and raising the men and boys were of the most unsafe and primitive description. Their security depended upon the tenacity with which they clung to the rope. The general practice, in ascending and descending, was for two men to sit each with a leg in the loop of the chain, and frequently five or six boys would cling to the rope, one above the other, trusting their lives to their capability of holding fast while the rope traversed a distance of 1,600 or even 1,800 feet. So inured were the boys to this hazardous mode of travelling, that it was regarded by them as 'fine fun'. Mr. Buddle relates an instance of a little fellow, who, in descending a pit 100 fathoms deep, fell fast asleep on the way, with his arms clasped round the rope, and in this condition was brought to the top again, when he was pulled off by the banksman, and even then was only awakened by a slap.[289]

John Buddle Snr, in his account of a miner who fell down a shaft, further describes the method of descending:

John Boys, a collier … going to his work very early one morning in 1763, and according to custom, on his turn to descend the shaft, in waiting to take the ascending hook, in order to his making a loop to introduce his thigh for that purpose …[290]

A verse of the *The Pitman's Pay* describes the impression made on young boys of winding at these speeds:

They popp'd us iv a jiffy down
Through smoke and styth* and swelt'rin' heat
And often spinnin' roun' and roun'
Just like a geuss upon a speet*

*styth = foul air, speet = spit[291]

The method continued until 1834 when the cage system was first brought into operation by Mr. T. Y. Hall.[292]

Curr claimed in *The Coal Viewer* (p. 12) that the speed of winding was

John Curr

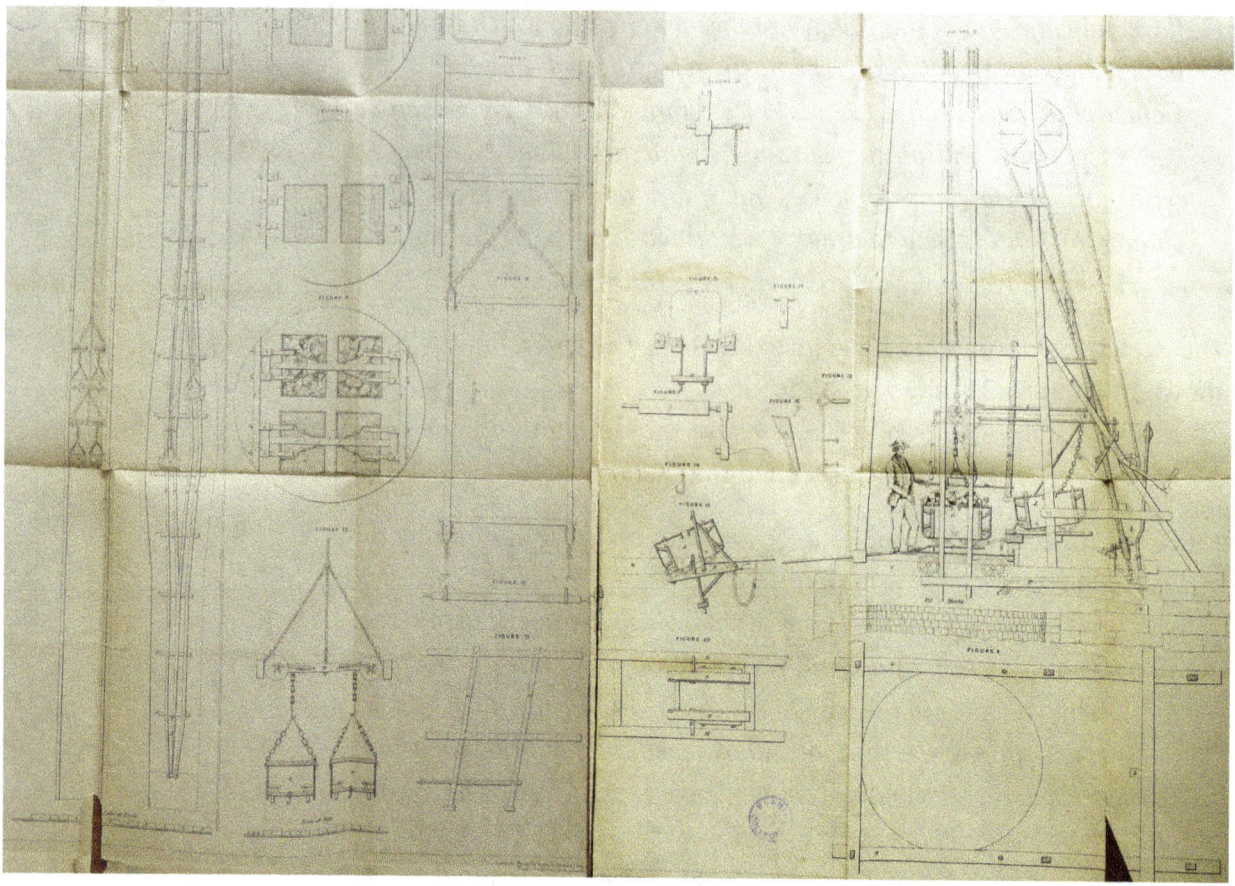

Figure 8.23. Photograph of the plate for patent no. 1660 (the print is 65 x 95 cm in size). From: SA/BC/9/8.

'*even through a space of one hundred and forty yards in half a minute*', a rate (840 feet per minute) unlikely in his day. In a letter to John Buddle Jnr dated 26th May 1803, John Curr III wrote:

> *…the Coal Engines have seldom been at a speed of less than 250 feet per minute our Engine at the Foundry does not perform its work if it goes less than 300 or 350 and in Staffordshire there are several engines which work upwards of 600 feet per minute.*[293]

Unlike for his plate rails and corf, he took out a patent for his conductors. This was the first of three which he took out during his time at Sheffield in the service of the duke.

Patent no. 1660 was published on 12 August 1788 [see Figure 8.23]: *Raising coals, lead, tin &c. out of mines [so as to prevent the corves running foul of each other: platform for landing and delivering]*.

The other two concerned rope and were: no. 1924, published 17th December 1792: *Method of working ropes and pulleys in mining operations &c.*, and no. 2270, published 17th November 1798: *Manufacture of flat rope*.

The inclined plane and the jinney

As part of these arrangements Curr made use of techniques which were, seemingly, rarely used at the time. C. F. Dendy Marshall tells us that:

> *The first appearance of the idea of causing full waggons in their descent to draw empty ones back is in Michael Meinzies' patent no.*

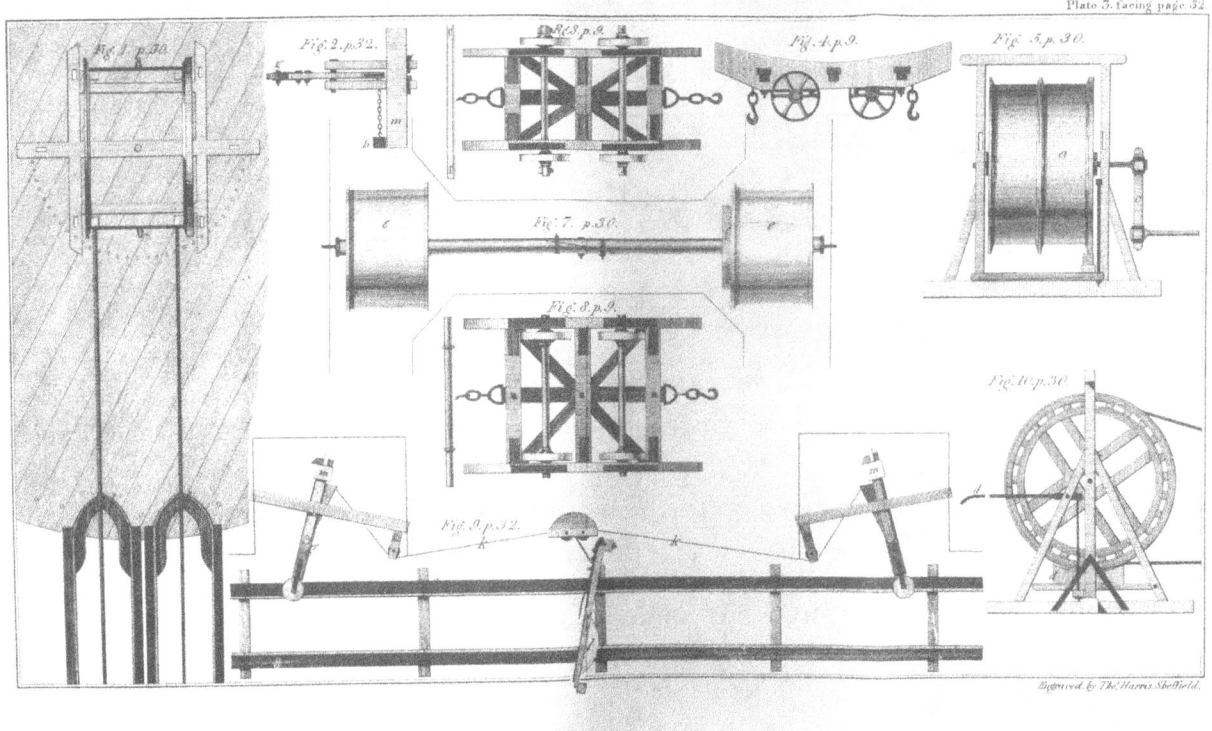

Figure 8.24. Plate 3, from *The Coal Viewer*, showing the jinney, at Figs. 5, 7 & 10.

653 of 1750 ...The beauty of the idea does not seem to have appealed to any one at the time; various people are given the credit in old books, Meinzies being apparently quite forgotten by the time his method of working inclined planes was put into practice.[294]

Galloway notes that:

A number of other improvements were introduced by Mr. Curr at the collieries under his care. Among these was the use of self-acting incline planes, whereby a train full of carriages in descending was made to pull up an empty train at the same time, the two being connected by means of a rope passing round a sheave, or pulley-wheel, at the top. By the employment of cast-iron rails these incline planes could be employed whenever the fall of the road amounted to three inches per yard.[295]

In *The Coal Viewer* Curr supplied instructions (p. 30) and estimates of expenses (p. 31) for the construction and use of jinneys. He describes two types: one for conveying coals above ground; the other below. He estimates the cost of the former to be £9 0s 4d, the latter £3 17s 6d.

Curr's instructions for conveying coals above ground illustrate his adoption of the inclined plane (p. 30):

Plate 3, fig. 5 [Figure 8.24], gives a front view of the jinney, the barrel (a) of which is 4 feet 6 inches in diameter. Fig. 10 shews the side view of it, and fig. 1, shews the platform of it, which turns upon a pin in the centre, and points to any required direction. A part of the planking for the corves to turn upon, and the points of the plates (bb) is also shewn. The ground on which the coals are stacked has a descent of

John Curr

Figure 8.25. Self-acting incline on Middleton Colliery waggonway. Plate from W. Strickland's *Reports on railways, roads etc., etc.* (1826), found in Tomlinson 1915.

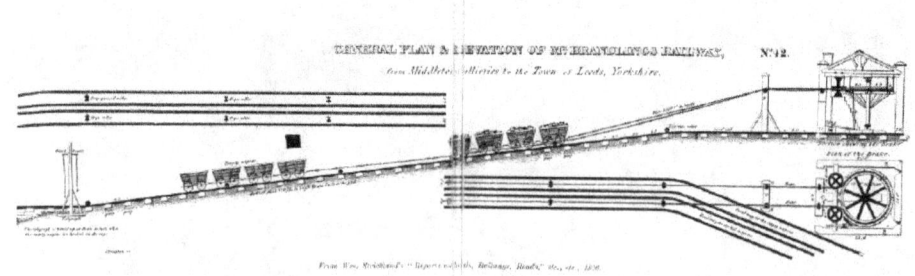

about 3 inches in the yard from the jinney, *and the momentum of the full corves going down the inclined plane, with the assistance of the communicating ropes, takes the empty corves back to the* jinney. *If power is wanted, there is a handle (c) to assist the jinney, and if it has too much velocity, there is a* brake *(dd) to retard its progress.*

The instructions continue for the underground version (p. 30):

Plate 3, fig. 7 [Figure 8.24], *gives a front view of the* jinney; *the two rope barrels (ee) are fixed in two inclining* board gates, *on which the corves pass, which are divided by a pillar of coal 4 yards thick. The ropes communicate round the barrel, and work upon the same principle as the* jinney *above ground, before described, and the narrow wheel (f) at one side of the rope wheel, is to retard its motion by the application of a* brake.[296]

Usually, the system employed was one by which the corves were attached to a rope wrapped around a wooden drum at the top of a gate. A crucial component of the device was the brake, necessary to control the speed of descent of the loaded corves. The system was applied both above and below ground.

The image shown in Figure 8.25 is an example of the use of an inclined plane.

In a letter dated 16th July 1800, Curr advised Buddle Jnr:

With respect to holding your Corves Down a descending Road where the jinny can't be introduced the best method you can take is to put the Horse behind the Corves and let him hold back with a Breechband and if he is still over forward you must Lock some of the Wheels.[297]

Mining transformed

Charles Hatchett's diary provides a description of Curr's innovations: his railroad; his corves; his conductors; and his tipplers in operation together. Hatchett relates that on 13 June 1796, after breakfasting with Curr, they went to Sheffield Park Colliery:

There is a level or opening by which persons may enter this work on foot but we were (to save time and trouble) taken to the mouth of the shaft by which the coal drawn up. …

Upon our arrival at the bottom we found some Boys and a Man who received the empty corves as they came down and hooked on those which were loaded in their room. The loaded corves are brought from the places which are worked for coal (and which are from the bottom of this shaft from 1 mile to 1 & ½ mile distant) these corves I say are brought along iron rail roads by a horse guided by a boy, and at every 250 yards, this boy and horse are met by another boy and horse either taking back the empty corves or bringing loaded ones. These boys make an exchange of corves so that each Boy and Horse goes and returns 250 yards in a stated place like a limited stage. Although each corf weighs about 250 [sic MS – 2½] cwt and when loaded contains between 5 and 6 cwt of coals yet by Mr. Curr's invention of iron rail roads by the rollers or wheels of the corves and by the passages being nearly perfectly level, one Horse is able to draw without difficulty from 12 to 14 loaded corves at one time, whereas before the Iron rail roads were used only two corves could be drawn at once. By this means 200 Tons of coals have been raised at this colliery in one day by the aid of 9 horses, and the machine above ground above mentioned. The corves are fixed one behind another by means of a Hook and Chain. As they are at present not full of work they only raise 150 Tons per Day. The places where the corves meet and exchange horses and drivers are called Pass By's because in these places only the corves can pass each other.

After we had walked a short distance from the bottom of the shaft we each got into a corf which being linked with others were drawn forwards in the manner already mentioned thro the passages which were just wide enough to receive the rail road for the corves and were thus transferred at the different Pass By's (to the number of nine) till we arrived at the working places about 1 mile and ¼ from the shaft where we had started. The height of these passages is about 5 feet and a half or near 6 feet.[298]

Here we see Curr's brilliance as an engineer and his significance in the history of mining. In stringing together his 'inventions' he revolutionised the transport of coal both above and below ground. Unfortunately, as will be seen, the Duke of Norfolk and his steward, Vincent Eyre III, who had allowed Curr such a free rein and provided financial support for his endeavours, were to benefit little from their introduction.

Underground canals

In John Buddle Senior's report of 7 April 1787,[299] he approved as *'peculiarly favourable'* a scheme of Curr's to move coal underground via canal at a proposed new colliery at Crook's Croft. However, there is no evidence to suggest that the scheme was adopted.

Earlier in 1787 Curr had visited the Duke of Bridgewater's Worsley Colliery in Lancashire, where canals were in use. Presumably enthused by this, Curr envisaged hoisting the wheel corves onto barges, or tub-boats, each containing twelve corves on an underground canal.

Ventilation

Another invention of Curr's, which is to be found on page 32 of *The Coal Viewer*,[300] was his *'Machine for opening doors underground'*. The benefit of this machine was to allow the automatic operation of ventilation doors, thereby dispensing with the need for trap door boys. As with Curr's scheme for underground canals, there is no evidence to suggest that such a machine was introduced.

John Buddle Senior's 1773 report refers to firelamps burning coal and level drifts for air and water.[301] Sheffield Park Colliery accounts for December 1783 show payments for trap door *'tenters'*. According to Hatchett, who visited in 1796:

> *Air shafts are sunk every ½ mile nearly and two are worked at a time, a fire being made at the bottom of one to keep up a current of air. Near the Working Place was a continuation of a passage into which we were cautioned not to go with candles as the fire Damp was there for want of proper circulation.*[302]

Medlicott notes:

> *The Norfolk collieries appear to have had few casualties as a result of gas explosions according to the number of payments ... recorded in the accounts ... Until 1805 candles and oil lamps were provided free ...*[303]

The 'fire engine'

From 1787, when the first steam pumping engine was installed at Attercliffe Common Colliery, Curr sought to improve their efficiency. The account and details that follow relate to the third engine, installed in 1789/90 at Attercliffe.

The longest section, over half, of *The Coal Viewer* is devoted to the *'fire*

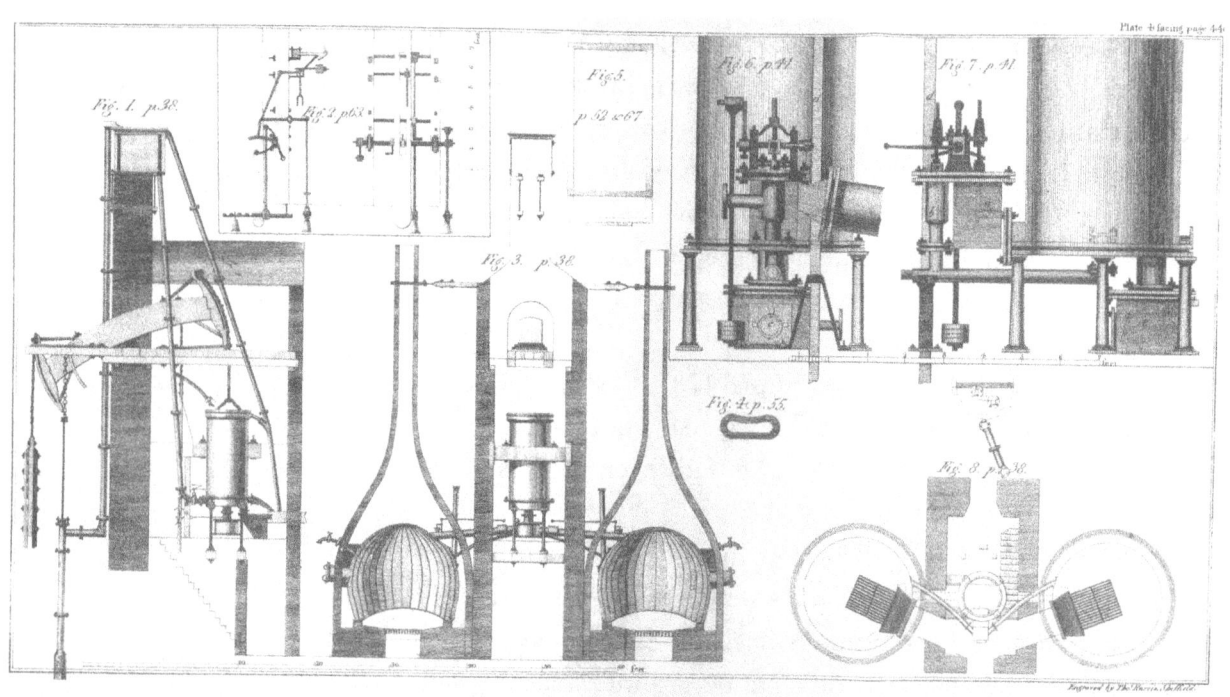

Figure 8.26. Plate 4: the complete engine in its engine house, with twin external boilers. From: Curr 1797 (1970), *The Coal Viewer*.

engine'. The section is composed of 59 pages, 26 ½ of which are tables. Curr describes the construction of a steam engine providing details, *inter alia*, of cylinders, pumps, pistons, regulator beams and pipes. The tables show weights and dimensions etc. of these.

There are also two plates. Figure 8.26 shows the complete engine in its engine house, with twin external boilers. Plate 5 [Figure 8.27] shows the individual components.

The entry for Curr in *A Biographical Dictionary of Civil Engineers* regarding *The Coal Viewer* states:

> *Although best known for its account of plate rails it also details the atmosphere or Newcomen-style steam engine in its most advanced form and it provides an early detailed account of boiler-making, discussing manufacture. Curr was a leading engine builder of the time...*[304]

In the preamble to the section on the *fire engine* Curr, who was keen to promote his expertise as a builder of common engines, having established his iron foundry at Sheffield Park, claims:

> *The various applications to which the fire engines under my care are adapted, have afforded me the opportunity of making several observations in this most useful of all machines, which I conceive to be of too great importance, to pass unnoticed, and have been so fortunate as to hit upon some deviations from the general rule in certain engines I have erected, which have produced an effect far exceeding my expectations, and which I flatter myself will be deemed worthy of engineers, as I can inform them, I have obtained a considerable addition of power, without any increase of fuel.*[305]

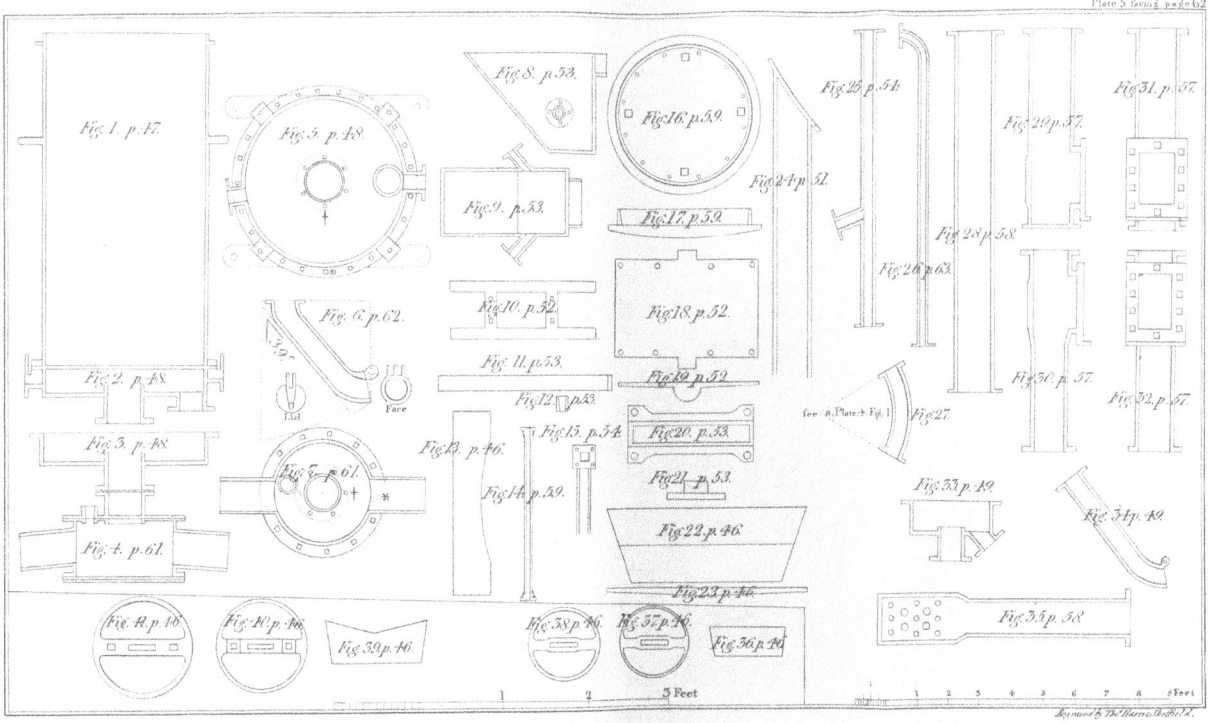

Figure 8.27. Plate 5: individual components of the engine. From: Curr (1797 and 1970), *The Coal Viewer*.

Two features, raising the injection cistern higher above the cylinder than heretofore, and the removal of the boiler from the engine house, are regarded as Curr's particular innovations. These are described by John Hunter:

> *The haystack-type boilers were positioned symmetrically on either side of the engine house, mounted snugly inside individual, elegant, brick-built 'houses' that combined the functions of boiler seat, flue and chimney in a single, decanter-shaped structure. The resultant improved efficiency of this had its downside. Boilers of that period were apt to fail within two or three years. There was even a danger of explosion should the interior water level fall too far. 'The Coal Viewer' offers no solution for servicing the boilers. Of necessity, servicing may have required dismantling the 'houses'. Such expense and inconvenience may have resulted in the boilers of a later era being uncovered, as shown in photographs, loss of heat being sacrificed for ease of repair. Another characteristic feature of the design is the greater elevation of the cold water cistern, which produced a better condensing spray. The engine house design improved the thermal efficiency of the boilers, but required lateral steam pipes to connect them to a common receiver beneath the cylinder.*[306]

John Farey, who saw the third engine at Attercliffe in operation in 1809,

> *seemed to regard Curr as the natural successor to John Smeaton [FRS; 1724–1792] for he published Curr's modified Smeaton tables on the different components of steam engines for cylinders of different size.*[307]

Farey's book also includes a plate showing the same sectional diagrams as in *The Coal Viewer* slightly redrawn, which is shown in Figure 8.28.

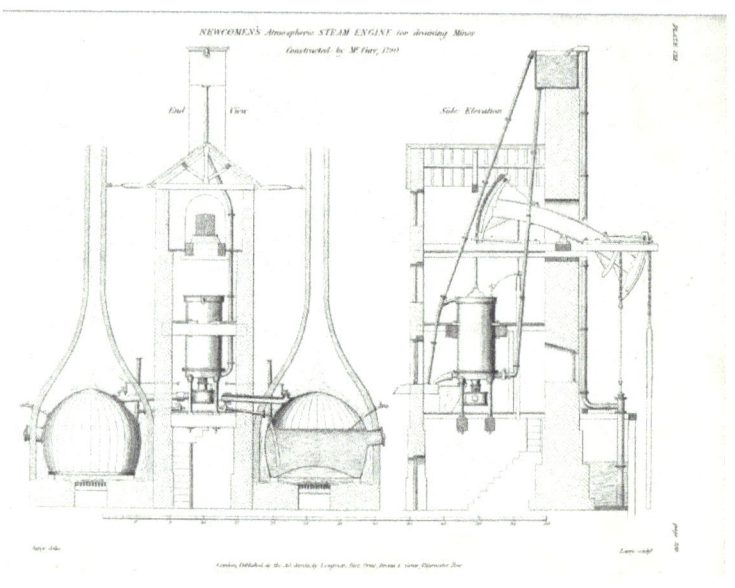

Figure 8.28. Newcomen's Atmospheric Steam Engine for draining mines; constructed by Mr Curr, 1790. From: Farey 1827 (1971), plate VIII, p. 205.

Farey's description (in part) follows:

> *The great working lever is composed of a single beam of oak timber, instead of being framed of several pieces, as was practised by Mr.*

Smeaton; …The injection-cistern is placed at the top of the lever wall, over the centre of the great lever, but at a greater height than in Mr. Smeaton's engines.

The diameter of the cylinder wall was 61 inches … The piston was capable of making 9 feet stroke, but usually worked at 8½ feet, and made 12 strokes per minute …

It worked five sets of pumps [(1) 13 in. dia., 24½ fathoms lift, weight of column of water 8480 lbs. (2) 13⅛ in. dia., 23½ fath., 8290 lbs.(3) 15¼ in. dia., 2302 lbs., (4) 15 in. dia., 5¼ fath., 2500 lbs.]… the injection-pump was 9 in. dia., 10 fathoms lift, weight 1657 lbs. suspended at 8 ft. from the centre of motion. The weight of all these columns combined, was equivalent to 20 434 lbs. resistance to the motion of the piston, or 9.12 tons, and at the rate of very nearly 7 lbs. per square inch.

The mechanical power exerted by this engine was more than 63 horse-power …

The engine had two boilers, on Mr. Curr's plan, of 14½ ft. diam., the furnaces of which consumed 10 hundred weight (=1120 lbs) of sleck [sic.], or small coals per hour,… and supplied the engine very fully with steam.

The boiler is surrounded by a dome of brick-work, built in the form of a bottle or decanter, with a tall neck rising from it, over the centre of the boiler to form a chimney. This brick-work does not touch the boiler in any part, but leaves a space of 9 inches all round between the boiler and the brick-work, for the flame to act in, as it rises up through the intervals between the 12 brick pillars from beneath the bottom, and the flame ascends on the outside of the boiler to the chimney, which is over the centre of the boiler. The internal diameter of the chimney is 22 inches and the height to the top is about 48 feet above the fire-grate.

In this way, the flame acts on the boiler in ascending, instead of circulating horizontally around it; and its action is not confined to the lower part of the boiler, containing the water, but the upper part, containing the steam, will also receive a share of the heat.

The injection-cistern is placed 36 feet high above the top of the cylinder, in order to give a greater velocity to the injection.[308]

Rise and fall

The years from 1786, when work began on a new colliery at Attercliffe Common, to 1801 saw the peak of investment in the duke's Sheffield collieries. They also saw the events which were to lead to Curr's dismissal in 1801.

John Hunter presented a paper in 2021 at the 2nd International Early Engines Conference which delves into the history of Attercliffe Common Colliery in great detail. The text and images in this section are taken, with

John Curr

Figure 8.29. Map showing the relationship between coal seam outcrops, the Rivers Don and Rother, and the manorial/township boundaries in the area from Attercliffe to Orgreave. From: Hunter 2021, Figure 1, p. 54; © John Hunter.

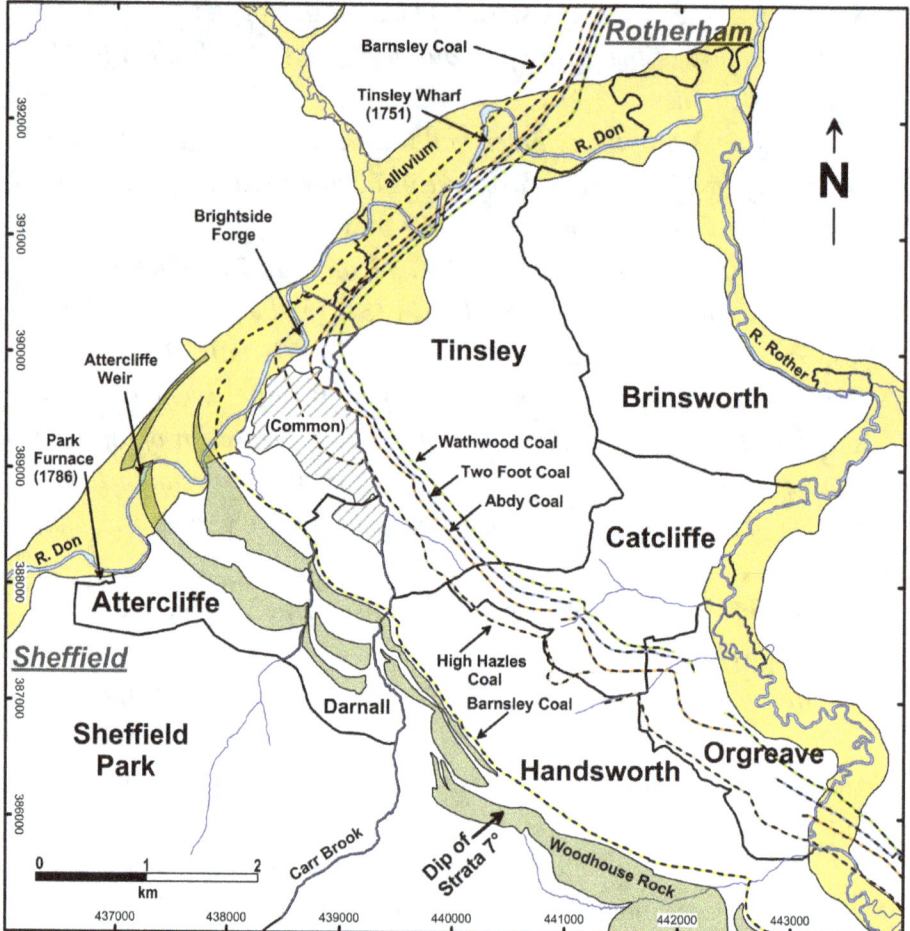

Figure 8.30. Map showing coal working in the Attercliffe–Darnall Coalfield, 1740s–1760s, in relation to the manorial boundaries, the limits of the commons, and the location of selected freehold property. From: Hunter 2021, Figure 2, p. 58; © John Hunter.

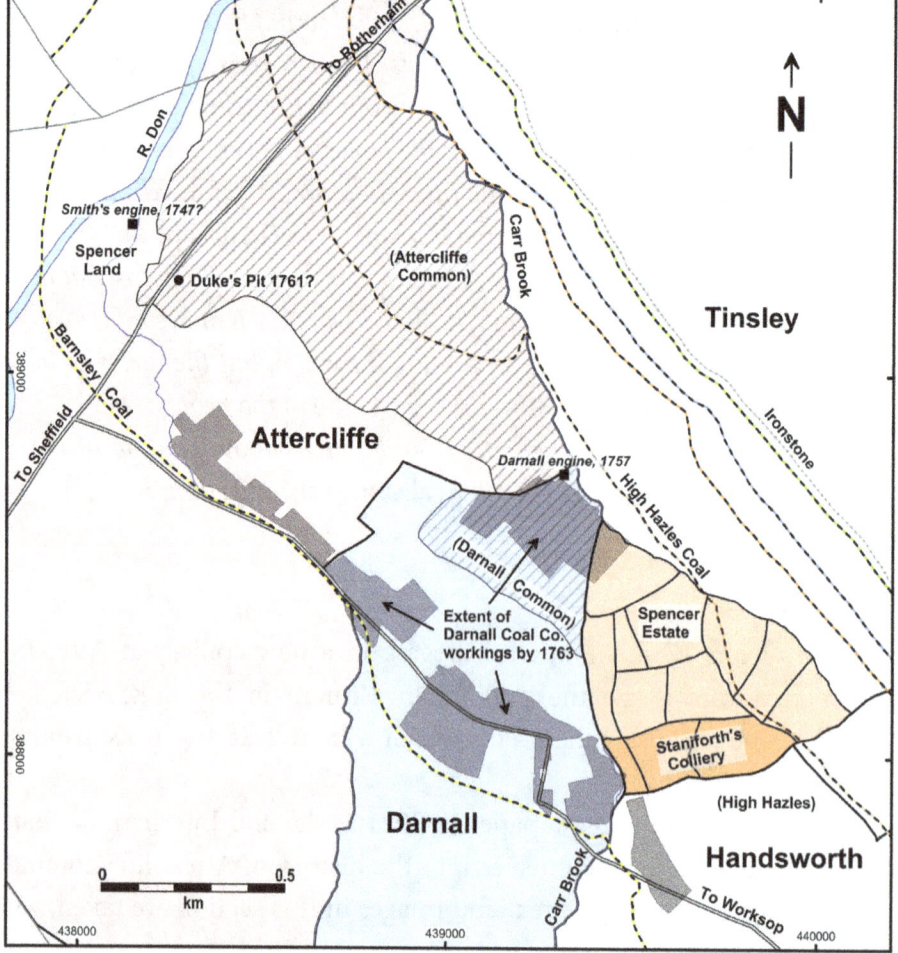

permission, from Hunter's 2021 IEEC paper. It is mostly quoted verbatim, with some minor paraphrasing, selections or edits (page numbers are included after quoted texts). Any exceptions are noted in the text.[309]

Hunter prepared two maps for this paper, showing the coalfield and surrounds [see Figures 8.29 & 8.30]. He summarised the problems encountered in the field as follows:

> *The history of the Attercliffe-Darnall Coalfield ... illustrates the consequence of uncoordinated development of adjacent early coal workings, opened in the same seam, by separate partnerships acting independently. The Darnall Coal Company initiated a problem of surface water infiltration by connecting shallow workings near the outcrop to a deep pumping shaft. When the pumps in that colliery shut down, the workings quickly became waterlogged and the boundary barriers of intact coal were of insufficient width and strength to prevent water migrating laterally into Attercliffe Common Colliery. Possible deliberate robbing of the coal barriers on both sides of Thomas Staniforth's small mine resulted in a similar problem in the other direction. The relentless inflow of water into the mine workings required far more steam pumping capacity than had originally been anticipated.*
>
> *John Curr's new 'model' colliery at Attercliffe Common, fully-equipped with his expensive inventions for efficient mining, was the principal victim, having been forced into premature closure with the loss of reserves and revenue. The unfortunate 11th Duke of Norfolk and his steward, Vincent Eyre, were the principal financial victims. John Curr was dismissed by the duke after the debacle. [from p. 73]*

Development of the duke's collieries, 1786–1801

> *In Curr's report of 1784*[310] *he noted that Clay & Co, whose workings at Darnall abutted those at Attercliffe Common, were still in production and that their coal was delivered directly to customers in the town of Sheffield, i.e. avoiding the duke's coal stage in Sheffield Park. ... By 1785, steel furnace owners in Sheffield were complaining about shortages of hard (Barnsley) coal, while coal carriers also complained about the hours wasted while they waited for their carts to be filled at Darnall Colliery or, even worse, returning empty to Sheffield.*
>
> *Several requests were made to the duke by third parties for a lease of the coal that he owned beneath Attercliffe Common, so that a new colliery could be opened. He was advised, presumably by Vincent Eyre III,*[311] *to reject these approaches because the terms being offered would not compensate for the competition his colliery in Sheffield Park would face from an increase in the supply of the by-product 'small coal'. It was proposed instead that the duke should consider opening his own colliery on the Common, and legal advice was sought about his right, as Lord of the Manor, to occupy part of the common waste*

land used by local residents to erect his fire engines etc.[312] *The duke was eventually persuaded to make the investment, and work began in June 1786. [The expenditure on the colliery between December 1786 and June 1790 was approximately £13,822 16s 11d.*[313]*] [from pages 61–62]*

1786–1789: John Curr's first two engines at Attercliffe Common Colliery

John Curr's account book for the Attercliffe Colliery new winning (mid-1786 to mid-1790) shows the sums of money paid for materials, engine components and labour, but unfortunately there is no description of how the work was undertaken.[314] *The sinking of two new shafts (the engine and bye shafts) began immediately in June 1786.*

When shaft sinking began, gin-tubs had been sufficient to control water seepage, but as the depth increased, so did the flow rate, and a steam pumping engine was needed to control it. In January 1787, a sum of £300 was paid to Walkers & Co. of Masbrough, while in April, £592 2s 6d was paid to Hartop & Co. of Brightside. These two unspecified payments to two local iron-founding companies are the largest individual items of expenditure that were incurred during the four-year construction period, and they are probably associated with the first engine purchase. Two boilers, made from 120 cwt of wrought iron supplied by Hartop & Co., costing £210 14s, were paid for in August 1787. An invoice for £383 for an engine house was paid in January 1788.

The new engine had not been in operation for very long before a second engine had to be ordered to cope with increasing volumes of water as shaft sinking progressed. This occurred during 1788, but its component parts cannot be easily identified among the regular orders for ironwork in the accounts.

Vincent Eyre invited John Buddle to visit Sheffield in April 1787 to advise on adopting Curr's innovative ideas at a proposed new winning at The Ponds, in Sheffield. Buddle also visited the Attercliffe winning, but sinking operations were still ongoing.[315] *He made a repeat visit in April 1789, when he recorded some details about the 'old engine'. It had a cylinder diameter of 50 inches, a piston stroke of 7 feet 5 inches, and it operated two 12 inch pump sets, each 48 yards in length, giving a shaft depth (including the sump) of 96 yards. Steam was supplied by two boilers, of 13 and 12 feet diameter, respectively. The engine also worked a jack head pump, drawing condensing water from a bore, 12 yards deep, and two additional pumps for drawing water from separate bores to power a water wheel used for coal winding [an example is shown in Figure 8.31].*[316] *The 'new engine' had a cylinder diameter of 46 ½ inches, with a piston stroke of 6 feet 10 inches. It also operated*

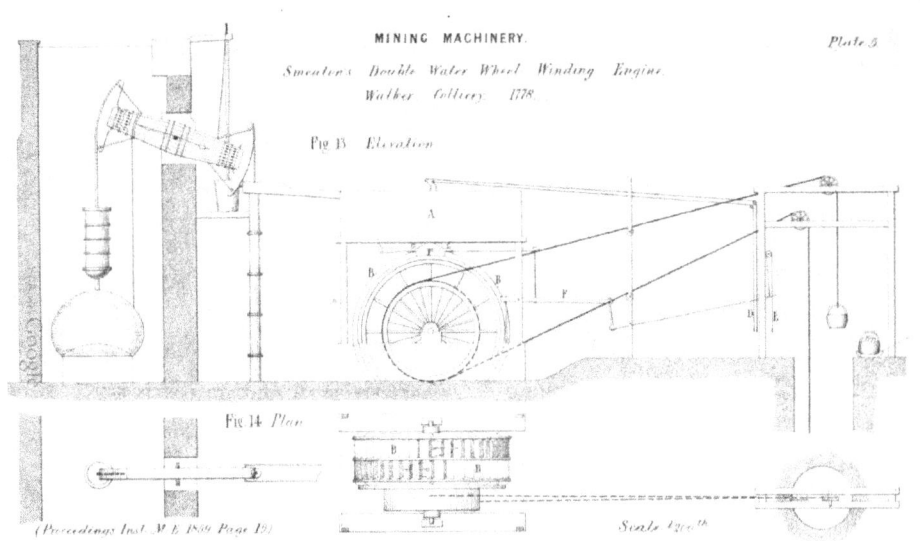

Figure 8.31. Example of a water-wheel winding machine from 1778. From: Taylor 1859, p. 14.

two 12-inch pump sets extending to the same depth as the first pair, plus three additional short pumps. Its boilers were both 12 feet in diameter.

Both the Attercliffe engine houses had twin external boilers and are shown in plan-view on several maps drawn by Fairbank, facing each other over a common pumping shaft [see Figure 8.32].[317] No contemporary drawing or painting of these engines (in perspective view) is known to exist. The sectional diagrams published in John Curr's book in 1797 and also in John Farey's book in 1827 belong to the third Attercliffe engine, although it has to be assumed that the first two engines were of a similar design.[318]

...

Except for the single payment made to Walkers & Co. in January 1787, all the routine purchases of manufactured metal items (including those made from brass and lead) were supplied by Messrs Binks, Booth and Hartop. These included other engine components, boilers, pump barrels (bored), buckets, windbores, clacks, spear plates, common pipes, steam pipes, chains, cog wheels, balance weights and large quantities of cast iron rails and corf wheels. Charges for new sand mould templates were common.

The introduction of Curr's innovations, the shaft winding system with conductors and corf by-passing, wheeled corves running on cast-iron plates, and counterweight jinneys and tipplers, involved a large financial investment. Capital repayments and interest charges, coupled with heavy pumping costs, made the venture economically vulnerable. [pp. 62–64]

1789–1790: John Curr's third engine at Attercliffe Common Colliery

By April 1789, Vincent Eyre had become concerned about the expenditure at Attercliffe Colliery, and also about the risk posed by an over-supply of small coal. This would depress the sale price in Sheffield, and therefore affect the profitability of all the pits. He invited

John Buddle to come to Sheffield again for consultation, accompanied by John Stephenson, a respected mining engineer from Kimberworth, near Rotherham. He needed independent advice to help decide whether further capital spending at Attercliffe Colliery was justified, or whether the venture should be abandoned and the investment written off. To continue in production, making only minimal profits, would squander the value of the duke's mineral estate. To abandon the colliery, however, would restrict the steelmakers' business in Sheffield, and would affect the whole economy of the town, which would have a consequent impact on the duke's income from property rents.

...

Buddle inspected both the Darnall and the Attercliffe collieries. He found that the roof was weaker and leakier at Attercliffe, requiring the top coal to be sacrificed to keep it sealed. He also noted that rainwater collected in crown holes above collapsed basset workings, and percolated directly into the seam. Heavy rainstorms caused water to rise quickly in the sumps at the pumping pits, requiring the engines to work harder. Joseph Deakin had already been forced to invest in a second engine at Darnall Colliery.

...

Buddle and Stephenson understood the critical financial situation facing Vincent Eyre and the duke, but were reluctant to recommend abandoning Attercliffe Colliery. They proposed that investing in a third, much larger engine to increase the total pumping capacity was justified, and estimated that an engine with a 70 inch cylinder, in a new house, fed by twin 17 ft. diameter boilers, would cost £2,100. The cheaper option of replacing the existing 46½ inch cylinder engine with a 70 inch cylinder machine was also costed, with the smaller engine being moved to Sheffield Park. The recommendation to purchase a third pumping engine was accepted and the cost estimate, including sinking a new shaft, was revised to £3,000.[319] John Curr's accounts show that payments were made for engine components and an engine house between August and November 1789. The cylinder (74 cwt) cost £93 3s, its separate bottom (24 cwt) cost £30 2s, and the piston (21 cwt) cost £15 2s, all supplied by Binks & Co.

Sectional diagrams of this engine were published in John Curr's book [The Coal Viewer] in 1797 [see Figure 8.26]. Instead of 70 inches, the cylinder diameter had been revised to 61 inches, the piston stroke was 8 feet 6 inches and steam was supplied by the usual pair of external boilers. When operating two primary pump sets, 13 and 13½ inches respectively, plus three additional short pumps, it worked at a rate of 12 strokes per minute. The combined weight of water lifted during each stroke was just over 9 tons, powered by 7 psi of effective (atmospheric) pressure on the piston. It delivered 34 gallons of water per stroke. Additional technical details are provided in Curr's book.

[…] A plan of underground workings at Attercliffe Colliery produced by Fairbank in 1790 shows the three engines located together in a group, pumping from two shafts [see Figure 8.32].³²⁰

Appendix 5 shows Newcomen-type engines built in the Norfolk collieries during the period 1747 to 1820.

1790–1797: Staniforth's, High Hazles and Dore House collieries

*Weekly accounts from 1790 to 1795 indicate that Attercliffe Common Colliery quickly became an established coal producer, but there is no information about the performance of the pumping engines.*³²¹ *Contrary to Vincent Eyre's fears, demand for coal in Sheffield began to increase in the early 1790s, but this only served to attract new*

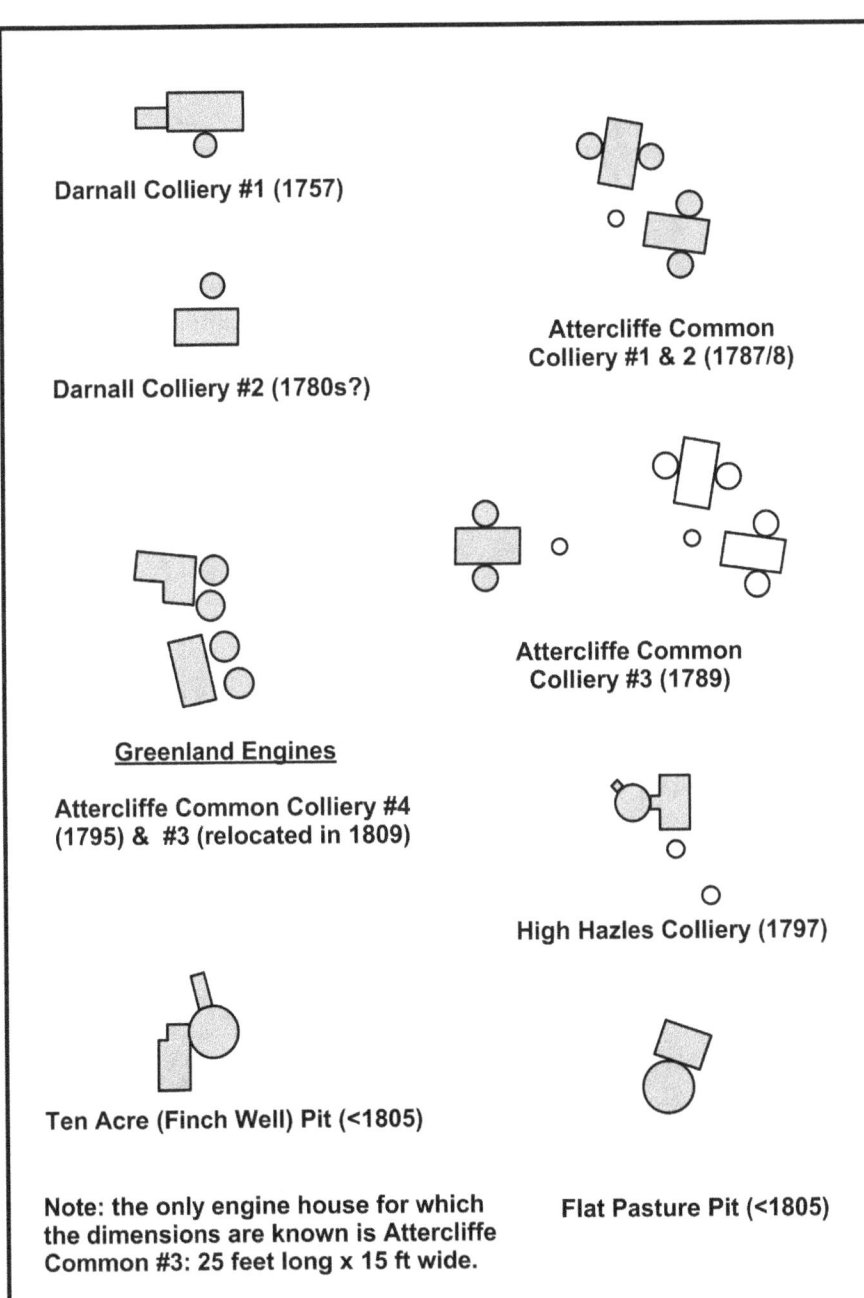

Figure 8.32. Plan view of the colliery engine houses (with external boilers) installed in the Attercliffe-Darnall Coalfield, as depicted on various maps prepared by the Fairbank family of surveyors. None of the sketches are scaled, and no other images of these buildings are known to exist. From: Hunter 2021, Figure 8, p. 73; © John Hunter.

John Curr

competitors to the Darnall coalfield. Surviving legal documents provide much detail about this activity, but the exact sequence and timing of the main events is still unclear. Unfortunately for Attercliffe Colliery, the uncoordinated development of new underground workings exacerbated its drainage problems.

The colliery workings of Messrs Clay, Deakin and Phipps had already been extended beneath land held by the Spencer family in the late 1780s.[322] Their deepest workings were contiguous with the old Darnall Colliery, but further expansion eastwards into the adjacent High Hazles Farm was blocked by a 16 acre, wedge-shaped, parcel of land that separated the larger properties [see Figure 8.33]. It was owned by a merchant called Thomas Staniforth. […] Refusing to sell his coal estate to Clay and Co., he decided to employ his own colliers to work it for himself. Having some slightly shallower coal, he had hoped that a horse-gin would be sufficient to manage the water, but rising ground away from the basset meant that his deepest pit reached 96 yards in depth, similar to that of Clay and Co.'s.[323]

Sinking the deep pit had involved excavating through a fractured sandstone bed which, unfortunately, was a strong aquifer. The single horse-gin was used to raise coal during the day, and could only be applied to lifting water tubs in the early mornings and the evenings, making it inadequate for keeping the mine dry. Staniforth's miners

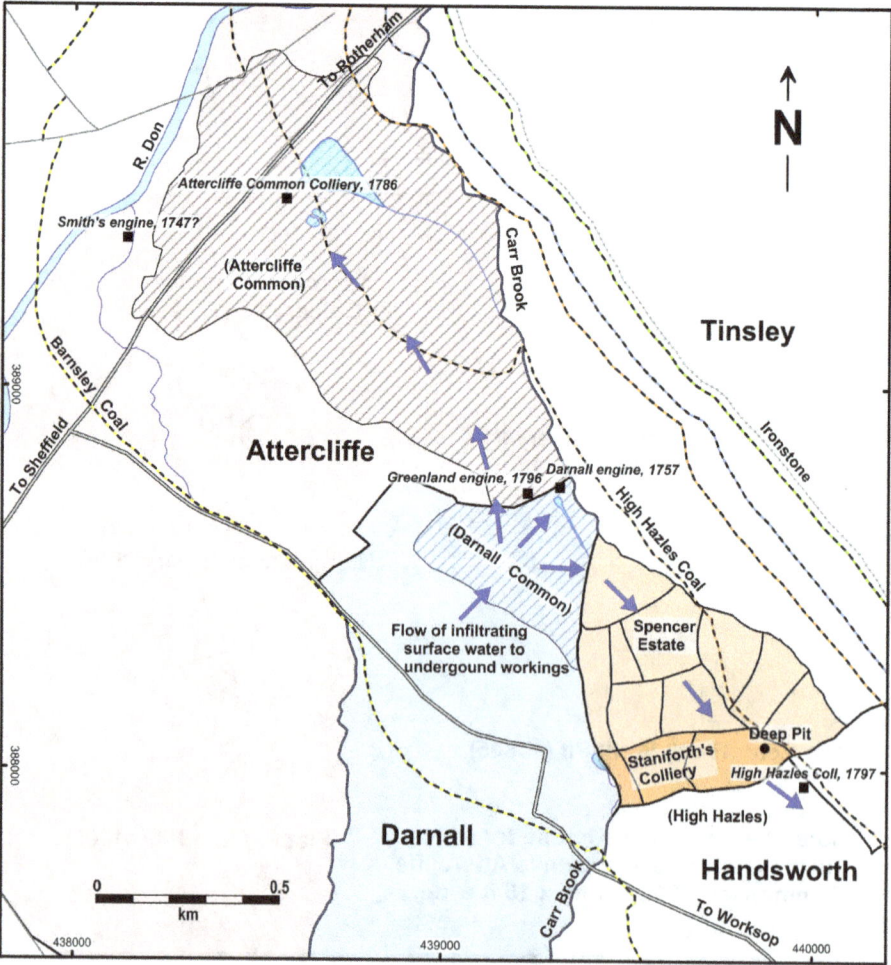

Figure 8.33. Map showing the underground flow directions of surface water that infiltrated the interconnected coal workings in the Attercliffe-Darnall Coalfield, 1790s–1820s. From: Hunter 2021, Figure 5, p. 67; © John Hunter.

should have retained a 13 yard-wide perimeter barrier of unworked coal around their workings. Whether intentionally, or through negligence, they robbed parts of the barrier, and holed through it in two places, enabling water to conveniently drain out into the adjacent colliery workings, which happened to be served by pumping engines.

In 1792, a new consortium comprised of steel refiners and cutlers, joined together to open their own colliery on Dore House Farm, on the east side of High Hazles, next to the Orgreave estate.[324] Its drainage system has not been documented, but being located much closer to the River Rother than the Don, it would have made sense to discharge mine water in the direction of the Rother Valley via a separate sough. [...] By October 1795 [...] the Dore House colliers were struggling to control inflowing water.[325]

A memo written by John Curr in July 1795 reported that Messrs Clay & Co intended to abandon Darnall Colliery. If their engines stopped, the mine would quickly flood, and rising water would exert pressure against the coal barrier separating it from Attercliffe Colliery.[326] In November 1795 William Dunn commented [...] that Deakin's men were 'drowned out', implying that pumping had definitely ceased at Darnall Colliery.

In August 1796, Dunn wrote that Joseph Deakin had begun to sink an engine shaft in Fitzwilliam's land at High Hazles.[327]
[...]
The High Hazles shafts reached the Barnsley Coal at a depth of 98 yards in March 1797, and production began immediately. Shortly afterwards, Deakin offered to sell the pit, together with all the unworked coal he held under lease, to the Duke of Norfolk. The duke agreed to buy it for a sum approaching £10,000, providing he could be assured of the integrity of the coal barrier surrounding Staniforth's Colliery.[328] This assurance was apparently given, presumably by John Curr. It was around this time, in the spring of 1797, when the first breech in Staniforth's barriers was discovered. None of the subsequent legal depositions give any indication that Deakin knew about the breach prior to the sale, but the timing is a suspicious coincidence. A steam engine was erected at the High Hazles shaft.

1795–1799: John Curr's fourth engine at Attercliffe Common Colliery (the first Greenland engine)

Darnall Colliery began to fill with water as soon as its engines were shut down, in the second half of 1795. As predicted, increasing pressure caused leaks through a weakened coal barrier into Attercliffe Colliery. Joseph Deakin had previously offered to sell Darnall Colliery to the duke for £4,300, including all its machinery and the residual unworked coal, but Curr advised against this because its shafts and engines were in poor condition and the latter were unlikely to cope with the pumping demands of the coming winter. He proposed instead that a new shaft should be sunk at the boundary of Attercliffe Common, just inside the coal barrier, 70 yards from the Darnall engine shaft. A new engine erected there, with a 50 inch diameter cylinder, would cost £2,000, including the shaft, although the exact cost would depend upon the flow of water encountered during sinking. A single engine, operating a pump-set with a 14 inch working barrel, would be able to intercept water at a depth of 63 yards and prevent it from descending to the 93 yard deep level in Attercliffe Colliery. To draw an equal amount of water at Attercliffe Colliery would require two additional engines and cost considerably more to pump it back to sough level.[329] The scheme was approved and the account books show that work began in August 1795.[330]

The new engine was located at a site that subsequently became known as the 'Greenland Engine', from its proximity to an eponymous farm, even though the old Darnall engines were closer to the farmhouse [see Figure 8.33]. As far as Attercliffe Colliery was concerned, it was referred to as the 'new', or 'fourth' engine. Few details of this engine were recorded ... [from page 68]

1799: Abandonment of Staniforth's Colliery and the sale of Dore House Colliery

Water accumulating in Darnall Colliery also began to flow in the opposite direction, running through Staniforth's mine into High Hazles Colliery which, by that time, belonged to the Duke of Norfolk.[331] A wooden 'frame dam' was built to block the flow, but it could not be kept watertight. Water squirted through the coal and 'rained' through the roof. Compounding these problems, the River Don flooded again in 1797, to 'an amazing height', indicating torrential rainfall. The river flooded once more in August 1799.[332]

The second breach in Staniforth's barrier was discovered in December 1799, when it was realised that wheeled coal corves, which normally rolled continuously along a surface tram-road, began to stall at a particular location, where almost imperceptible subsidence had affected the gradient. The underground dam was strengthened by backfilling the workings with compacted earth and stone. Unworked coal surrounding the breach had to be abandoned because of the hazard.

The water level in the Attercliffe shafts also began to rise soon after the discovery of the second breach, despite the efforts of a fourth pumping engine. Faced with the prospect of continually-increasing costs, Curr recommended that the Attercliffe engines should be restricted to raising water from a depth of only 40 yards, probably by disconnecting the lower lift of pumps. This would result in the

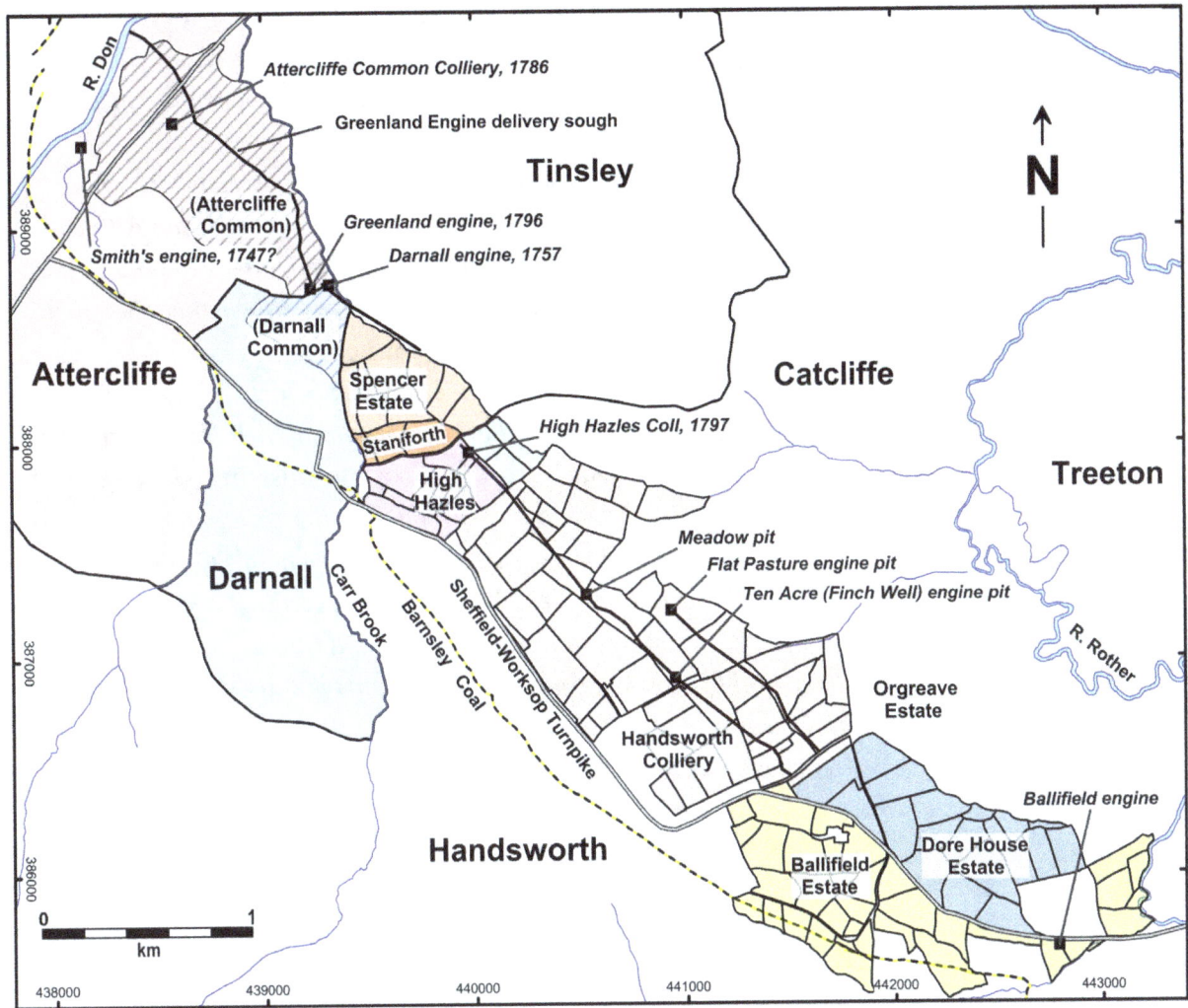

Figure 8.34. Map showing the extension of coal working from Attercliffe to Orgreave, early 19th century, the location of known pumping (or winding) engines, and the distribution of related freehold estate property. From: Hunter 2021, Figure 6, p. 69; © John Hunter.

abandonment of 20 to 30 acres of unworked deep coal, amounting to a commercial loss of between 6,000 and 12,000 guineas. One quarter of this sum was royalty owing to the duke, who had already suffered losses from the stolen and abandoned coal surrounding Staniforth's Colliery.

Recovery of the Attercliffe deep coal was prioritised until the decision was made to remove the pumps, beginning in March 1800.[333] *In April Curr changed his mind about the integrity [of] the High Hazles frame dam and recommended an independent investigation. This involved William Dunn who, in a private letter to his son, blamed John Curr for the current state of affairs.*[334] *In January 1801, John Jeffcock, who was William Dunn's son-in-law, was appointed the Duke of Norfolk's new agent for High Hazles Colliery.*

The proprietors of Dore House Colliery were also struggling to manage their water and they decided to invite Vincent Eyre to buy their investment for £10,000, but the offer was declined.[335] *Their colliery was separated by a leaky coal barrier [...] having a supposed width of 60 yards [see Figure 8.34]. In February 1800 Eyre instructed John Curr to inspect the colliery and estimate its value. Curr was critical of the poor condition of the workings and advised that further capital expenditure was essential.*

[...] the proprietors eventually accepted Eyre's counter-offer of £5,000 for the colliery and the residue of an 80 year lease on 110 acres of coal.[336] *[...] Vincent Eyre [...] died [...] in April, aged 57. His son, Vincent Henry Eyre, became the duke's new steward [...]*

Staniforth was forced to abandon his small colliery, probably around September 1800 [...] It appears, therefore, that the Attercliffe engines had been shut down completely by then, after little more than a decade of operation.[337] *It is not clear, however, whether the fourth engine at Greenland remained in operation. [...]*

In April 1801 John Buddle came back to Sheffield to assess whether the leaky barrier separating the Staniforth and High Hazles collieries could be strengthened. [...] The duke, by this time, had become suspicious of Curr's proposal for repairing the frame dam and also with his ambitious engineering schemes in general. On May 1, the duke and Vincent Eyre's heirs began legal proceedings against Thomas Staniforth, claiming compensation for additional pumping costs, stolen and abandoned coal, and other damage caused to their colliery by the continual leakage of water.[338] *On 14th October 1801, Vincent Henry Eyre dismissed John Curr. Dore House Colliery closed in 1802.*

The dispute between the duke, the Eyres and Staniforth continued into 1803, when the parties agreed to a settlement by arbitration. These proceedings concluded in November and, after reviewing the evidence provided by Staniforth, the outcome was that no damages or costs were to be awarded to the plaintiffs. The duke and the Eyres were understandably outraged and submitted a legal challenge at the York assizes on 24th December.[339] *Unfortunately for them, Thomas Staniforth died on 15th December, aged 68, and the case could not be transferred to his heirs. [from pages 69–70]*

Dismissal

Curr's dismissal in late 1801 'without any reason being offered',[340] came as a surprise to him.

Why was Curr dismissed? Significant capital investment was required to solve the problems arising from rising water levels at Attercliffe Colliery and the leaky barrier separating Staniforth and High Hazles collieries. The duke, resistant to this expenditure, was presumably aware that Curr would find it hard to accept an overall cut in capital investment and the implied end to his quest for increased efficiency in mining. Perhaps, too, the duke and Vincent Eyre III's heirs (his widow, Catherine, and his brother, Thomas) were at the end of their tether and wished to see some return on their investment. Profits at the duke's Sheffield collieries had reached a peak in 1793 but thereafter declined. Indeed, Attercliffe worked at a loss between 1794 and 1797. The colliery did not return to profitability until 1800/01. Profits at Sheffield Park and Manor collieries fared little better.[341]

Another possible explanation may be that Curr and Vincent Eyre, who had worked together for fifteen years, had established a good relationship. Eyre died in April 1801 and it can be speculated whether, had he lived, Curr would have been dismissed so abruptly later that year.

Whichever, after Curr's dismissal the partnership withdrew £21,000 from the collieries between 1804 and 1805.[342] By 1805 they had extricated themselves altogether. Hesley colliery was sold to Richard Swallow for £2,500 in 1804,[343] and in 1805 the Sheffield collieries were leased to a consortium of local businessmen together with a payment of £72,500.[344]

While stressing that Curr was a highly proficient mining engineer both Medlicott and Lock cast doubt upon his abilities as a manager. In particular:

> *By 1800, and notwithstanding Curr's innovations, the Sheffield collieries were again operating at a loss ... Curr's abilities as a manager were limited. Faced with rising costs and falling profits from 1791, he had failed to balance investment with financial return, focusing too much on the experimentation and implementation of his new technologies. This said, Curr's inventions significantly extended the working lives and productivity of the mines in Sheffield ...*'[345]

This may well be true but, in Curr's defence, allowances must be made for the effects of flooding and underground skulduggery. These were beyond his control and may well have defeated the most able of managers.

Curr's letter and the duke's response

Curr's immediate reaction was to write a long letter of self-justification, dated 23 October, to the duke: *Report of Inventions introduced by John Curr in the Sheffield and Attercliffe collieries; and reasons why collieries of late have been unproductive.*[346] His letter 'vigorously protests' the duke's decision, 'unconscious as I was of any cause for so unexpected a dismissal'.

In his *Report* Curr was at pains to underline the advantages accruing to the duke from the introduction of his inventions and improvements:

> *I need say no more to recommend Rail Roads, than about 16 Collieries out of 20 have introduced this mode of carrying coals in the County's of York, Lancaster, Salop, Derby, Stafford, Warwick, and great part of Wales, and is now adopting near London, and in the neighbourhood of Newcastle upon Tyne it has been adopting very much for 2 or 3 years past. ... those who live 10 or 15 years will probably see my Rail Roads all over this Kingdom, notwithstanding 12 years passed over before they were much imitated.*

Mott wrote that Curr went on to say:

> *that his [patented] system of using guides to enable corves to be wound up a pit shaft, and the use of tipplers, had been adopted at the Sheffield, Manor, Hesley Wood and Attercliffe collieries of the Duke without benefit to him [Curr], though he had received awards from other coal*

owners who had applied the system. He added that the use of double ropes, flat ropes, and inclined planes had proved beneficial.[347]

There was, too, competition from rival collieries sunk nearby at Dore House and Intake, which depressed the price of coal. In relation to Dore House colliery Curr added:

The loss to themselves in Interests and money sunk has been about £8,000 but the loss sustained (in being deprived of their consumption, and in keeping down the price of Coals) has been 3 times as much to the Duke's Collieries.

In Curr's opinion:

... bad as the Collieries have lately proved, they would have been worse if I had not made the improvements I have stated... Here, my Ingenuity has been buried.

Inflation, too, had played its part in increasing overheads at the collieries. The last two decades of the eighteenth century saw a period of inflation, which was particularly aggravated between 1790 and 1800.[348] Curr's *Report* draws attention to this:

The high prices of hay and Corn, Workmens wages in acct. of the high price of Provisions, Punch Wood and leading, Deal Timber, Powder, Ropes, Candles, Iron, Cast Iron, & Oyl & have for 2 years past been distressing.

Curr was concerned that dismissal would reflect badly on his judgement and integrity. The duke's reply was dated 27 October 1801:

In answer to your letter of expostulation on being dismissed from the management of my collieries in the neighbourhood I have to say that the want of success in concerns so important to myself & the trade of Sheffield has appeared to me a sufficient reason for placing the management of them in other hands, to try whether different measures may not produce better consequences.

Should these alterations succeed, I do not see why you are to consider your character arraigned as the past deficiency may have been from the error of your judgement without impeachment to your integrity. If they do not succeed you will *be clear from public censure.*

Events were to prove that Curr must be considered 'free from public censure' for, as has been seen, in 1805 management of the duke's Sheffield collieries was passed to a consortium of local businessmen.

It appears that the duke and Curr continued to maintain a cordial relationship. In his letter the duke said that his dismissal was not due to personal dislike, and that he would appreciate any advice Curr may like to offer in the future.

Curr did not look on his salary as generous in relation to the responsibilities that went with the position, and the expenses he had to pay. After his dismissal he wrote to the duke:

> *If I had charged a less price some years back for my good & taken the same pains in manufacturing them as I have done, I must have made a forfeit of as much if not more than the salary I received for my Agency to the Collieries; having for all my trouble and Inventions, and including the Risque of Moneys deposited in my hands no more than £190 per annum: and when I have deducted an assistant Clerks salary, whom I was obliged to Keep; the Keep of only ½ my Riding Horse, House Tax, Coals, Candles for Office, and a sufficient Agents House to live in, all of which are in general found by the proprietors of collieries; I have not pocketed towards the maintenance of myself and family more than £80 or £90 per annum.*[349]

Despite his protestations, Curr had other sources of income than the £190 per annum for his salary as Superintendent. As well as his work as a consultant, in his *Report* to the duke he acknowledges that:

> *I have received something handsome for the Patent Rights from sundry Proprietors of Collieries.*

Astutely, too, he had established, in his own name, a foundry at Sheffield Park. The foundry made such items as cast-iron rails, boilers and engine parts. Originally supplied by Booth & Co., Curr convinced Vincent Eyre that the duke's collieries would benefit were these items supplied by him.

Nine

AFTER DISMISSAL

Chronology

1802–3 In Ireland as consultant (actual dates unknown).

1805 Applied stationary steam engine to line from Birtley to Black Fell.

Patent no. 2891 published (manufacture of ropes – laying of ropes).

1806 Patent no. 2914 published (spinning hemp for the manufacture of ropes or cordage).

Patent no. 2947 published (ships' cables and capstans).

Patent no. 2960 published (laying and twisting yarns for making ropes).

1808 Patent no. 3157 published (apparatus for towing vessels, catching or detaining whales, &c.).

1811 Patent no. 3502 published (manufacture of ropes – laying ropes).

1813 Established ropery at Sheffield Park.

Patent no. 3711 published (flat ropes for mining purposes).

1820 Sold ropery.

Left for Paris.

1821 Returned to Sheffield.

1823 Death.

After dismissal Curr turned his attention, in particular, to the development of rope, and to the improvement and promotion of the products of his foundry.

His consultancy work continued and, in 1802/3, he visited the Castle Comer coal field in County Kilkenny, Ireland. The collieries had been advertised for let, but such was the degree of disorganisation and corruption that no taker had been found.

> *Mr. Curr was brought over from England to endeavour to re-arrange matters after a proper form; but it does not appear that much was*

effected, the master colliers having so great a hold of the property.[350]

An advertisement in the *Derby Mercury* for Thursday, 13 April 1809 shows that, as well as his foundry and consultancy work, Curr had other interests:

> *A COLLIERY to be either Let or Sold. A Very Valuable COLLIERY to be either Let on Lease or Sold, in the centre of the Kingdom, and within 100 yards of an inland Navigation communicating with the Metropolis. – An Engine for drawing Coals is erected on this Colliery, Weighing Machine, &c – Coals are now turning at one Pit; other Pits may be sunk at a very moderate expence, and from the excellence of the Coals, and the Colliery being the nearest in the Kingdom to the London Market, a Sale to a very great extent may reasonably be expected. Beds of Coal are about 26 feet thick, and the present Coal Pit only 70 yards deep. – An iron rail Road from the Colliery to the Navigation is already laid. A Steam Engine for raising Water is at present unnecessary; but should it hereafter be thought advisable to erect one, the present proprietors will make ample allowance towards the expence. – A Tenant may have the option of a Farm of Arable, Meadow, and Pasture land, containing about 170 Acres, with an excellent House upon it, within half a mile of the Colliery. Every encouragement will be given to a good tenant – Further particulars apply to Mr. JOHN CURR, No. 10, George Yard, Lombard Street, London; or to Mr. John Woodhouse, Bedworth, near Coventry.*

John Buddle Junior

At this time Curr collaborated greatly with John Buddle Jnr, the son of his old mentor. Not only did young Buddle provide orders for Curr's rails, engine parts, ropes and other items for collieries, but he played

> *a significant part in testing the inventions of others and in disseminating knowledge of recent improvements – the Davy lamp; Chapman's locomotive, John Curr's flat ropes, cast-iron rails and shaft conductors …*[351]

John Buddle Junior (1773–1843), Curr's 'close friend',[352] was born at West Kyo in Co. Durham, near the Curr's home at East Kyo. Considered *'the most celebrated mining engineer of his day'*[353] he was to become a leading figure in the development of the North East coalfield, known in the region as 'King of the Coal Trade'. When his father became viewer at Wallsend colliery in 1792 he served as his assistant.

From about 1801 Buddle *'was receiving salaries and fees for colliery viewing independent of his father'.*[354] On his father's death in 1806 he succeeded him as viewer at Wallsend. He managed Tanfield Moor and also held salaried posts as viewer at other collieries: Percy Main (1802),

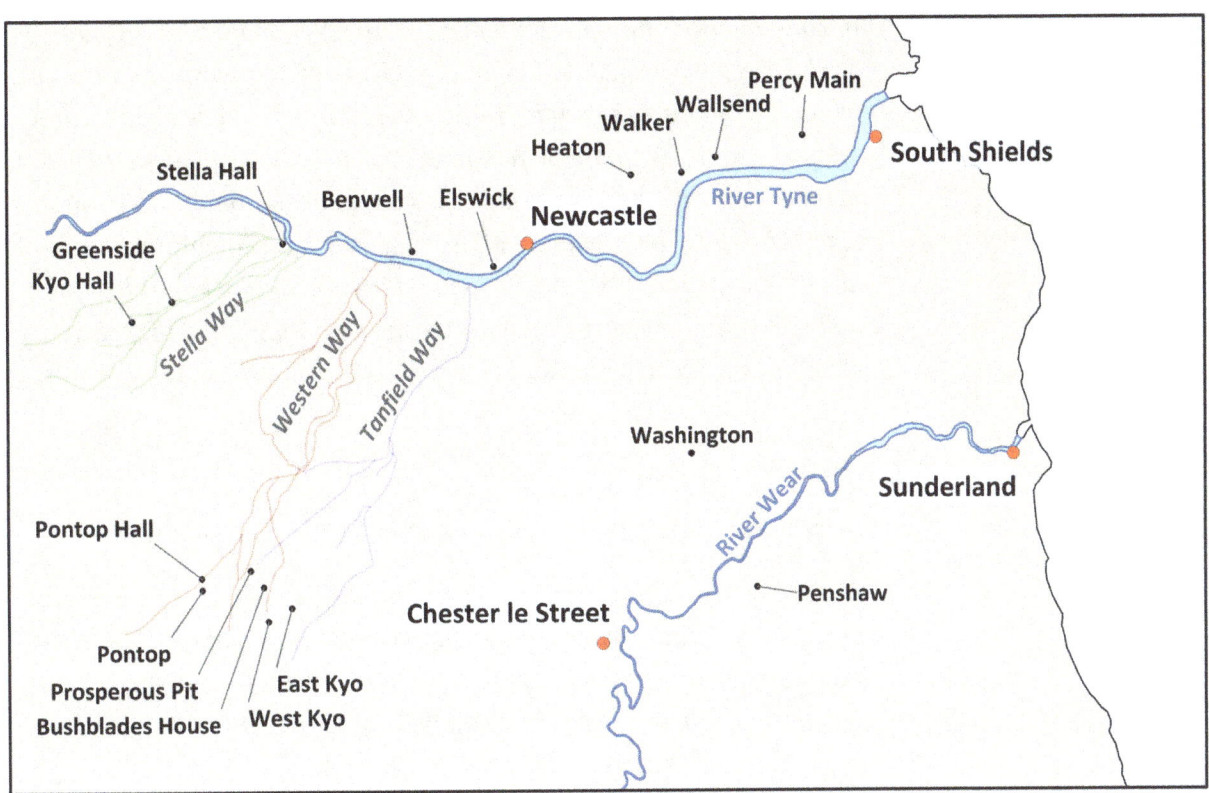

Figure 9.1. Tyneside and Durham collieries mentioned in the Curr–Buddle letters. © John Hunter. All rights reserved.

Hebburn (1803), Sheriff Hill (1804), Elswick (1804), Heaton (1807) and Jarrow (1811). The map in Figure 9.1 shows a selection of contemporary collieries in the Co. Durham and Tyne & Wear areas.

A substantial number of letters between Curr, his son John Curr III and Buddle Jnr are held by the Durham Record Office (DRO)[355] and by the North of England Institute of Mining and Mechanical Engineers (NEIMME).[356] In the main they cover the period 1800–3 (DRO) and 1802–11 (NEIMME). There is a short correspondence between Curr III and Buddle Jnr, concerning steamboats, from 1814 (DRO).[357]

Some of these letters precede Curr's dismissal, but the bulk are after it. They are primarily concerned with products manufactured by Curr at his foundry and the development of rope. The letters also show that it fell to Buddle to collect monies due to Curr for his products in North East England.

The correspondence also shows that Buddle Jnr, as well as assisting Curr in the development of his inventions, enlisted his help with schemes of his own. In February 1804 they began a correspondence relating to Wykin colliery, near Coventry.[358] One Mr Inge had offered his shares to Buddle for the sum of £60,000. In a letter, dated 12 February, Buddle said that '... *it is a Business of that magnitude as to require a personal interview which the pressure of my business will not allow of at present ...*', and requested Curr to prepare a report on the colliery: '*in Short I should be glad to be favoured with your opinion of Mr Inge's last offer*'. On receipt of Curr's report, which was dated 19 March, subsequent letters asked for more detail.

In another letter, dated 27 February that same year, Buddle asked Curr for his assistance in another matter:

In consequence of an advertisement in the Newspapers of a Colliery on the River Burry in S. Wales to sell a Gentleman of my acquaintance who is inclinable to purchase in case the thing should be eligible, and allow me to take a share on advantageous terms, has requested me to make inquiry of any person likely to give information on the subject, whether there is a probability of its being worth his attention; and as I have some idea that it is the Colliery which you had in hands last year will esteem it a favour if you will give me your opinion of the business as soon as you can conveniently ...[359]

Foundry

In Curr's letter to the Duke of Norfolk after dismissal he continues:

About the year 1792 Fire Engines for drawing Coals were introduced and got fairly under way in the neighbourhood of Coalbrookdale, when I took an opportunity of informing Mr Eyre, that as these Articles were certain to be of great use in our Collieries; to take away the necessity of employing so many horses ...[360]

Curr asked Vincent Eyre for *'leave at my expense to establish a foundry for Rail Roads and Engines'*.[361] Leave was granted and a one-acre plot of land, off Duke Street in Sheffield Park, was leased from the duke's estate where the foundry was established [see Figures 9.2 and 9.3 overleaf]. For this enterprise Curr took on a partner, Joseph Adkins, but he was to leave in December 1799, after which Curr continued alone. His son, John Curr III, was employed there.[362]

The site was located less than 1 km from Booth & Co's foundry. Booth & Co, in its various guises, had been the main supplier of cast-iron goods to the duke's Sheffield and Attercliffe collieries before the establishment of Curr's foundry. Curr placed many orders with Booth & Co for iron pigs which were re-melted and cast as rail-plates, corf wheels and the like as well as steam engines (with cylinder diameters up to 32 inches). In an advertisement Curr claimed to be melting 10–12 tons of pig iron per week.[363]

Between March 1792 and March 1801 the bulk of cast-iron goods for the duke's Sheffield collieries were ordered from the foundry. During this period castings were supplied to Sheffield and Manor collieries to the value of £5,643 1s 0½d and malleable iron totalling £159 6s 1½d. In 1794 an engine for drawing coals was supplied to Manor colliery for £200. Over the same period castings were supplied to Attercliffe Colliery to the value of £7,711 12s 10d and, in December 1795, an engine for £370.[364] Together this amounted to £14,069.[365]

Besides the duke's Sheffield collieries the foundry supplied other customers, but, except for collieries in the north-east coalfield, evidence is sparse.

In the case of Wallsend colliery the evidence is found in the correspondence with Buddle.³⁶⁶ On 17 February 1800 Curr wrote to John Buddle Jnr recommending his rails:

> *As a great deal of the mettle your Founders use comes out of this Country I don't doubt but I can serve you cheaper than they do. Rails about ½ an inch thick (such as is described in the Coal Viewer) I would lay you down at £13 per ton 6 months Credit, Corf Wheels un[flan?]ged, 18s per plate. The Common Plate described ... the 23d Page of the Coal Viewer being so generally used I have seldom less than 15 or 20 tons by me and if they are what you want could send you some immediately, and if you gain direction and choose to make any Alteration it shall be punctually attended to.*

He wrote again on 11 March: 'Hoping to hear from you soon on the subject of Rail Roads ...'

Success came in a reply from Buddle dated 25 March:

> *With regard to the Rail Road, your common Plate described in the 23 Page of the Coal Viewer will suit our Purpose exactly, only I think they ought not to be longer than 4 feet, as the Weight of one of our Full Corves, with the Carriage is abt. 10 Cwt – you will be so good as forward us 500 yds of Way (which will be abt. 10 or 12 Tons) as soon as convenient, they must be straight Plates as we must have the turn of Siding Plates made for particular Situations as they occur, must therefore have them cast at N.Castle ...*³⁶⁷

Curr wrote on 12th May to say that the order had been shipped.³⁶⁸

These letters show that when placing his orders, it seems that Buddle consulted *The Coal Viewer*. Two years later, extra copies were requested: 'Pray send me 2 copies of your Coal Viewer by the Waggon as soon as convenient'.³⁶⁹

Buddle had little choice in ordering cast-iron goods from elsewhere. At the time Northumberland and Durham were very badly off for iron foundries.³⁷⁰

Because the common plate ordered by Buddle was 4 feet long, rather than 6 feet as recommended in *The Coal Viewer*, Curr had to adapt. He wrote:

> 6 June: *As the Article of Castings (being a little shorter than I commonly cast them and raised a little in the middle of the Margin) which I have sent you seem to me so extremely Strong and useful at a moderate weight have no doubt but they will be more call'd for than the others; on what Act. I have given orders to my Fireman to cast a Quantity immediately, so that you may be supplied I presume almost immediately when you order the Goods. If you want turns or Pass Byes of any description and will give me a Scetch I will make the proper Molds and serve you charging only 10sh per ton Extra as these are more troublesome than the common Rails.*

John Curr

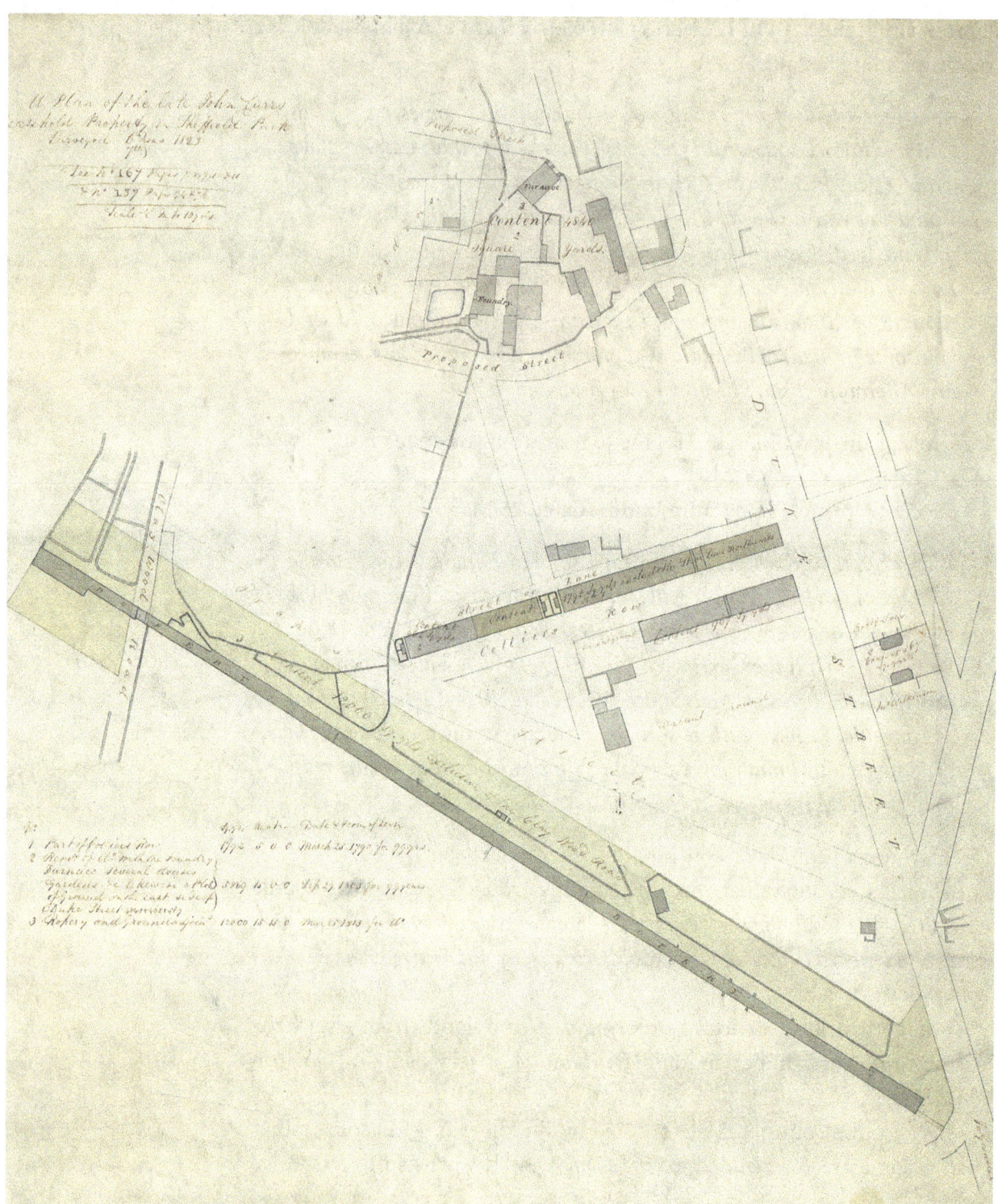

Figure 9.2. 1823 plan of John Curr's leasehold property at Sheffield Park. From: Sheffield City Archive: FC/P/SheS/800L (picturesheffield: arc04185).

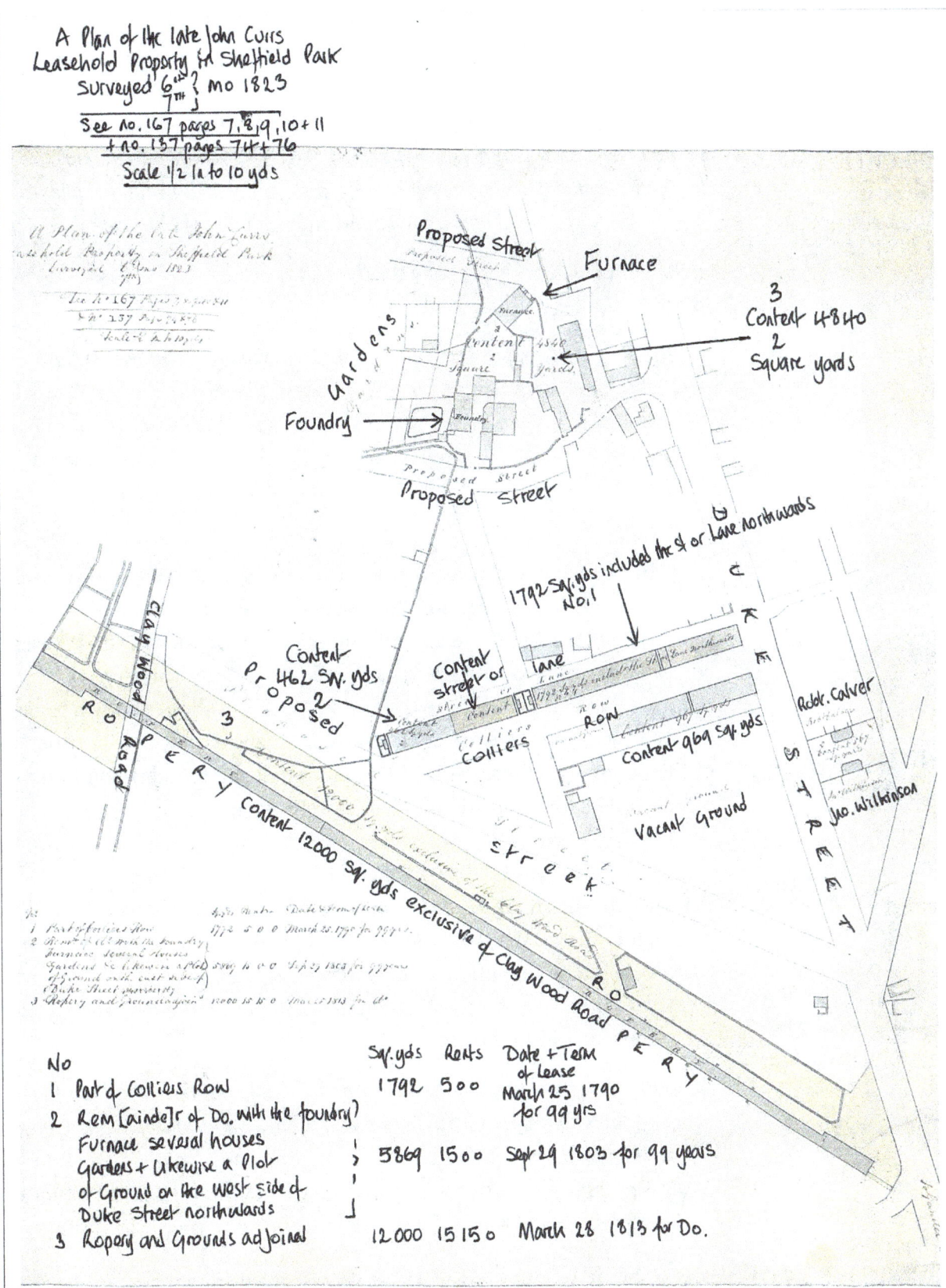

Figure 9.3. The 1823 plan annotated by the author to improve legibility.

Again, on 24 June: *I hope you will ere this have received your Rail Roads and that they will answer your expectation and shall be happy to execute your future orders in that way. You will find them I expect to be in your opinion pretty well executed, but a discovery has lately been made here that enables the founders to make their Castings smoother on the surface than they have heretofore been enabled to you, which I shall be able to give you a specimen of in your second orders if more is wanted.*[371]

In September Curr wrote to say that he had also adapted the pass bye plates to meet Buddle's requirements:

The Viewers Companion will show you how to lay down the Pass Bye Plates, the Plates sent you varying only in the length of plain Plates that are in the Coal Viewer about 7 ½ feet long which we have cut into two lengths in order to strengthen them, and should advise you to lay these down upon the Top of the Grounds before they are sent down into the Pit.[372]

In all, by October 1800, it seems that Curr's foundry had supplied Wallsend colliery with 200 rails 4-feet long and 26 pass bye plates, weighing nearly 3 ½ tons, for £40 13s 4d.[373] The quantity supplied increased. In 1805 Wallsend colliery was supplied with 6 ½ tons of rail plates (340 each 4 ft 8 ins long, 4 switch & 2 siding plates) for £98 15s 6d.[374] Other collieries known to have ordered rails were Flatworth, Hebburn, Whitefield and Elswick. The last of the surviving letters, from 1807, shows that the foundry continued to receive orders for rail plates until, at least, that year.

Steam engines and other iron parts

Evidence for the supply of steam engines and other iron parts elsewhere is sparse. It is known that an 18-inch cylinder engine was supplied to Joseph Clay at Hell Rake lead mine, near Bradwell, in 1795.[375]

A very brief quotation for an engine is found in a letter dated 16 July 1800 from Curr to Buddle Jnr:

With respect to the Engine for drawing Coals 60 Tons per Day the depth of 50 fathoms I would lay down all the Materials of Wood (the leading Bords excepted) Iron, Cast Iron, Brass and Lead at Stockwith or Newcastle for the sum of £350, and the Bricks and setting up would not cost above £50 more I Presume and should give your Friend a very compleat Engine.[376]

In a letter about Wykin colliery, dated 11 April 1804, Buddle asked Curr:

You will also please to say if 53 Cwt of your best Coal will work a 14 feet boiler, fairly driven for 24 hours? Will not your slack work your steam Engine what quantity will work a 14 feet boiler as above, and what would it cost you per Ton from your own Colliery?[377]

More evidence of other common engines built by Curr is found in contemporary advertisements. A 6-hp engine was advertised for sale in the *Leeds Intelligencer* for Monday 25 February 1805:

CARPET MACHINERY.
To be SOLD by AUCTION
by Mr. BARDWELL

Upon the premises at the Carpet Works in Coulston Croft, Sheffield, belonging to the late Mr. Wildsmith, ...

THE MACHINERY consists of a STEAM-ENGINE of Six Horse Power, erected on a Wooden Frame by Mr. Curr, and calculated to work Machinery of any Kind, or to draw Coals, to which purpose it is well adapted, excepting that the Rope Wheel Apparatus will require to be annexed, if applied to that Purpose; also ...

**** For Particulars reflecting the Engine, apply to Mr. John Curr ...* '

Another engine was advertised for sale in the *Leeds Intelligencer* for 8 May 1819 [Figure 9.4].[378]

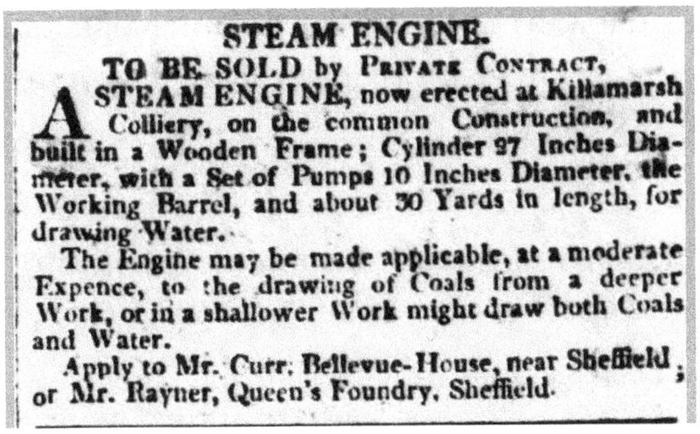
STEAM ENGINE.
TO BE SOLD by PRIVATE CONTRACT,
A STEAM ENGINE, now erected at Killamarsh Colliery, on the common Construction, and built in a Wooden Frame; Cylinder 27 Inches Diameter, with a Set of Pumps 10 Inches Diameter, the Working Barrel, and about 30 Yards in length, for drawing Water.
The Engine may be made applicable, at a moderate Expence, to the drawing of Coals from a deeper Work, or in a shallower Work might draw both Coals and Water.
Apply to Mr. Curr, Bellevue-House, near Sheffield; or Mr. Rayner, Queen's Foundry, Sheffield.

Figure 9.4. Ad for sale of steam engine, *Leeds Intelligencer*, 8 May 1819.

Cast-iron props, an invention which did not catch on, were advertised in the *Derby Mercury* for Thursday, 3 June 1802. For the full text of this lengthy advertisement, see Appendix 6.

Evidence that the foundry also supplied pipes[379] is found in letters written in 1804 from Buddle Jnr to Curr III:

27th January: *I am duly favoured with yours of 22nd Inst. offering to deliver 50 to 60 Fath[oms] 8 in. Pipe in Newcastle at 16s/- per Cwt. to be delivered in two months from receipt of Order which I accept, and beg you to cart 60 Fat[hom]s of that Size for Wm. Russell Esq. & Co (Walls end Colliery) and forward them as soon as possible you will please not to make them too heavy and the Spigot Joints must be made for wedging. I hope you will be able to deliver in much less time than two months ...*
7 March: *Pray let us have the 8 in Pipes as soon as possible.*[380]

Haydock colliery's cash book for February 1810 has an entry:

'Paid John Curr for Pulley Wheels for Haydock Colliery – £7 7s 4d.'

Improved corf design

Curr continued to tinker with the design of his corf and wrote on 17 February 1800 to Buddle to tell him about the advances he had made:

> *I ought to inform you that since I was with you have made a great improvement in the Corf Wheel by lengthening the Axis to 4 ½ Ins and leaving 2 ¾ Inches of the inner part of it hollow, 3/8s of an Inch deep, so as it never touches the axletree. The hollow we fill with grease and the Corf runs 30 or 40 Miles without greasing, and in the other case we was obliged to Oyl every 3 Miles and one greasing is abt. equal in Expence to one Oyling. Our Wheels now are on the Outside of the Corf in order that they may be taken of[f] to Grease. In the Corf you see in the Coal Viewer we are obliged to oyl them all, which is done with great difficulty, and we have paid for that oyl about £250 per ann[um].*
>
> *The lengthening of the Axis of The Corf Wheel gives it a better direction than a short Axis does and the Wheel works much longer before it gets too large in the Axis.*[381]

Buddle responded to Curr in a letter dated 25 March, noting that he'd been at Durham 'where I have been electioneering in the course of 8 days past', placing an order for 500 yards of rail road, among other things, and added:

> *… and also 1 doz. of your improved Corf Wheels, as the saving of Grease is an Object of consideration to us; the Grease used by our Underg[roun]d Carriages amounts to £410 per Annum.*[382]

In a letter to Buddle dated 6 June, Curr noted:

> *The Hollow long method Wheels which I have sent you a specimen of continues to show itself a great improvement in the saving of Grease and greatly eases the Burden of the Horse. They work a whole Week without Greasing, and the Wheels are pulled up to take of[f] in order to Grease them, in which case we fill up the hollowed part and when the Axletree warms it helps itself to the Grease. To facilitate the wheels taking of[f] we put a cotter (which is defended by a thin ring) through the Axletree end, and the Cotter having a small piece of Leather drawn through a Hole at the end of it prevents it Loosing on the Roads.*[383]

In September, Curr wrote to Buddle with news of a further advance:

> *I have lately been so fortunate as to hit upon an improvement in the Waggon for Conveying Coals on a Rail Road upon the Top of the Ground that, might in my opinion be well applied in the situation when you are making your New Winning as I should call it by the side of the Tyne which you took me to when I was with you last Summer.*
>
> *As I presume you won't bring the Coals drawn at that Pit up to your present Staith you will of course have to build a new Staith*

in which case the Carriage I have invented will save nearly ½ the Expence in the first erection of the Staith having only occasion for 1 Tier of Geers(?) or Upright Frames running Parallel to the River nor will you with this Carriage have any expence in the Article of Trimming Coals. The Principle of this Carriage is not to let out its load through the Bottom as your Waggon does, nor to let it out from the end of the Carriage as a Cart does but it unloads on both sides of itself and on both sides the Rail Roads and delivers its load to the breadth of about 4 feet from the side of the Rail Road, and when you are laying up a Bank of Coals in the Staith it only requires the Rail Road shifting side ways upon the Bank it last made to give the accommodation of forming a Bank of any extent. I have applied the scheme to use here about 2 Months ago, and some of our ablest Mechanicks pronounce it the most compleat method of conveying and Banking up of Coals they have ever seen. My motive in mentioning this to you now is to prevent you ... any progress in the Building of your Staith until I see you or of engaging your Timber for it.[384]

No evidence has been found as to whether Buddle followed Curr's advice on this occasion.

Shipping

How did Curr transport his goods to customers who were not in the vicinity of Sheffield? There were two routes, neither ideal. There was a packhorse road from Sheffield to Bawtry which could be difficult to negotiate in wet weather, or even impassable, especially for the horse-drawn carts which would have been required to move heavy cast-iron goods. At Bawtry goods were loaded onto vessels on the river Idle and thence to the Trent. The other route was by river. In the first half of the eighteenth century the river Don had undergone improvements which made it navigable as far upstream as Tinsley, a mere three miles from the centre of Sheffield, where a wharf was established.[385] In this case the route was to Hull via the Ouse from where the Don joins it just below its junction with the Aire. In 1789 a consortium of owners of shallow-draft vessels, with limited capacity, had begun to advertise weekly sailings to Hull.[386] In 1802 this route was greatly improved by the opening of the Stainforth–Keadby cut which linked the Don directly with the Trent. Transport by river could be interrupted by low water level in summer or by frozen rivers in winter. See Figure 9.5 for further details.

Curr's rope was manufactured until 1813 by William Bourn, roper, at Gainsborough and loaded onto vessels on the River Trent.

The cost of carriage added much to the price of the goods and shipment was, not infrequently, subject to delay. Both are illustrated in letters written by Curr to Buddle Jnr in 1800 (except that of 3 June, which was from Buddle to Curr).[387]

An indication of the cost of freight is found in a letter dated 16 July:

John Curr

I am very glad you did not pay this very exhorbitant charge made by Captain Beecham for Rail Roads as I am well assured the whole charge of Cast Iron to Newcastle from Tinsley does not exceed 20 shillings or 1 Guinea per Ton, and it is a thing well known as a great deal of Mettle is sent from here to your Newcastle Founders of whom you might easily be informed of the Price and what ever is customary I expect to pay and not more, but the freight from Tinsley to Gainsborough I can pay here which is 9s per ton the Price from Gainsborough to Newcastle I expect is 11 or 12 shillings per ton. When I informed you of my Terms for Goods Delivered I expected the carriage would cost a Pound or a Guinea per Ton.

The following are excerpts from the correspondence regarding delay:

3 June: *I presume the delay has taken place at Hull, as the Vessels which take in Goods there, frequently lie a Fortnight or three Weeks before they receive their full Loading.*

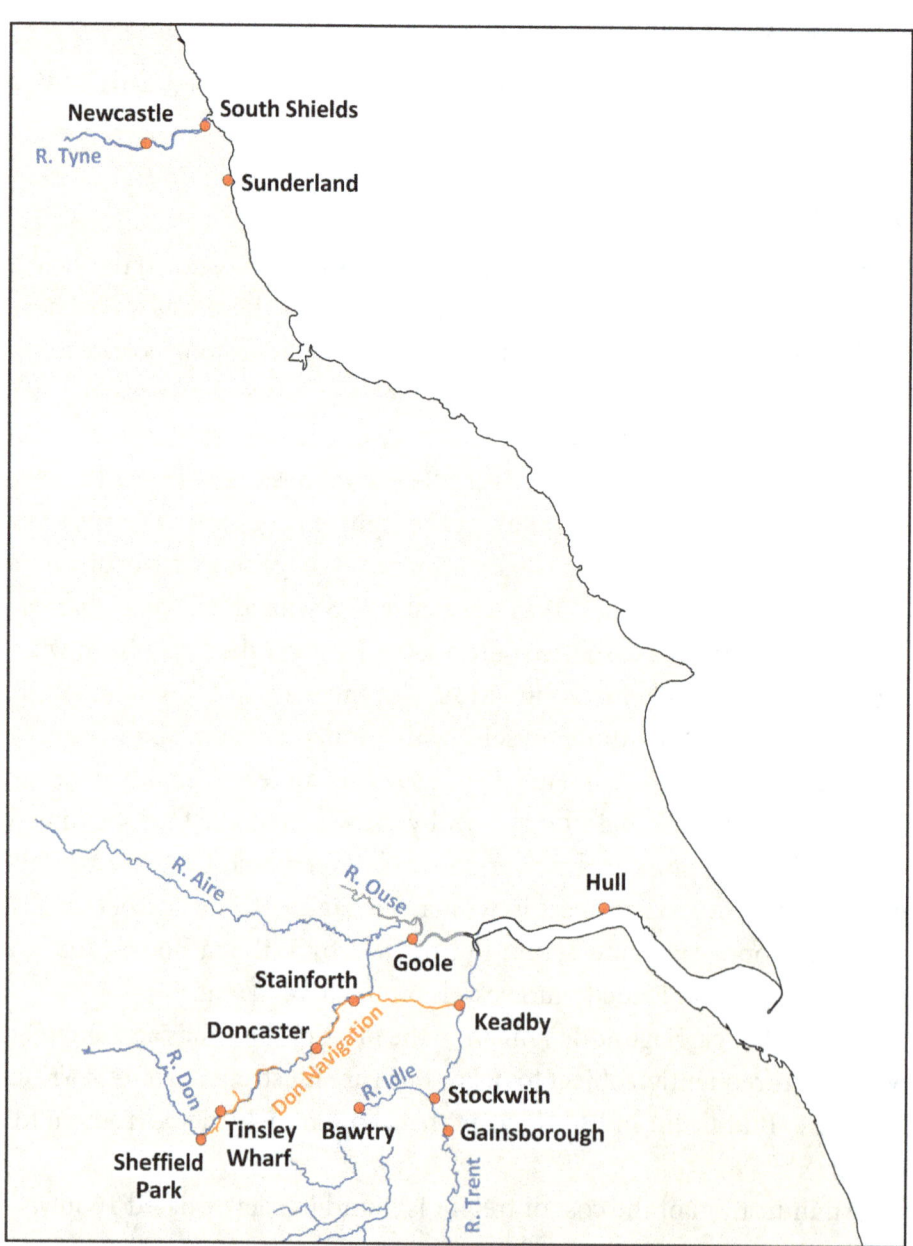

Figure 9.5. Shipping routes from Sheffield to Newcastle. © John Hunter. All rights reserved.

> 6 June: ... *am sorry your Castings has been detain'd. It unfortunately happened that they was laid fast by the failure of a Lock before this had proceeded no more than 20 miles, but the Boatman told me more than a fortnight since they had got at liberty.*
>
> 13 August: *If I find when the flatt Ropes are ready that there is any delay in our Navigations (which is very common and Water is very low) had I not better send them over land.*
>
> 4 October: *Your Rails are at Mr. Brightmore's Wharf Gainsborough who informs me by this days post he has two Vessels now on their passage from London and as soon as they arrive one of them goes to N.Castle.*

On 4 October 1800 Curr wrote to Buddle Jnr to say that a shipper from Shields proposed direct sailings from Tinsley to Newcastle every 5 or 6 weeks:

> *A Mr. Graham of Shields has bought a small Vessel in this River which he proposed trading regularly from Tinsley near Sheffield to Newcastle and which he expects will make a voyage every 5 or 6 Weeks and what will be pleasant the Goods will go forward in one Bottom by which means I hope in future we shall have less Delay and trouble on our Conveyance. This Vessell I believe is just now arrived here and shall send off by her in a Week or 10 Days the Hebburn Ropes and the remainder of your order for Rails which you will receive perhaps as soon as those that were sent of[f] 3 or 4 Weeks sooner.*

Curr wrote again on 17 October:

> *The Hebburn Ropes and a Pair of Pulleys went down to Tinsley this Morning together with the remainder of your order for Plates and as they are gone by the New Trading Vessel (the Blessing of Shields) and as it is not to be put aboard any other Vessel am in hopes they will arrive in less than a fortnight.*

This appears to have been an improvement because letters written by Buddle to the Currs after this date make no further mention of delay.

Rope

Background

In 2015 Edward Sargent gave a talk to the Docklands History Group, 'A history of rope making'. Although about maritime rope, his talk is a useful introduction:

> *Traditionally rope is made of 3 strands (hawser-laid) although sometimes it is 4 strands (shroud-laid). The strands are made by twisting yarns together. A cable consists of 3 hawser-laid ropes twisted together. ... rope in England was, until the middle of the nineteenth century, normally made from hemp ...*

The fibres had to be removed from the plant and cleaned and then they were beaten but later a breaker was used. Fibres were taken to the rope works for hatchelling which meant they were dragged through spikes to draw them out straight and remove seeds, roots and other matter. The fibres were then spun by hooking some on a wheel and the spinner walked backwards paying out the fibres as the hooks turned. The yarns were often tarred in a vat of hot tar. After this the yarns to form a strand were pulled out along a rope walk. They were then twisted in the opposite direction to the twist of the yarn to form the strands. Three strands were then connected to a hook on a sledge at one end and at the other to three hooks. The strands were then twisted while the single hook at the other end was twisted in the opposite direction to close the rope. The strands were kept apart until they came together to close the rope by a conical piece of wood with grooves in called a top.

There were technical advances at the end of the 18th century and beginning of the 19th century when machines were devised which could form strands and lay rope while winding them on although they did not improve the quality of the rope. ... In 1793 Captain Huddart, a master mariner, invented the register plate which has concentric holes for the yarns to pass through and a forcing tube to compress them together as the strands were formed. Ropes made using this system were twice as strong as conventionally made ropes. Several inventors

Figure 9.6. Ropery at Chatham Historic Dockyard. Photo credit: John Hunter.

patented similar designs later in the decade. ... Chapman [William, see below], later in the same year, took out another patent which had a modified sledge that would pull itself along a ground rope as the whirls were turned. This provided a direct relationship between the speed of the whirls and that of the sledge but also gave a means of powering the process using horse or steam power.

Rope for mines was sewn together.

The Trust at Chatham still operates the rope-walk at the Dockyard which is now the last working mechanised rope-walk probably in the world.[388]

For a comprehensive description of rope making see Appendix 7. The image in Figure 9.6 shows the Ropery at Chatham Historic Dockyard.

At the end of the eighteenth century, the mining industry faced a variety of problems concerning rope. Galloway outlined one of them:

As the depth of mines increased, the weight of the ropes employed in raising the coals became of itself a considerable load upon the winding machine. In addition to the weight of coals, the engine had to raise the whole weight of the rope hanging in the pit at the commencement of its run, the other rope at this point being wound upon the drum and rendering no assistance. Several expedients were brought forward in the latter part of the eighteenth century. Smeaton applied conical drums, others used counterbalance weights of various kinds. With the same object Mr. Curr invented the flat rope. It consisted of several small round ropes stitched together, and was made to lap itself on winding. Thus at the commencement of a run the loaded rope began to coil upon a small diameter gradually increasing, while the empty rope began to coil off a large diameter gradually decreasing – an arrangement which rendered great assistance to the winding engine.[389]

Flinn described other problems, and their solution:

In the early eighteenth century hemp rope was exclusively used in winding. It was strong, flexible, and fairly durable. The winding rope used by the Butterley Company in Derbyshire at the end of the eighteenth century was generally about two inches in diameter and was purchased from a Newcastle supplier. But hemp rope was expensive, and tended to twist to some degree under tension, so that corves, or miners who simply put a leg through a loop of rope to ascend or descend, spun round in the shaft. Unless replaced in good time, sooner or later, the ropes snapped, with danger to the workers at the shaft bottom as corves fell down the shaft, or loss of life if miners were using the rope.

An invention by John Curr, ..., in 1798 solved the problem of spinning. His flat rope was made by stitching together two or more ropes with hemp or linen thread, or with brass or iron wire. The flat rope was wound on itself on the drum, thus reducing the speed of winding as the corf neared the bottom of the shaft. The use of Curr's flat ropes spread slowly but steadily to other coalfields.

Curr's system offered two economic advantages: by eliminating the continual twisting and untwisting of the ropes it reduced wear on it and prolonged its life ...[390]

The invention and implementation of improvements to rope making was a crowded field. For instance, William Chapman (1749–1832), civil engineer,[391] and his brother Edward (1762–1847), founded a successful ropery, managed by Edward, at Willington quay, near Newcastle, in 1789. They filed a number of patents for improvements to rope-making machinery. William Chapman wrote a book on cordage, published in 1808, which gave a good account of the early development of rope-making machinery (see Appendix 8 for further details).[392]

Curr and rope

Curr, in his capacity as 'Superintendent of the Coal Works' of the Norfolk estate collieries at Sheffield, was aware of the problems outlined above

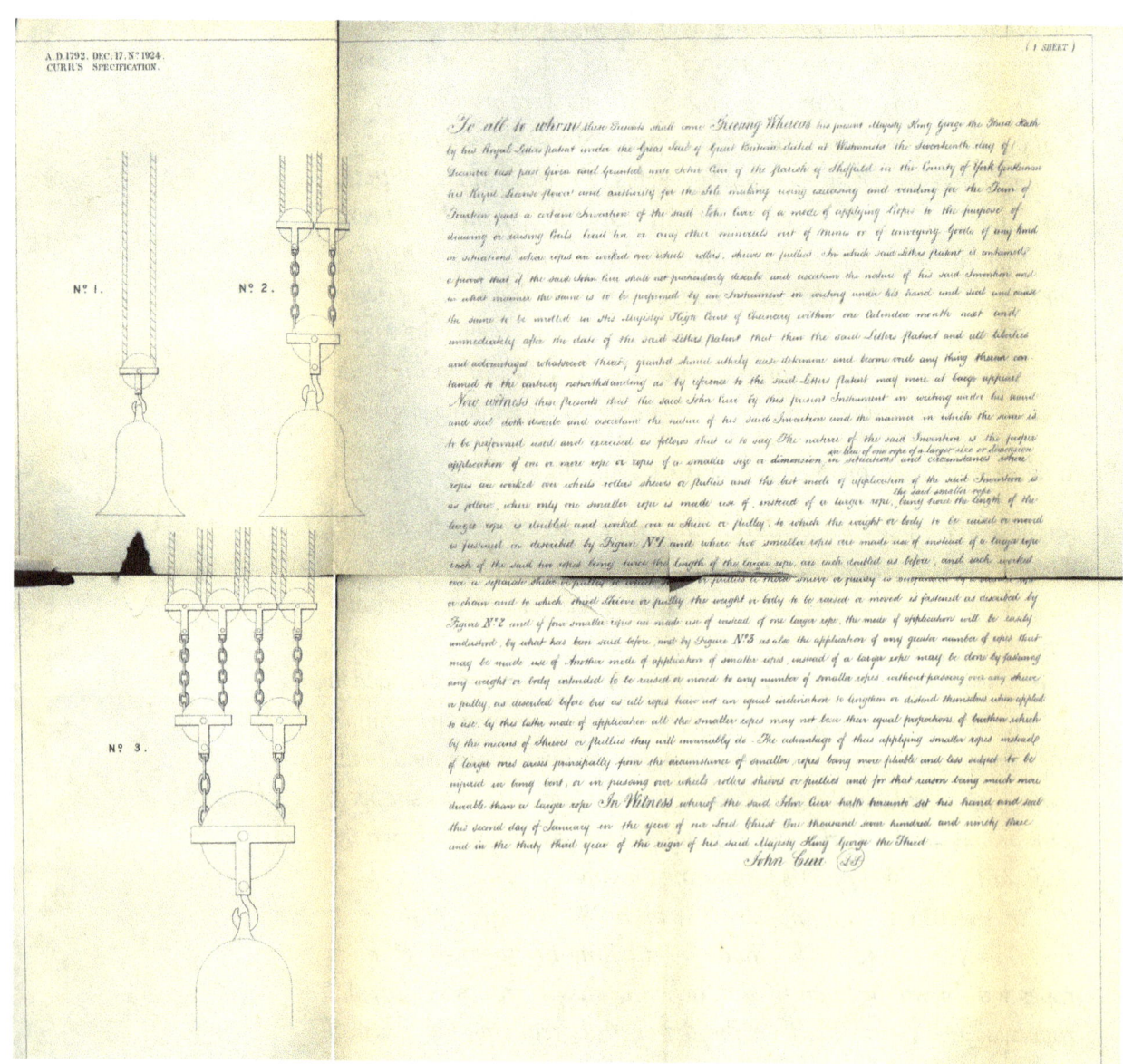

Figure 9.7. Plate from patent no. 1924. From the author's collection.

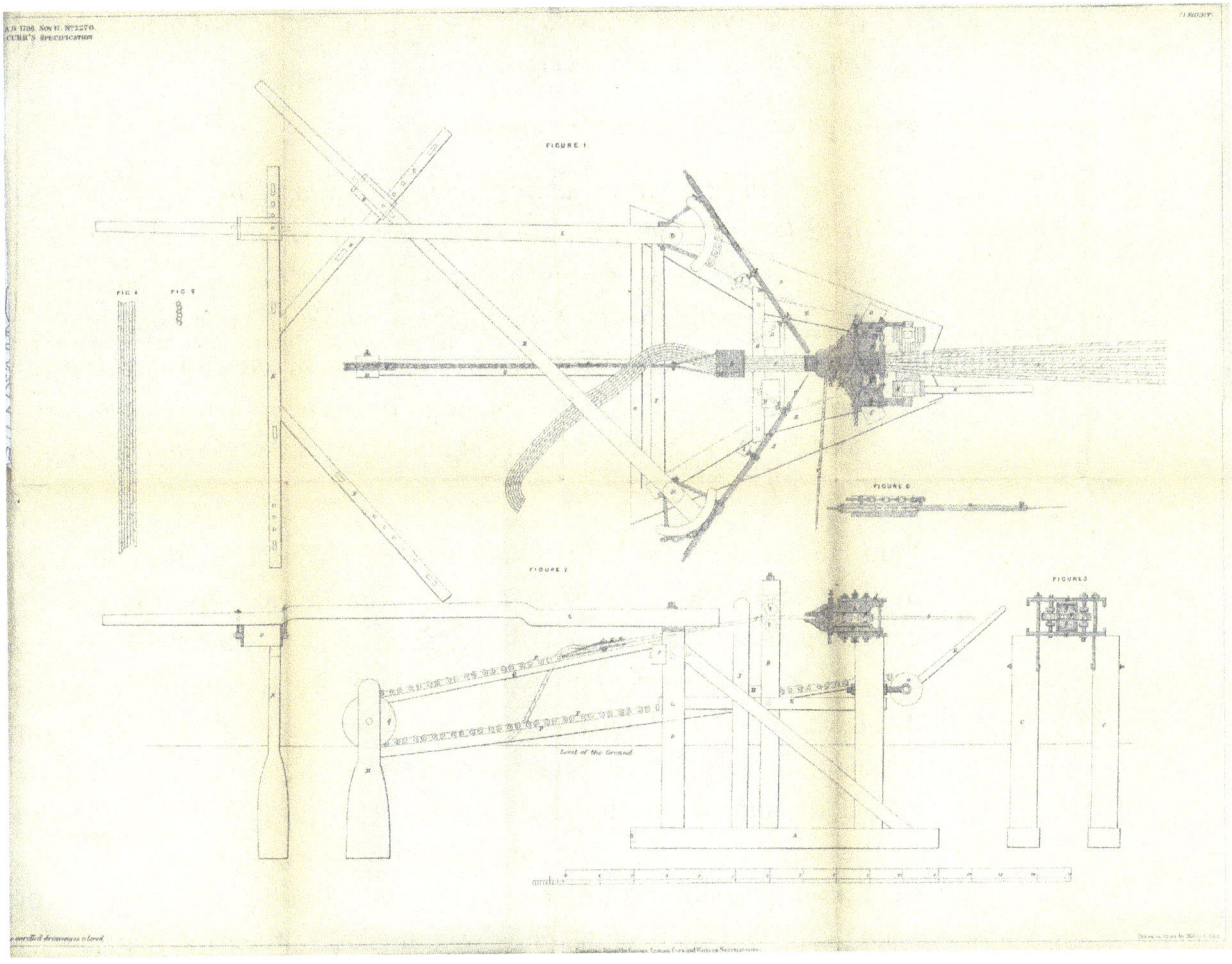

Figure 9.8. Plate from patent no. 2270. From the author's collection.

and set his mind to solving them. His first attempt was his invention of 'double rope', for which he took out a patent in 1792. This was patent no. 1924, '*Method of working ropes and pulleys in mining operations &c.*', published on 17th December. The plate from the specification published at the Great Seal Patent Office, London (1856) appears in Figure 9.7.

In his letter of protest to the Duke of Norfolk after dismissal he claimed of this:

The double Rope which I introduced at the Collieries 10 Years ago, produced a great annual saving as the Ropes of the same weight drew more than double the weight of Coals drawn by the single rope.[393]

The double rope proved unsatisfactory for, as noted, Curr next invented the '*flat rope*'. For this he took out patent no. 2270, '*Manufacture of flat rope*', which was published on 17th November 1798. The plate from the 1856 printing of the specification appears in Figure 9.8.

In the same letter to the duke, noted above, Curr further claimed:

[The] *invention of the Flat Rope (3 years ago) has drawn in one instance 5 times as much as the com.n Round Rope, in another instance 3 ½ times, and in others 3 and 4 times as much, but the Ropes now in use, being more perfectly manufactured will do 6 or 7 times as much work*

> *as the Common Ropes, and thereby save 3 or 4 hundred pounds, per annum in the Duke's Collieries.*

Flat rope had its disadvantages. Galloway noted:

> *Though the ingenious arrangements introduced by Mr. Curr at Sheffield, in the latter part of the eighteenth century, for raising the coal in carriages, were found to answer very well in shallow collieries, they were not sufficiently matured to admit of their application in deep ones. The carriage being suspended at the end of a rope, it was scarcely possible to raise more than one at a time; and the mechanism for landing the load at the surface was somewhat clumsy and tedious.*[394]

John Hunter explains that in shallow collieries, such as those managed by Curr at Sheffield, flat rope made a contribution towards counter-balancing the ascending and descending cages during shaft winding. Because it was designed to lap upon itself on the spool, its lapped diameter increased during winding. Therefore, when the engine began to lift the loaded cage from the shaft bottom, spool diameter was at its minimum, and the 'moment' (i.e. torque) was minimised. At the same time, the rope for the empty descending cage was fully wound on its spool (on the same axle), and, although the weight of the empty cage and its hanging rope was much less than the loaded cage, the 'moment' was increased by the radius of the fully wound rope on the spool.

This effect may be demonstrated by noting the difference between holding a heavy weight at arm's length and holding the same weight close to one's chest. The 'moment' of the former, which requires greater effort (strength) to hold the weight stationary, is greater than that of the latter.

Although the partial counterbalancing aided in reduction of effort by the engine when winding in shallow collieries, its assistance was much less in the deeper collieries in north-east England. As a result, more effective methods were developed, including *'the staple pit and chain'* at Monkwearmouth colliery, which is described by Galloway, amongst others.

John Hunter also notes that, although stronger, flat rope had other disadvantages: its expense, excessive weight and tendency to chafe and wear on the edges unless the reels and pulleys were perfectly aligned.

How much Curr benefited financially from the flat rope was questioned by Matthias Dunn:

> *He also invented the flat-rope, which he manufactured himself; but the difficulty of stretching all the four strands alike, with the then machinery, was so great that his patent yielded him little or no advantage; especially as the numerous improvements of the round rope were brought into powerful competition.*[395]

Was Curr the inventor of the flat rope? Most are agreed that he was, but an interesting obituary was published in the *Sheffield Independent* for 27 September 1828:

> *On Sunday last, aged 76, Mr. Isaac Penniston, of Harvest-lane, of the firm of Penniston, Wheatcroft, and Singleton, edge tool makers. He was a man of strict integrity and superior natural abilities, and before he became a partner in the above firm, he was in the service of the late William Curr, Esq. of Belle Vue* [mistaken names abound in newspapers], *rope manufacturer. He invented the flat rope, so generally adopted in coal pits, for which Mr. Curr got a patent. ...*

Ropery

The satisfactory development of flat rope was a long process. Initial development was made in conjunction with William Bourn, ropemaker, of Gainsborough, Lincolnshire. On Curr's retirement in 1820 Bourn purchased much of his rope-making machinery and published an open letter 'To the Proprietors of Coal and other Mines' in which he states:

> *... After frequent Letters and many Consultations between us on the subject, I made him four Ropes, in such a manner as I thought would form a Flat Rope. – They were found to answer the purpose; and I continued making them for MR. CURR for about eight years, until the Flat Rope had established its credit; at this time MR. CURR and I could not agree as to the terms of my manufacturing Ropes for him, and the consequence was that we parted. As I had allowed him at all times to have free access to my Manufactory, (he having been three weeks or a month at my house at a time,) an opportunity was afforded him of making what observation he chose, as to my mode of making the Ropes, he then fixed his Manufactory at Sheffield, where he continued until he declined business in toto, and retired into France ...*[396]

Curr leased a further 12,000 square yards at Sheffield Park from the Norfolk Estate in March 1813 and established there 'The Sheffield Patent Ropery', which employed his machine for lacing.

John Hunter has overlaid the 1823 plan of Curr's leasehold property at Sheffield Park on the 1850s OS map [Figure 9.9]. Shown, from north to south, are furnace and foundry, Colliers Row and ropery. Part of the 1774 waggonway has also been added. These, except the waggonway, are also shown overlaid on the present-day OS map [Figure 9.10].[397]

According to H. W. Dickinson[398] a rope-walk could be as much as 440 yards long to accommodate what subsequently became the British standard length of rope, viz. 120 fathoms. The length of Curr's was about 410 yards.

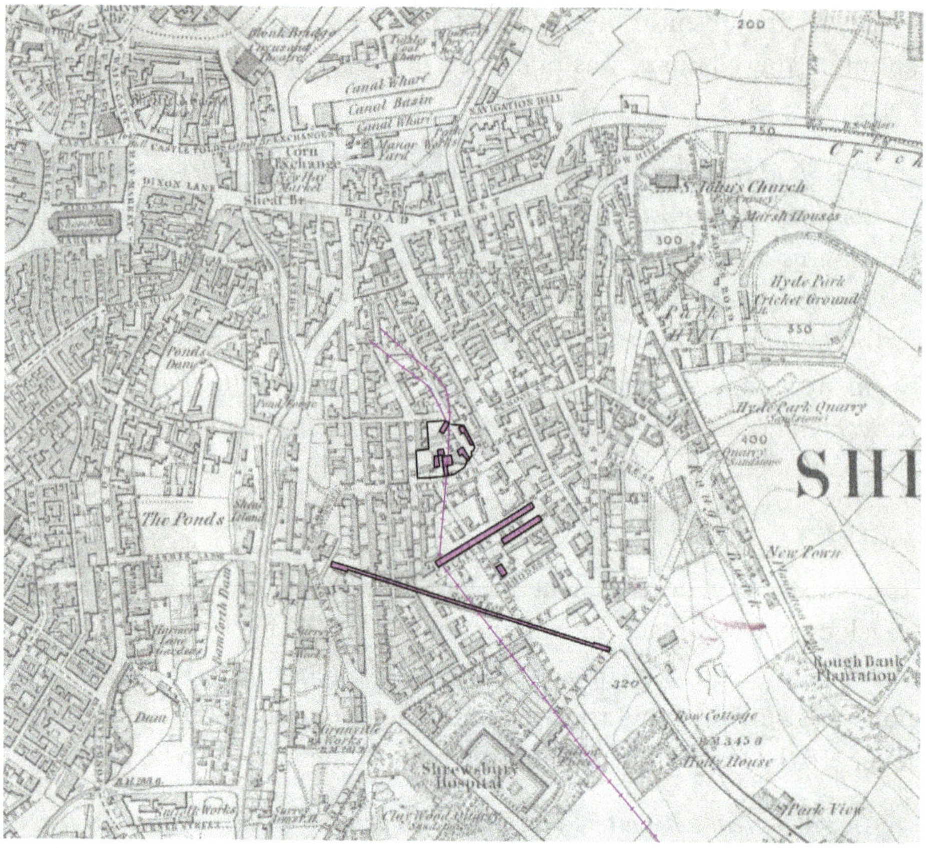

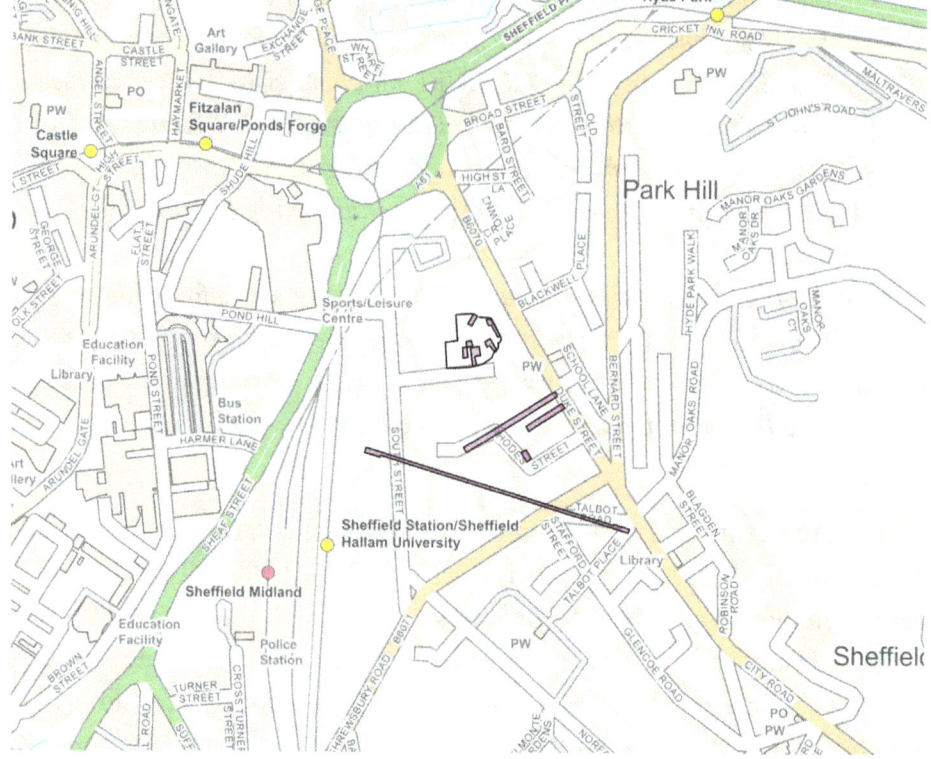

Figure 9.9. Route of 1774 waggonway and (from 1823 plan) foundry, Colliers Row and rope-walk overlaid on 1850s OS map.

Figure 9.10. Foundry, Colliers Row and rope-walk (from 1823 plan) overlaid on present-day OS map.

For both: Images contain OS data© Crown copyright and database right (2017). © John Hunter. All rights reserved.

Twisting

A major problem besetting the development of flat rope, twisting, proved hard to resolve. An insight into its solution is found in correspondence between Curr and John Buddle Jnr at Wallsend colliery. Buddle had installed flat ropes at Wallsend prior to early 1800, for, in a letter to him dated 7th February that year, Curr wrote:

> *I have just received your favour of the 3rd and am much concerned to hear your new flatt Rope inclin'd to contract so bad a habit in its infancy as you state of turning upon the Pulley Wheel. I was in great hopes as the Rope seem'd so strong that it would not easily bend so as to make the turn, and why these do not resist that inclination more than the former ones leaves me quite unable to account for. A Week or 10 Days will determine what these Ropes will do, but if they continue their bad habit it will certainly be advisable to take them off before they are materially injured, and if on the receipt of your next I find them in no better state than your present letter informs me, I feel much inclined to go down by the Coach and consult with you and your Father what is best to be done.*[399]

On 11th March Curr claimed:

> *The temptation to Collieries seems great as ours that have now been working a year at 400 Tons per Week seem much the same as they have been for 6 Months past and which I am extremely happy to say I hope soon to give you a sight of.*
>
> *It occurs to me that a regular system of making & twisting Ropes (such as Mr Chapman's I conceive must be) would be of great importance to the flatt Rope as one Rope then probably would not have more inclination to distend than other and could wish to have a tryal made of them.*
>
> *Be so good as enquire if there is a possibility of getting a Pair of Ropes made by Mr Chapman or the Ropers at Shields that work by his Patent.*[400]

In reply Buddle wrote on 25th March:

> *I'm very glad to hear that your flat Ropes continue to answer so well & I must again observe that it must surely be owing to our greater Depth of Speed that they are not equally successful here … There is no doubt but if the Ropes which are laced together for a flat Rope were made by Mr Chapman's Machine that it might be an Improvement but not unless all the Ropes which are to be laced together have an equal Number of twists in a given length & in that Case it is natural to suppose they would be considerably superior to the common Ropes; There is not any Roper except Mr. Hood who has agreed with Mr Chapman in this Neighbourhood for liberty to make Ropes by his Method, and I have no Doubt but that you may have Ropes of any Size made there … As a 6 In. Rope of Mr Chapman's is nearly as strong as a common 9 In. Rope, of course the Ropes for a flat Rope need not be so thick; If you wish to make Tryal of Mr C's Ropes for a pair of flat Ropes, after you fix the Size I can very easily get them put forward at Mr Hood's; you will also consider of the cheapest mode of getting them laced, whether by sending the Ropes to Sheffield, or your lacing Mach[ine] to N.Castle.*[401]

After a further exchange, Curr wrote to Buddle on 6th June to say:

I would like to make a Tryal of 4 Ropes manufactured by Mr Chapman's Patent which must come here of course to be laced. My Flatt Rope continues to do such wonders that it would certainly be a loss to the Publick as well as myself to sit down tamely under our late disappointments. Experience has shown me that several of the Ropes I have ordered the last 6 Months past, wherein I expected to have made considerable improvements, have turned out quite to the contrary, and I doubt not now but I can make a flatt Rope work even your pace, but should like at the time I try another of my own to have one of Mr Chapman's (made flatt) to try them together. I find that it is only regularity in the Manufacture of the Rope that is wanted without which the separate ones that compose the flatt ones cannot take a fair lift and gives them the Turning inclination, and in the Spinning 2 of the Ropes the contrary way the Men cannot make so even a thread and those Ropes are never twisted regular as the others are. Another incon[venience] is in the Tapering as the turn is not so regular upon it as it is in the even Rope. The effect of some of our Ropes now at work here on Act. of the above reasons has been that one of the 4 Ropes has absolutely Broken at several times on which act. they have been obliged to be cutt and spliced again. To avoid this in future I shall order all the threads twisted the common way and no more tapered ones, and shall have the 2 Ropes Cable laid which gives the contrary twist only observing that there must be a yarn or 2 in each strand less in the Cable laid Ropes than the others in order to get them of the exact same Circumference as the Stranded laid Ropes are without which they cannot be laced properly. I shall also in future let the Ropes all lie a Week or 10 days to stretch so that if one has an inclination to distend more than the other it will do so before it is laced. The inference you have had of the Ropes I first sent you, with these observations I have now made will enable you to form an opinion on the flatt Ropes which I doubt not will corroborate with mine.

My Ropes being so small I don't expect much mischief(?) from Mr Chapman's Patent in the Yarns taking all their regular lift, but I conceive in other respects Mr Chapman's Ropes being laid by Machinery it is done regular, and the Common Rope Maker has not regular rule which in my opinion is the only reason why you see a strand disappear in the common Rope when it is nearly wrought out.

He continues:

Be so good as order me if you please 4 Ropes to be sent with all possible dispatch here each 2 ⅞ Inches Circumference and 245 Yards long in which case if they don't suit you at Walls end they will cut into 2 for sending places that I serve. 2 of these of course will be Cable laid with a yarn or 2 in each strand less than the others in them that their circumference may be exactly the same as the others are. Don't let them be laid harder than common, if any thing a little softer but perhaps they cannot easily alter that. I have now great reason to believe the flatt Rope here will draw 5 times as much as the common Rope draws here, and compared with yours at Newcastle our Apparatus and Corves being so much heavier will draw 30 or 35 S[under]l[an]d Tons of Coals, and am also satisfied that the twisting of the Rope with you has been owing to the irregular manufacture of the separate Ropes.[402]

Buddle replied on 16th June:

In compliance with yours of the 6th Inst I called upon Mr Chapman's or rather(?) Mr Hood's People respecting your Order for the 4⅔ In. Ropes; they inform me they make none so small by the new Mode, particularly Cable laid Ropes, their smallest size being 5 In for Cable & 3 In for shroud laid Ropes, but if you would wish to have 4 of the latter, they will make them to the Size you wish.

I'm glad to find you continue so warm in the Cause of the flat Rope, and I hope your laudable Perseverance will in the End be crowned with Success. The great Desideration in the flat Ropes seems to be the Ropes which compose them not acting exactly together, or not bearing each an equal Share of the Burthen, and I think your Scheme of stretching them for a length of Time, before you join them, must be a great step towards effecting that Purpose.[403]

On 24th June Curr wrote to Buddle:

I am now more & more satisfied that our failure there was owing entirely to putting the Rope makers out of their way in twisting the yarns the contrary way and tapering the Ropes by which means the Ropes have different inclinations to distend and that is the only cause of the twisting which the Rope exhibits when in use, and that was the sole cause of the failure in your second flatt Rope. In some later experiments I have made it appears evident to me that I have not the least necessity of having one half of the Ropes made to twist the contrary way to the other half in order to make the Rope retain its flatness, which will be one very grand point gained, as I may then hope to have to have the Ropes made so as they will take a more regular lift. In a fortnight I hope to try a Rope made in this manner and shall put it on the Pit where we have no Conductors and will inform you of its success. I must now beg you'll order me 4 shroud laid Ropes made by Mr Hood 2⅞ circumference and sent of[f] with all possible dispatch.[404]

And on 16th July:

I have now got a Rope tried on our Pit without Conductors and have the satisfaction to say it is working without turning even once round in the Depth of 90 Yards and goes about 7 feet per second! The Rope is composed of 3 and it appears that the Number gives a better Byas than 4.[405]

On 13th August Curr wrote:

I informed you in my last I was in hopes of making a perfect flatt Rope or in other Words a flatt Rope that has little or no inclination to turn, in which I have now the satisfaction to say I have succeeded to my wishes, having got one working here without Conductors and doing very well. The whole inconvenience I have discover'd has arose from the irregularity of the Rope makers in twisting their Ropes, which requires to be done to the greatest exactness every Rope when a Weight is put to having an inclination to untwist, and if they are all twisted to an equal degree of hardness the two shroud laid acts against the 2 Cable laid and the flatt Rope remains either with a greater or lesser Weight upon it always straight and flatt and has not the least inclination to untwist or turn. I hope to have you a Pair finished in a fortnight and shall be glad if you will be so obliging as put them on as soon as they arrive. I have lost much time of great value both to the Country and myself which you will be inclined to believe when I tell you that the Old Flatt Rope here continues working, and that I am now assured my flatt Rope will do 6 times the common Rope.[406]

Finally, on 30th August he wrote:

I finished the Lacing of one of your Ropes on Saturday last and have had it at work on a Pit of ours near 100 yd Deep without Conductors for 4 days in order to try it and have the satisfaction to say when it was first put on that it twisted in that depth only 1 ½ times which was coming full as near perfection as I expected and which in the case of the greatest speed of drawing can be of little or no inconvenience, but when we took it of[f] it twisted only ½ a time in that depth, and had it staid [sic] on 3 or 4 days more I expect it would not have twisted at all. If the Ropes don't serve you 9 Months I shall be a little surprised. We have now finished the Ropes at Attercliff Coll'y which had been working 70 Weeks all but 2 days and drawing including the Corves and Chains &c to which the Corves are suspended and including the empty Corf &c going down as near as I can calculate 1114 tons per Week, so that these Pair of Ropes have drawn nearly on the whole 80 S[under]l[an]d tons. The Account you gave me of your Common Ropes was 719 scores in 6 weeks and allowing as above for the Corf and Chain both up & down as I have calculated above they drew on the average about 1500 tons per week and on the whole less than 9 S[under]l[an]d tons, and the above flatt Rope was composed of 3 only of 3 In. circumference each and our Corf, Coals and Apparatus weighs 8 ¼ cwt and I expect yours weigh only 9 cwt, so that if you compare

Figure 9.11. Flat rope (made of iron wire) from Hemingfield colliery. (Iron wire was introduced in England around 1835). Image © John Hunter.

all these points together it certainly leaves a flattering prospect for the flatt Rope for you which is composed of 4 Ropes.

Your 2nd flatt Rope will be finished on Wednesday next and will endeavour to send of[f] by a Boat the Ropes and Rail Roads by the end of next week if possible and shall be glad if you will get the flatt Rope to work as soon as you can that no more time may be lost.

A later example of flat rope is illustrated in Figure 9.11.

Curr was sufficiently confident in his rope and that the problem of twisting had been overcome that the letter continues:

I think there is now little doubt of a machine for lacing being wanted at N. Castle and think the lower part of the Quay side would be favourable especially as I propose to have a Warehouse for Rail Roads which my Agent would be able to look after also. If you could be so good as to be making some enquirys before I see you it would oblige me much.[407]

Progress

Curr's confidence was justified for he heard from Buddle in a letter written from Wallsend colliery on 15th September 1802:

The William & Ann has arrived this Tide with Mr. Wade's Engine etc. etc. and I will have every Thing sent to their respective places of Destination with all convenient Speed. The Cowpen Flat Ropes are not yet set to work the alteration in the Shaft for new Meetings which I informed you of in my last, not being completed yet, but it must soon be done of <u>Necessity</u> as the last pair of Dicky's round Ropes are at work, and one of them is a very bad one. At Hebburn I may probably get the new 4 laid on, on condition of drawing the Work as quick as the round Ropes, as soon as the present pair fail, but we cannot possibly attempt with one of the 6's and one of the 4's in the present bustle of the Trade both Pits being at work double Shift as hard as ever they can go, and the very idea of losing a single Score is excruciating to us.

Buddle continued on the subject of payment:

> *I was with Major Wade on Monday, and mentioned the Bill for the flat Ropes to him, and he promised to do something for me next Monday. I expect to see Mr Russell tomorrow and will ask him to settle your Balance for Walls end and will endeavour to remit you as soon as possible ...*[408]

Payment is a recurring theme, as seen from a letter dated 28th September 1802:

> *Balance of Acct. for Ropes at Walls end, as settled when you were here £29 14s 2d.*
>
> ...
>
> *Equating the Weight, Price & Quantity of work drawn at Hebburn by the Flat Ropes to Grimshaw's I make £102.9.6 the Sum due for them, but the Major would only pay £80 on Acct. and it was with considerable Difficulty I obtained that Sum from him, as he alleges, that the Ropers have given twelve Months Credit, and that you had told him you would be at the Expense of putting up the Pulleys, shifting & adjusting the Hogs etc. I do not know anything of this Business, but you will recollect how it was to be adjusted, but as this is not a final Settlement you may probably be here again, before it may be necessary to adjust the Acct.*

By now flat rope was being employed at other collieries. The letter continues:

> *I have been at Cowpen this Morning where everything is nearly ready for laying on the Flat Ropes, whether we shall be able to put them upon the Machine on Saturday first or Saturday Week I know not, but will not be longer than the latter Day at any rate. Mr Watkins sends for his ropes tomorrow, or Thursday, and will set them to work as soon afterwards as possible. We lay the old Pit at Hebburn Tub on Saturday next for the purpose of repairing the Tub, and I expect the Major will allow the new Pair of flat Ropes to be set to work there as soon after as the present pair of round ones fail, so that ere long I expect we shall be satisfied as to the Merit of the flat Ropes and sincerely wish it may succeed as well here as it has done in the Southern part of the Kingdom. I am rather afraid of the narrowness of Cowpen Pit being unfavourable to us. You have not mentioned Mr Peareth's flat Ropes in your Letters since I mentioned them to you, but expect you have put them forward, as he was enquiring after them yesterday.*[409]

Success

A series of letters from Buddle to the Currs confirms that the flat rope was performing well:

24th May 1803: *I have just come from Cowpen where the Splice is going on exceeding well. The Hebburn Flat Ropes are at work and going on as well as possible, as are Mr Watkin's also which have not undergone any Material change since Mr Curr saw them.*

12th June 1803: *The flat Ropes everywhere are doing well; I saw Mr Watkins' Agent yesterday who says he does not perceive the smallest alteration in the Pensher Ropes. One of the Cowpen Ropes (the spliced one) finished its mortal Career last Friday Week after having drawn 2278 Scores 9 Corves.*

17th July 1803: *P.S. You will please put the following Order in execution as soon as possible. Viz.*

A Pair flat ropes	*120 Fathoms long for Messrs Fenwick & Co. Lambton.*
A pair ditto	*110 Fathoms long for Mr Watkin at Harraton Colliery*

No change in the Pensher Ropes yet.

12th February 1804: *P.S. The flat ropes are going on well everywhere but Money is very Scarce I expect to remit something in a few Days but cannot get the principal Sums I am afraid till April. I will however remit as soon as I receive anything …*

7th March 1804: *I should be glad to have a few more Flat Ropes for 4" sent when you have the opportunity …*

11th April 1804: *The flat ropes are doing well everywhere.*[410]

By 1804 Curr's confidence in flat rope had grown to the extent that he had printed (by Akenheads, printers, Newcastle-upon-Tyne) an open letter for circulation to colliery owners in the north-east coalfield (see Appendix 9).[411] The letter, printed in pamphlet form, has two parts.

The first, written by Curr, makes claims of the superiority of flat over round rope and the resultant savings:

*As ropes are become expensive articles in collieries, I beg leave to hand you a fair and impartial statement of my flat ropes introduced by Mr Sober Watkin, at Painsher Colliery, on the river Wear, which was set to work on the 30th day of January, 1803. … On the 12th December, 1803, the flat rope was taken off on the east side, when it had done nearly the work of seven of the common round ropes … It will be proper to state also, that a saving of at least seven shillings per week in the article of grease has been made … and, of course, by the flat rope not twisting round, the corves and pit sides are very much preserved, and the pitmen like them better than round ropes, conveying them so steadily, and more particularly in not being liable to the dangerous circumstance of **GLUEING**.*

Curr also states that pairs of flat ropes have been laid on by Mr Watkin and Co. at Fatfield Colliery and by Mr Fenwick on the E Pit at Lambton Colliery, where *'the depth of shaft is 93 fathoms'* (558 ft.).

The letter concludes:

> *... and may be met with at the Three Indian Kings, Quay-side, Newcastle, on Saturday the 16th of June Inst.*
>
> *Ropes may be supplied in two or three months from the date of ordering.*

The second is the *'fair and impartial statement'* mentioned above, a copy of a letter to Curr written by William Anstice, Successor to the late W. Reynolds, Esq., lauding the merits of flat rope installed at Madely Wood Colliery, near Shiffnal, Shropshire. The letter concludes:

> *P. S. We have for a few weeks past had a pair of your flat ropes at Blest's Hill, where they promise as well as those already tried; and we expect to apply another pair in our colliery before long. W.A.*

A long letter written by Buddle to Curr from Wallsend colliery on 23rd February 1806 goes into great detail regarding the performance of flat rope at Washington D Pit, Lambton and Pensher (see Appendix 10).[412]

Advertisements

Curr advertised the products of his ropery extensively. Examples of his advertisements appear in newspapers nationwide from as early as 1803 [see Figure 9.12] until at least as late as 1814. Their text is verbose by today's standards (see the advertisement in Appendix 11 for the *The Hull Advertiser*, 5 December 1807, which runs to over 1,250 words). In the modern way, Curr offered interest-free credit and warranties as inducements to the prospective purchaser (see the advertisement in Appendix 11 for the *Chester Chronicle*, 9 November 1810).

Figure 9.12. Ad for patent flat ropes. Aris's *Birmingham Gazette*, 13 June 1803.

> **PATENT FLAT ROPES.**
> THE Patent Flat Ropes, invented and manufactured by Mr. Curr, of Sheffield, for the drawing of Coals and Minerals out of Mines, being tried and much approved of by the Coal Masters for their Duration, as well as by the Colliers for preventing the Breakage of Coals in drawing up the Shaft—this is to inform the Public, that for the immediate Accommodation of the Collieries in this Neighbourhood and at Coalbrook-dale, and to prevent extra Expences and Damages in conveying the Ropes over Land, Mr. Curr has engaged a Warehouse at Mr. Pitchfork's, of Tipton, near Dudley, where Ropes of sundry Lengths may be had, by personal Application, or by Letter addressed to Mr. Pitchfork; as also proper Pullies and Niches.
> Price of Flat Ropes April 18, 1803, one Shilling per lb. six Month's Credit, and the Coal Masters pay the Water Carriage.
> The Flat Rope, if properly fixed up, will wear out from six to eight common Round Ropes of equal Weight.

Other patents

Such was Curr's attention to the development of rope that between 1798 and 1813 he took out a further seven patents.[413]

In 1805 Curr took out patent no. 2891, *Manufacture of ropes* [laying

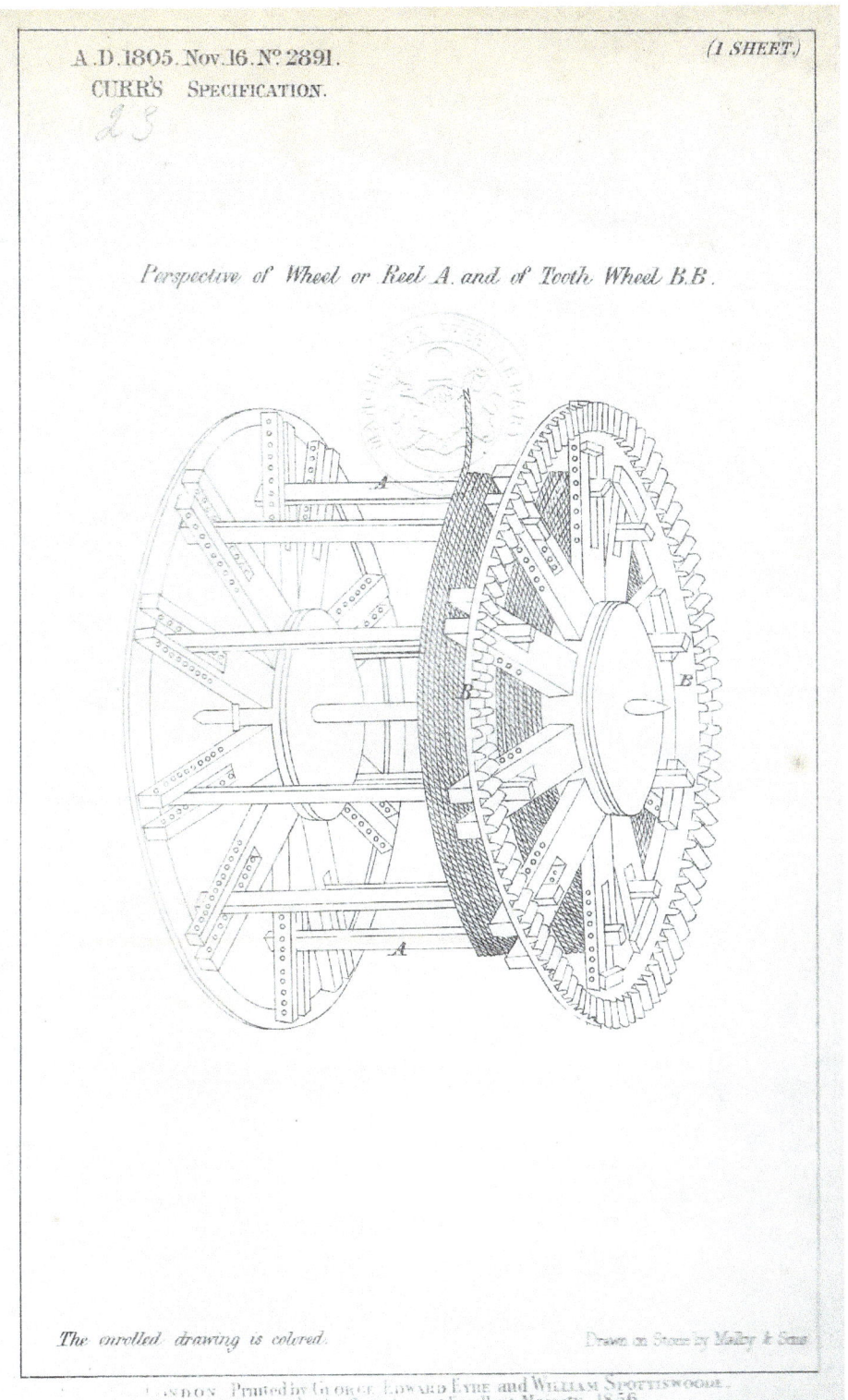

Figure 9.13. Plate from patent no. 2891. From the author's collection.

ropes], published on 16th November. The plate from the 1856 printing of the specification is shown in Figure 9.13.

This was followed in 1806 by patent no. 2914, *Spinning hemp for the manufacture of ropes and cordage*, published 8th March (see Appendix 8).

Not satisfied with applying his flat ropes to mining operations, Curr attempted to break into the market for ships' cables. So, in 1806, he took out patent no. 2947, *Ships' cables and capstans* (applying cables of ships and vessels upon the windlasses, capstans, or drums), published on 4th July. The plate from the 1856 printing of the specification appears in Figure 9.14.

John Curr

Figure 9.14. Plate from patent no. 2947. From the author's collection.

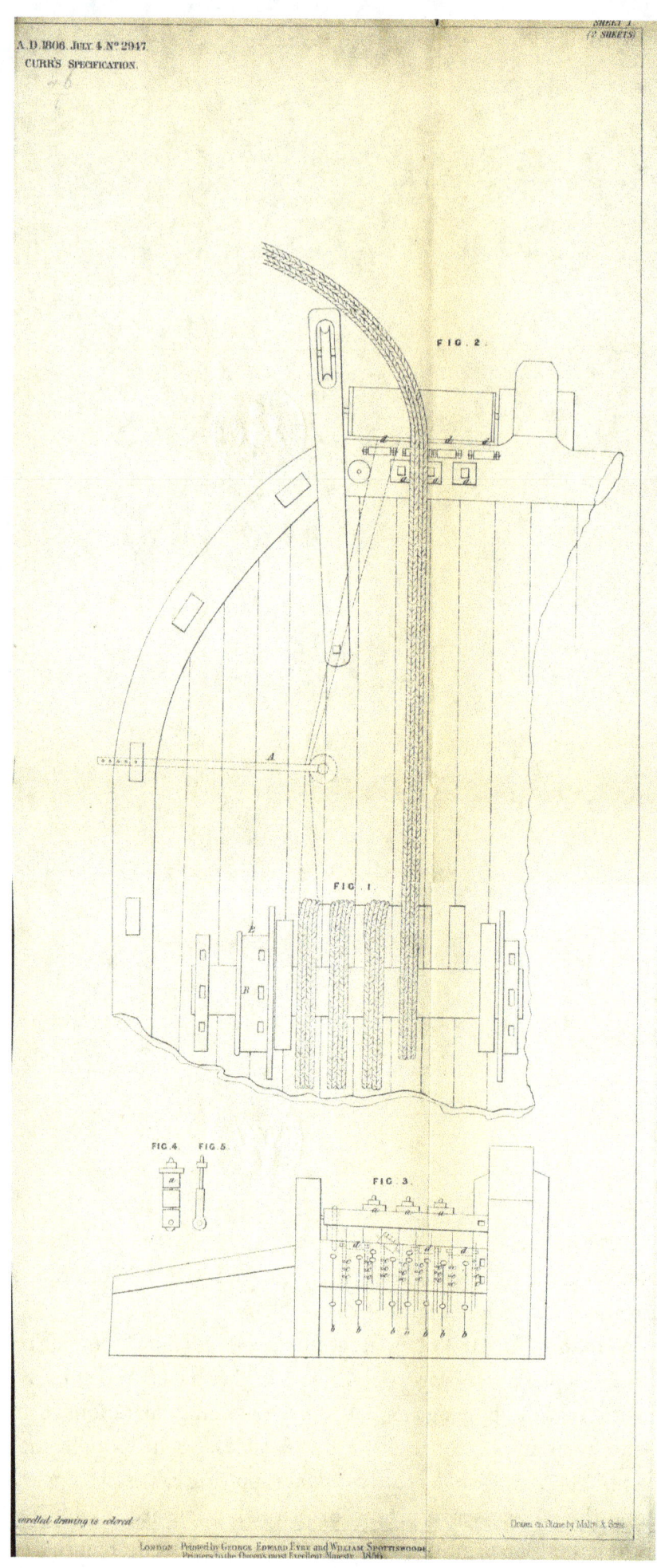

Curr clearly hoped that the Royal Navy would introduce his improved cables, but the Admiralty was not enthusiastic, as shown below in a note dated 6th September 1808 from William Wellesley-Pole:

> *The flat cable and machinery for working it invented by John Curr have been examined on board Ocean Transport and found to be unsuitable and not worthy of trial use on any class of ship.*[414]

In William Bourn's view the failure proved expensive. In May 1820 he wrote:

> *... I have lately purchased of MR. JOHN CURR, the large Engine with which he used to make the Patent Flat Rope. At the time it was constructed, MR. CURR had an idea that he should have succeeded in his endeavours to introduce the Flat Cable, both into the Navy and the East India Company's services; and although he was at a considerable expense, yet he was not able to accomplish his wishes. The Engine was sufficiently powerful to have made a Flat Cable large enough for a Ship of 1000 tons burthen ...*[415]

Curr was not alone in failing to convince the East India Company and the Admiralty of the benefits of his invention. Captain Joseph Huddart, F.R.S. (1741–1816), offered his patent ships' cable (No. 9512, 25 April 1793), after a further seven years' experimentation and satisfying himself that his ideas were sound, to the East India Company around 1800. The Company relegated his invention to the manufacturers they employed who, wedded to their old methods, declined to take it up. Huddart next offered his invention to the Admiralty, with much the same result.[416]

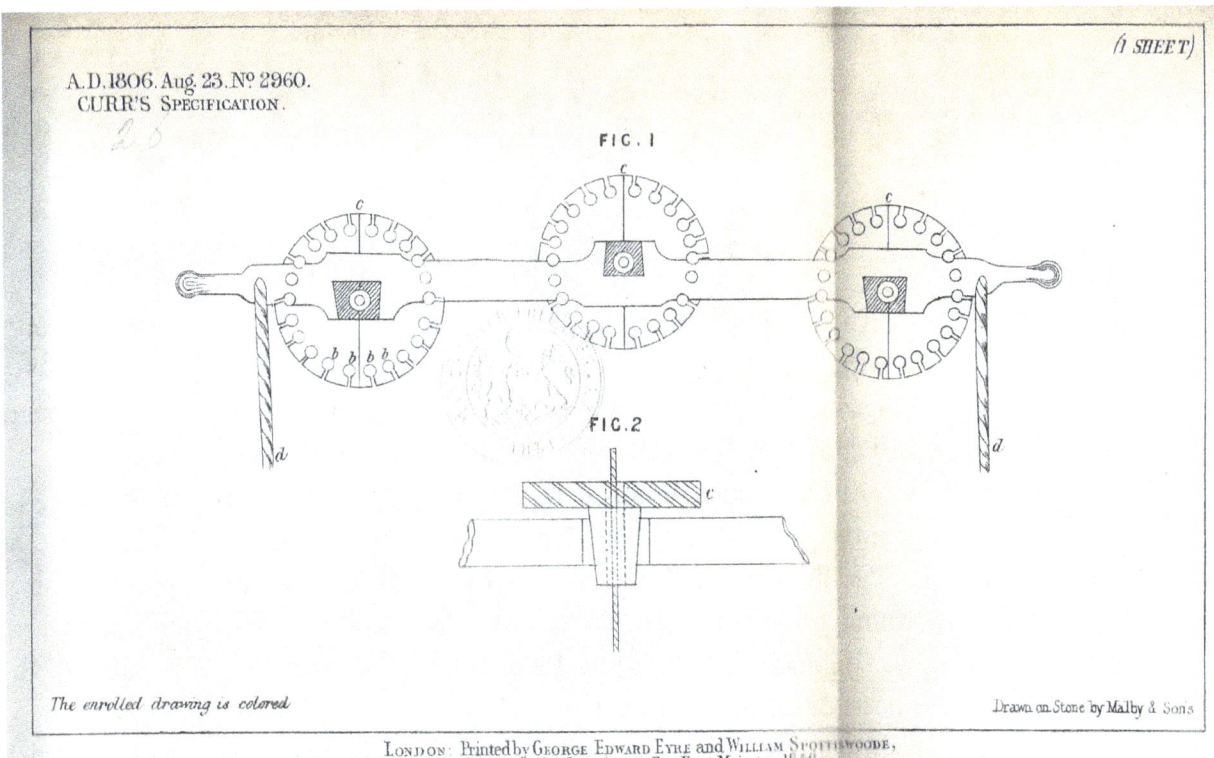

Figure 9.15. Plate from patent no. 2960. From the author's collection.

John Curr

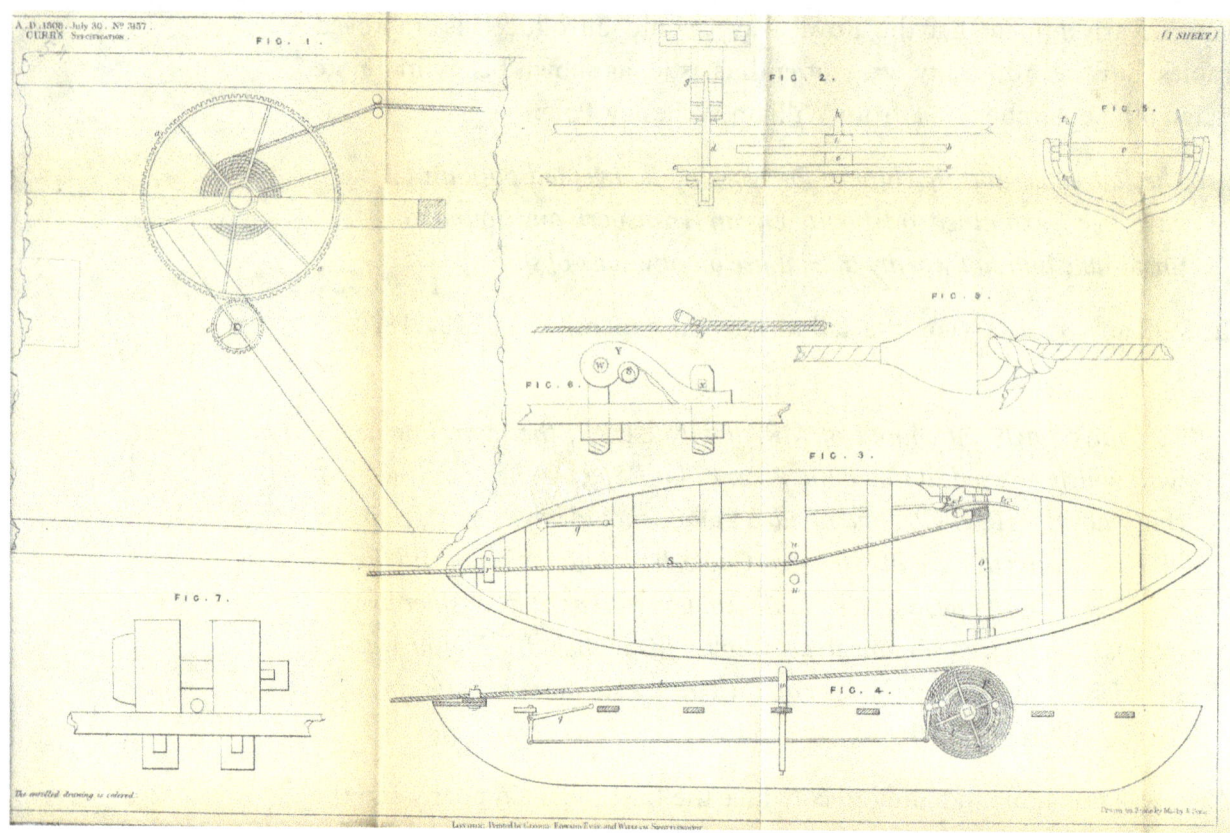

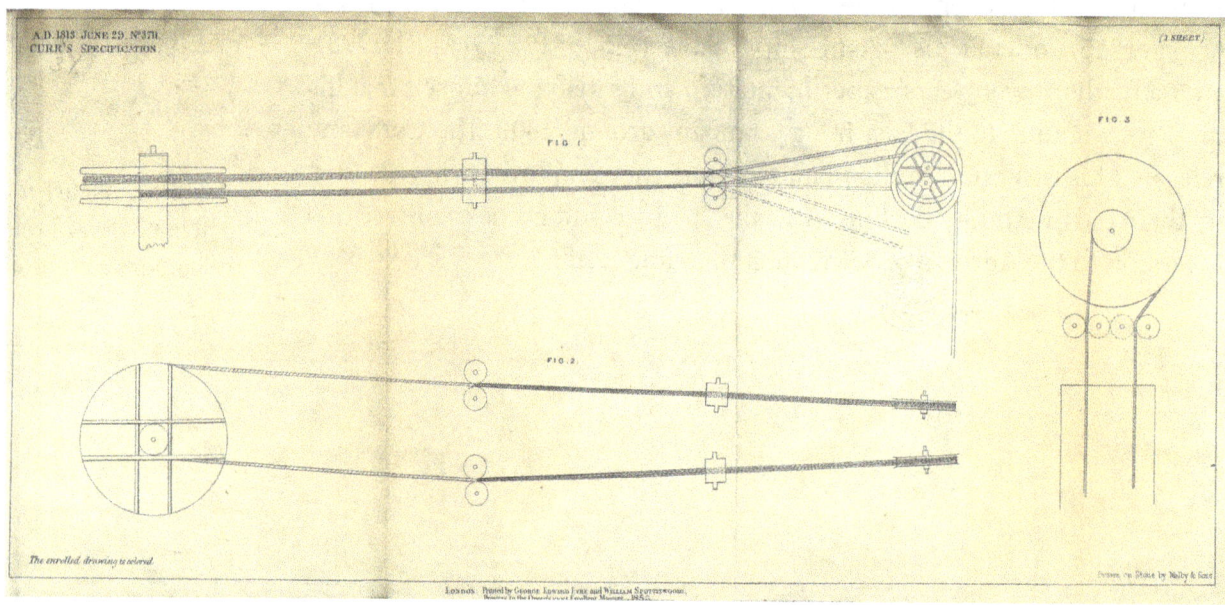

Figure 9.16 (top). Plate from patent no. 3157. From the author's collection.

Figure 9.17. Plate from patent no. 3711. From the author's collection.

Curr's final patent that year was no. 2960, *Manufacture of ropes* (laying and twisting yarns for making ropes), published on 23rd August. The plate from the 1856 printing of the specification appears in Figure 9.15 (see also Appendix 8).

Another patent, no. 3157, *Apparatus for towing vessels, catching or detaining whales, &c.* (applying flat ropes and flat bands or belts to capstans and windlasses of ships and vessels, for towing the same; applying flat or round ropes, lines, bands, or belts, for catching and detaining whales), was published on 30th July 1808. The plate from the 1856 printing of the specification appears in Figure 9.16.

A further two were to follow: no. 3502, *Manufacture of ropes*, published on 30th October 1811, and no. 3711, *Flat ropes for mining purposes*, published on 29th June 1813 [Figure 9.17].

An advertisement in *The Royal Cornwall Gazette* for 3rd December 1814 continues to make claims for the use of Curr's ropes, but now both in mines and on board ships (see Appendix 11).

A royal visit

The *Bath Chronicle & Weekly Gazette*, Thursday 23 November 1815, recorded:

> *Archdukes John & Louis, brothers to the Emperor of Austria, with their attendants, visited Sheffield, staying at the Tontine Inn. They visited the flat rope works in the Park, grinding wheels, screw mills, cutlery manufacturers, The National and Lancasterian schools, etc.*

Ropery sold

In his printed sheet dated 1st May 1820, William Bourn stated that, after Curr's retirement:

> *I have selected and purchased such part only of his Machinery as I know to be really useful and necessary to form a complete Flat Rope.*[417]

On the other hand, Medlicott noted:

> *Curr stopped rope making in 1818, sold some of his machines to R. & W. Furley & Co., Gainsborough, and provided them with his connections, in return for a percentage on all flat ropes sold. Colliery proprietors continued to use hemp and later wire flat ropes in the nineteenth century.*[418]

After Curr's retirement in 1820, a substantial quantity of materials from his foundry and ropery were advertised for sale by auction. The auction lots are listed in Appendix 12.

The stationary steam engine

> *At the very commencement of the [nineteenth] century a combination of circumstances occurred favourable to the application of the steam engine to the haulage of coal-waggons on railways. Watt's patent for the double-acting engine had then expired; the steam-engine entered on a new phase of its history when Trevithick discarded the condensing apparatus, and brought out his high-pressure engine; while to this was added the gradual improvement of railways, which was going forward in virtue of the substitution of iron for wood in their material. Thus it happens that almost simultaneously projects were started for applying the steam-engine to the haulage of railway waggons, both in the form of travelling engines, or locomotives, dragging the waggons behind*

them, and of fixed engines hauling with ropes: Trevithick being the first to attempt the former system in South Wales, and Curr the latter in the North of England.'[419]

In a letter of 1802 John Buddle Jnr had enquired of Curr:

Have you seen or erected any winding Engines for drawing Waggons along horizontally or up ascents. If so have the goodness to inform me to what Extent and if, with good Effect.[420]

The letter appears to have intrigued Curr, for, in 1805 he applied a stationary steam engine to haul coal waggons with ropes from the valley at Birtley to the higher grounds at Black Fell.[421] The system was a simple one; the rope was passed round a large wheel revolving in a horizontal plane to which ropes were attached. The practicability of this innovation was soon established.

Galloway wrote:

More success was achieved by at first applying the steam engine to haul with ropes, especially in cases where a double slope could be arranged for, as an engine placed at the summit served the double purpose of hauling up and lowering both the loaded and empty trains. The first to employ fixed engines in hauling waggons was Mr. Curr ... This was immediately followed by other schemes of a similar character ...

Mr. Hall also adopted edge rails and flanged wheels (Trans. N. E. Inst., vii., p. 6), in lieu of the tram plates and sharp wheels which were still in general use. Thus, he perfected the improvements which Mr. Curr had begun, and established the system of drawing and conveying coal in carriages, as universally practised at the present day.[422]

The consequence in the Durham coalfield was described by Matthias Dunn:

Since the year 1810, inclined planes and public railways have become so generally adopted, as facilitating the conveyance of coal, that many of the distant collieries have again been brought into operation, viz., Stanley, the country to the westward of Chester Burn, Marley Hill, Andrew's House, Greencroft, the new district to the westward on the river Wear above Brancepeth, Shincliffe and the parts east of Durham, Medomsley, Pontop Pike, &c. ...[423]

Here concludes the description of Curr's inventions and innovations.

Ten

CHILDREN

From 1782 to 1800 Hannah bore nine children to John. One child, Charles, died in infancy. Strong faith was a characteristic in common to the lives of all the surviving children. Biographical notes for each are below.

Elizabeth Mary (1782–1823)

Elizabeth was born on 18th March 1782. She was sent to be educated at Bar Convent, York in 1793.[424] She married John Bernard Furniss, also Roman Catholic. Listed as *merchant* in contemporary trade directories, he was one of the trio of local Catholics, alongside his father-in-law and John Smilter, who purchased the Lord's House from the Duke of Norfolk in 1814. Elizabeth bore five children.

Of note is John (1809–1865), a Roman Catholic priest, and a writer of religious pamphlets for children. According to Wikipedia:

> *His writings were assailed as 'infamous publications' by the rationalist historian William Edward Hartpole Lecky in his* History of European Morals, *chiefly on account of the somewhat lurid eschatology of the children's books, which today can seem traumatising, even as more than four millions of his booklets were sold. He warned the 'little child' reading* The Sight of Hell (1861),[425] *for example, that 'if you go to Hell, there will be a devil at your side to strike you. He will go on striking you every minute for ever and ever'. In the 'dungeons of hell', he shows his child readers a girl wearing a dress of fire that always burns her, but never kills her, because she went dancing instead of to church; a boy, who also went to 'dancing-houses' and theatres, endures an eternity of his blood boiling throughout his body.*

According to her nephew 'She [Elizabeth] was thrown out of her carriage and killed'. This accident took place on 28th August 1823.[426]

John III (also John Jnr or Curr III) (1783–1860)

> *Unlike the rest of the family he was a little man, but a good horseman. He painted well, and had studied engineering in which he was well read.*[427]

John Curr

The Currs' oldest son, John, was born in 1783. Nothing is known of his early life until, by his own account, it is learned that, from at least 1801, he worked at his father's iron foundry in Sheffield Park:

> ... at 18 years of age he was entrusted solely with the engineering department, both as to the plans and their execution, of what was then considered a rather extensive establishment in Sheffield Park, belonging to his father, and in which profession he continued about 16 years.[428]

Letters from 1803–4 show that he corresponded with John Buddle Jnr concerning orders for rails, flat ropes and pipes.[429]

His father showed faith in his abilities for, on leaving for a consultation in Ireland in 1802–3, he entrusted responsibility for the supply and installation of a common engine for Percy Main colliery[430] in Northumberland to his oldest son. John Buddle Jnr, who had been appointed viewer at Percy Main in 1802, does not appear to have shared that faith. In a letter dated 15 September 1802 to the elder Curr he asks:

> Yours of the 8th Inst. Informs me of your intention of taking a long Tour which I wish you pleasantly through, but you do not hint anything about the Time when I may look for you here, as I expect you still keep your intention of visiting us on your return from Ireland.[431]

All did not run smoothly. On 24 May 1803 Buddle wrote to Curr III expressing his concerns:

> Now I submit to your Consideration whether this is not too great a speed for the Engine to travel, as I am apprehensive that by its alternating so quick, too great a Strain will be laid upon the moving parts ... The Nut Wheel I am also of opinion is too weak ...
>
> I am really seriously apprehensive of the Axle breaking when the Engine goes to work unless it is previously strengthened, on which Account I should be very glad if you will consult Mr. Curr (who may probably have come home) on the best mode of securing it. ... The best way to enable the Engine to do her work with sufficient dispatch would be to enlarge the Nut Wheel ...

He adds that completion of the engine is running 3 or 4 weeks late (the time fixed was the 19th) and that he'll try to give warning:

> when the time draws near, and I should then be exceedingly glad to see you or Mr Curr ever before we make a Start in order that you may have everything properly adjusted as any accident to delay more time would be a very serious affair indeed, as I do not think there was ever a Coal Machine built of such Importance to the Parties concerned as this.[432]

Curr III replied on 26th May in a long letter in which he gives his calculations regarding the speed of work of the engine and the nut wheel, comparing it with one working on an engine at Attercliffe Colliery. He describes his experiments with the 'spur wheel axle', adding:

> I am afraid when you compare this axle with those you have broken you do not look at each of them fairly.

His letter concludes:

> If you cannot get a nut cast in time I would advise you to start with the present one ... You will see that the engine is capable of performing the work req[uire]d and even more. – My father instead of com[in]g home as I expected is in Ireland, he expected to be home in 3 weeks from this time & to return by Whitehaven – if so he will probably be with you at the starting.[433]

Buddle's next letter is dated 12th June:

> ... I'm glad to find that we are likely to have M[r]. C. here in a few days, as the P. Engine will be ready to turn around about Wednesday next and be set to Coal Work by tomorrow Week, by which time I expect there is a Chance of Mr C being with us. I think you must have mistaken my meaning with respect to making a new Nut, as I never thought of making one without your directions, ... As to the Axle and Nut Wheel I am glad to find that you have such perfect confidence in their sufficiency, and although you may have allayed you have not entirely cleared up my doubts as to that Point. As to the Axle being capable of supporting 40 tons suspended from its Centre ... You will however find that your principle of calculation is erroneous if you refer to Emmerson's [sic] Mechanics[434] 4 to 5th Edition Prop[osition] 73 where this problem is demonstrated ... and the mode of doing it is what I particularly wish to consult Mr Curr about.[435]

It proved that Buddle's concerns were not unfounded, for, in a singularly unapologetic letter dated 16th June, Curr III wrote to Buddle:

> I am sorry to hear by your last letter of the ill fate of the Percy Engine – it is now too late to remedy what is past and in preventing future ..., I fear I shall be of little service. ... Yet with an intention of rendering my best endeavours, and with a hope of reducing a little your Anxiety, and perhaps to allay your fears I propose being at Walls E[nd] on Tuesday next ...[436]

By now, it seems, Buddle had lost patience with young Curr for his letter of 17th July reads as follows:

> At a meeting of the Percy-main Company yesterday, in consequence of the unfortunate Accident and Delays which have attended your Engine, they have come to the resolution of ordering a new one of the best construction on Boulton & Watts Plan, to avoid possible similar Disappointments in future; on which Account they request that you

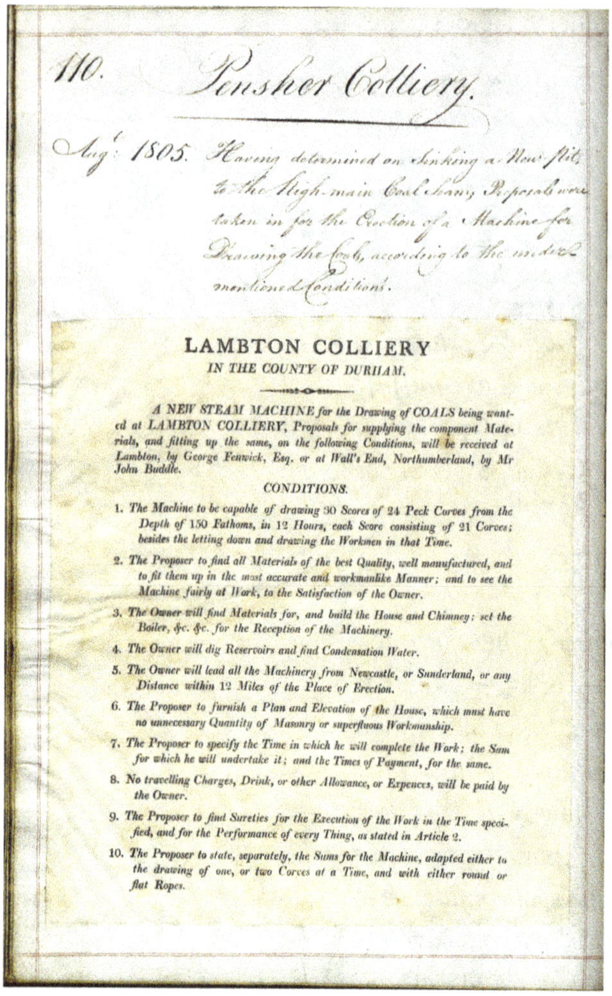

Figure 10.1 (left). Invitation to tender for engine at Pensher Colliery. From: TCR NEIMME/Bud/13/110.

Figure 10.2. John Curr III's proposal for engine at Pensher Colliery. From: TCR NEIMME/Bud/13/113.

may not proceed with the proposed alterations as fixed on the 5th and 6th Inst. As they intend to repair the fly Shaft again and endeavour to draw Coals with the single Corf, till the new Engine can be got ready for work. It is much to be regretted that Mr Curr was not present on this Occasion, as in that Case probably, something more advantageous to both Parties might have been suggested.[437]

The letter concludes with orders for flat ropes. Relations between Curr III and Buddle do not seem to have been damaged, or were repaired, for Buddle invited him, in a letter dated 7th March 1804, to tender for another engine:

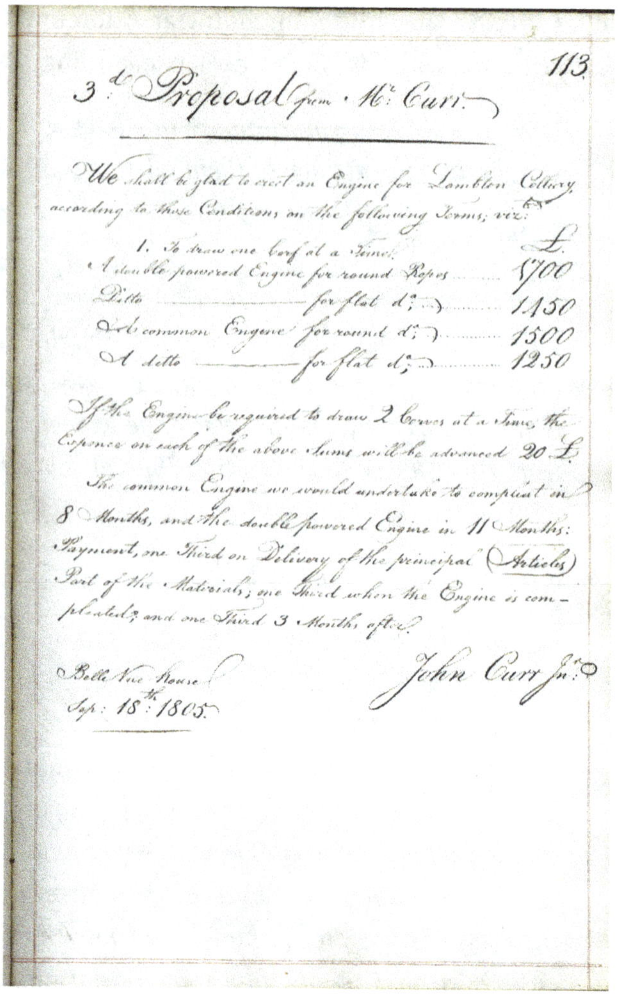

I shall be glad to know as soon as possible in what time and for what Sum you can erect an engine at Durham to draw 20 Score of 16 Peck Corves from 33 fathoms deep in 8 Hours, finding and fitting up all Materials except Foundation and Brick Work and exclusive of leading the Materials from Sunderland. The Engine must also be adapted for pumping Water out of the Pit to the depth of 16 Fathoms when done drawing Coals.

I should be glad to know the Price of both a common and double powered Engine for doing this Business.[438]

And again, in 1805, Curr III, among others, was invited to tender for a steam machine at Pensher colliery. The printed invitation to tender and

Curr's proposal are shown in Figures 10.1 and 10.2.[439] Curr's proposal is noticeably short on detail, particularly since he offered the choice of either a double powered (Watt type) or a common engine. It is not known whether his tender was accepted.

Steamboats

Provided, presumably, with capital from his father, Curr bought an interest in the Norwich Iron Foundry, an old and established iron and brass foundry, under the partnership of Aggs & Curr. The year the partnership was established is not known. The earliest evidence is found in an advertisement in the *Norfolk Chronicle* for Saturday, 5th January 1811.[440] The foundry's premises, in Faith's Lane, adjoined the navigable river Yarmouth and the partners became involved in building steamboats. At this time the steamboat was at a very early stage of development. According to Robert Thurston: 'In 1814 there were five steamers, all Scotch, regularly working in British waters; in 1820 there were 34…'[441]

In July 1814 John Curr III wrote to John Buddle Jnr from the Norwich Iron Foundry, '*understanding that you are interested in the Tyne Steam Packet*'.[442] The reason for Buddle's interest in steamboats was described by Matthias Dunn:

In the year 1818, the application of steam boats to the towing of vessels in and out of harbours, gave an extraordinary impulse to that part of the machinery of the trade; for by getting in and out of the harbour by all ordinary weather, they were enabled to set at defiance those stoppages which formerly affected them, and this led to the abolition of the practice of lying up during the two winter months, both with them and their collieries. Hence the immense increase to the supply in these particulars.[443]

In his letter Curr describes the *Orwell*, the first paddle steamer with which Aggs & Curr were involved, 'designed by Mr. Dodd junr', he states. According to Charles Dawson:

PS Orwell *presents one of the most fascinating histories of the early steamships, particularly in connection with the personality of one of the men responsible for her original engines, John Curr, junior …*[444]

The *Orwell* was built in 1813–14 by James R. Leppingwell, shipwright, of Great Yarmouth. Curr's letter provides brief details of her engine (but does not state who was responsible for building it), paddles, speed and tonnage. R. Malster claimed that Aggs & Curr's was '*undoubtedly the first marine steam engine manufactured in Norfolk.*'[445]

Buddle wrote to Curr on 25th July with some details about the Tyne Steam Packet and on 5th August in much greater detail.[446] Curr replied to Buddle on 8th August. In this letter he claims:

> *I have recd. your favour of the 5th instant and finding that on a comparison between your Engine and ours that ours seems to do much more work ...*

The letter continues:

> *It appears we are going to have a law suit about ours – we lay the fault on the Boat they on the Engine – they say it is only of 2 horse power. What we have to do is prove it is of 12 horses power, without any reference to the velocity of the Boat, which I expect we can easily do. Barring the twists of the law.*[447]

Curr's confidence was misplaced. The *Orwell's* maiden voyage was delayed while a trio of consultants, Maudslay, Donkin and Galloway adjudicated. *Orwell* was originally advertised as going into service in September 1813 and next in April 1814.[448] Sixteen months were to pass before a replacement engine, built by Maudslay and Field, was delivered, on 8th August 1815.[449] The *Ipswich Journal*[450] recorded that she arrived at Ipswich '*with the Union Jack flying at the stern*'. The *Orwell* operated on the river of that name between Ipswich and Harwich '*for a few weeks*'.[451] Her brief service was long enough for an artist to have made a quick pencil sketch shown in Figure 10.3.[452]

Figure 10.3. Sketch of the steamship *Orwell*. From: Dawson 2001, p. 215.

The partners' next steamboat was more successful. The *Defiance* was described in the *Journal of the Franklin Institute*:

> *The month of August [1815] had now arrived, and with it came the third steam vessel,*[453] *named the* Defiance, *built at Norwich, about 112 miles from London, by two scientific young engineers, under the firm of Aggs & Curr. She was seventy-three feet, by twelve, drew about three feet of water, was clinch built, and an excellent model for speed, in smooth water, or river way; had two ten inch horizontal cylinders, with three feet stroke, working into rectangular crank pins, on pinion wheels, gearing into large drift wheels on the paddle shaft; the paddle wheels were large in proportion, and the engines were worked by cylinder boilers, the whole of workmanship superior to anything that had then appeared afloat. This vessel, also, was put into the Margate trade, and had she been somewhat larger, and better fitted for the sea, she would*

> *have given the Thames some trouble; no one however would go in so low a vessel, when the lofty Thames was near, and besides this, the character the Thames had acquired was an additional aid to her, ... The* Defiance *had been sold to Holland, in fulfilment of a patent, granted to some Englishmen, to navigate the rivers of that country by steam; and she, during the year 1816, astonished the natives of that country, by her performances; but when, at the end of the year, an increase of accommodations became evidently requisite, a disagreement had arisen between the patentees, and on the vessel coming to England for repairs, she was suffered to lie and decay for several years, and the patent was eventually forfeited by non-user, ...*[454]

Whether Aggs & Curr built *Defiance* is open to question. Although the letter to the Franklin Institute gives them the credit, Charles Dawson, writing in *The Mariner's Mirror*, states that she was built by Richard Wright of Yarmouth with a double horizontal cylinder engine.[455] Curr, whose sense of injustice over his treatment with regard to the engine for *Orwell* had festered for many years and, quite possibly, a fantasist, claimed in his 1847 book, *Railway Locomotion*:

> *I, with the concurrence of my then partner, Henry Aggs ... built a vessel to receive the ejected engine, (the* Defiance*) ...*

He also claims her success in a race from near London Bridge to Gravesend against the paddle steamer *Margery* and adds:

> *On another occasion, the* Defiance, *in consequence of an affront offered by the* Prince Regent, *at Margate, issued through the town a printed placard, saying that she would lay at her moorings until the* Prince Regent *was under full way, that she would pursue and steam round her, and leave her miles astern in the first hour, all of which was accomplished to the very letter, and in a short time afterwards every new steamer which appeared on the Thames was supplied with an engine on the same principle as that of the* Defiance, *and of which the writer hereof is the inventor.*[456]

The tragedy of the *Courier*

Aggs & Curr built a third vessel, the *Courier*, which made her maiden voyage in November 1816.[457] On 4th April 1817 the *Courier's* boiler exploded causing the death of twelve persons on board.

The *Norfolk Chronicle* reported the disaster in all its gory detail on Saturday 5th April 1817:

> *... It is our painful task to state the occurrence of a most dreadful and fatal accident, which took place in this city yesterday morning. – About nine o'clock, as Wright's Norwich and Yarmouth steam packet had just started from the Foundery* [sic] *Bridge, the boiler of the engine*

> *burst, with a tremendous noise, and, by the irresistable [sic] force of the explosion, the vessel was literally blown to atoms. There were, at this unfortunate moment, 18 persons, adults of both sexes, and two children, on board. – Of this number, melancholy to add eight, (viz. five men and three women) were instantaneously killed. Six women, miserably wounded, were immediately conveyed to the Hospital. The remaining six persons, including two children, have, we hope, escaped without sustaining material injury. – Every possible assistance was promptly rendered to the survivors by those who happened to be near at the time; and among the most active in these humane exertions, it is due to mention Mr. Aggs, opposite whose premises this shocking and deplorable event happened.*
>
> *Not being able to procure an accurate list of the names of the unhappy victims, we abstain, for the present, from mentioning the few we have already learnt, as well as further particulars, which will appear in our next.*
>
> ** * Since writing the above, we have received the names of the six women, who were conveyed to the Hospital, viz. – Sarah Smith, (aged 24) simple fracture of the leg, – Mary Harrison, (50) fracture of the right arm, with wound in the face. – Susan Carr, (22) servant of Mrs. Crockett, in St. Simon's, and Martha Dewell, (58) with wounds of the scalp. – Father Welton, (50) from Yarmouth, fracture of the right leg, (amputated). – Mary Stephen (36) from Yarmouth, compound fracture of both legs, which were immediately amputated.*[458]

What caused the explosion? Evidence given to the Select Committee

The almost immediate result was the setting up of a Select Committee of the House of Commons to inquire into the accident.

Giving evidence to the Inquiry, George Dodd, civil engineer, designer of the *Orwell*, stated in 1817 that he *'had been on board and was well acquainted with twenty steam-boats; knows that there are more than forty in Great Britain'*.[459] News of the nascent technology was spreading. Dodd noted that he

> *had declined purchasing the Norwich steam-packet because it had a high-pressure engine. – Went with a party of German gentlemen from Bremen, who were anxious to make an immediate purchase of a steam-vessel; and they also declined to purchase that, or any of the boats upon the river Yare, solely because they had high-pressure steam engines on board.*[460]

A report on the German party's visit in August 1816, prior to the ship's launch, was detailed in a later article in the *Morning Chronicle*.[461] The observations noted by the visiting party proved prophetic.

From evidence given to the Select Committee it was established that:

The boiler of the Norwich steam-vessel was eight feet long, with a cylindrical boiler four feet two inches diameter; it was first made with an internal angle iron at one end, and an external angle iron at the other end. In consequence of the internal angle iron having given way, a cast iron end was substituted, which certainly was not accurately performed. It was originally intended to sustain a pressure of forty pounds to the inch. ....[462]

Mr Bryan Donkin, who had volunteered to visit Norwich in order to inquire into the cause,

Was of the opinion that the immediate cause of that explosion had been the use of steam of a very high expansive force; the approximate cause was a deficiency in strength of the end of the boiler. The boiler was cylindrical. The cylindrical part, and one end, was wrought iron; and the other end was cast iron. It appeared to have been previously of wrought iron, but, for some reason, the wrought iron end had been cut out, and a cast iron end substituted in its place. – Was of opinion that any high-pressure boiler so constructed was unsafe ... The strength of cast-iron boilers was extremely uncertain: cast iron was liable to contract in various degrees in different places, and therefore was liable to break. – Thought that all cast-iron boilers were dangerous when used for steam of high expansive force.[463]

After hearing the evidence the Select Committee's report, published in 1817, included four resolutions, summarised as follows:

1. Resolved, That it appears to this Committee, from the evidence of several experienced engineers, examined before them, that the explosion in the steam-packet at Norwich, was caused not only by the improper construction and materials of the boiler, but the safety-valve connected with it having been overloaded; by which the expansive force of the steam was raised to a degree of pressure, beyond that which the boiler was calculated to sustain.

...

3. Resolved, That it is the opinion of this Committee, that, for the prevention of such accidents in future, the means are simple and easy, and not likely to be attended with any inconveniences to the proprietors of steam-packets ...

...

4. Resolved, That the Chairman be directed to move the House, that leave be given to bring in a Bill for enforcing such regulations as may be necessary for the better management of steam-packets, and for the security of his Majesty's subjects who may be passengers therein.[464]

For the key features of the bill see below.

Slow uptake in the United States of America

In the years following that terrible accident, the subsequent lack of action and resultant loss of life from steamboat explosions in the United States prompted the following letter to the *Journal of the Franklin Institute* in August of 1834:

> SIR, – So many, and such very serious, accidents, have happened among steam vessels in the United States, that it appears high time that some legislative enactment should be adopted to protect the public from their dreadful consequences ...[465]

In December, the *Journal* noted that *'The Spring of 1817 was marked by the most disastrous steam packet accident in England'* and that the Select Committee had *'reported a bill prohibiting the use of high pressure steam engines, in vessels, and directing a severe test of the strength of boilers, used for such purposes'*.[466]

In the United States of America the first attempt to regulate steamboats by a law was not until 1838. The legislation proved inadequate as disasters increased in volume and severity. The Steamboat Act of 1852 tightened the legislation, after seven boiler explosions in the eight months preceding its passage claimed more than 700 lives.

Notably, though, the USA had seen the first successful commercial introduction of the steamboat; in 1811 the *Juliana*, the world's first steam-powered ferry service, was put into operation by John Stevens and, in 1816, Francis Ogden established the first steam paddle-boat service on the Mississippi River. Also noteworthy is that it is generally accepted that the non-condensing high-pressure engine was developed by an American, Oliver Evans.

Aftermath

The saga had its final flourish in 1847 when Curr, after returning to London from New South Wales, published his book *Railway Locomotion and Steam Navigation: their principles and practice*. According to his nephew:

> I think it was in 1849, after having elaborated a scheme by which steamers could do their voyages with only half the coal then used, that he went to England for the purpose of getting his views brought into use by shipowners. In this he failed.[467]

The book is eccentric. As well as an attack on the eminent mathematician Charles Hutton (1737–1823), the book is replete with strange claims, such as: *'For the purpose of railways* [in New South Wales] *wood is decidedly preferable to iron'*.[468]

The book was reviewed, not unkindly, in *Mechanics' Magazine*:

> *The circumstances under which this book has been written, the evident wish of the author to be useful, and we may add, the amusement some of his declamations have afforded us, induce us to give it a more*

> *extended notice than it would otherwise lay claim to. It appears that Curr, whilst an engineer in England, some thirty or forty years ago, had a dispute with his professional brethren (MR. Maudslay amongst others), and was so disgusted at 'finding ignorances silvered over with plausible appearances,' that he quitted the profession, and transferred himself and his acquirements to New South Wales; where, however, it does not appear that they were any better appreciated. …*
>
> *For such a man no generous mind can help feeling the deepest respect, however unsuccessful the effort. The author's case is by no means a solitary one. We have reason to know that there are hundreds of men at this moment – not exiled like Mr. Curr in such distant lands as Australia, but here in England, and in London, – who have wearied themselves for years in labours that* must *end in disappointment; because without any sound and rational foundation.*[469]

Curr was furious and wrote to the editor of the *Mechanics' Magazine*, who responded in print:

> *We have received a very long letter from Mr. Curr, on the subject of our review of his book. He finds fault with it, of course, for what bemauled author ever yet kissed the rod where he was chastised? 'I do not,' he says 'ask the insertion of this in your Magazine as a favour, but demand it as an act of justice;' but with the manifest purpose at the same time (for Mr. Curr is no simpleton, whatever else) of rendering compliance with his request impossible, he has interlaced every passage of his letter with terms of abuse too gross for insertion in the pages of any respectable journal. One or two of the things in the epistle, we must, nevertheless, in fairness to ourselves, and in despite of Mr. Curr's insolence, advert to in a general way. …*
>
> *He defends, too, his absurdities about the thermometer in a way which shows, if possible, more glaringly than his book, that he knows and understands absolutely nothing about it.*[470]

This editorial further provoked Curr, culminating in his rebuttal, *The Learned Donkeys*. This name is presumably a veiled pun on the names of Messrs Maudslay, Donkin and Galloway who, all those years ago in 1814, had adjudicated in the dispute over the power of the *Orwell*'s engine.

Curr claimed in *Railway Locomotion* that they had reported that

> *the engine instead of being of ten horse power was only of five. The cause assigned for their decision being, that the engine having two cylinders, the connecting rods applied to cranks at right angles, and without the accompaniment of a fly-wheel, had no more power than one cylinder, for it was like two men turning a grindstone, and they had no more power than one!! In what school they acquired such notions of mechanics it is not for me to say, but such is the substance of their report as expressed to myself by the lips of Mr. Donkin.*[471]

Fourteen pages long, *The Learned Donkeys* consists of letters to the editors of *Mechanics' Magazine*, the *Artisan* and the *Civil Engineer and Architects' Journal*. It concludes with a section 'TO THE PUBLIC'.[472] He advertised this pamphlet on the front page of the *Morning Chronicle* [Figure 10.4.].

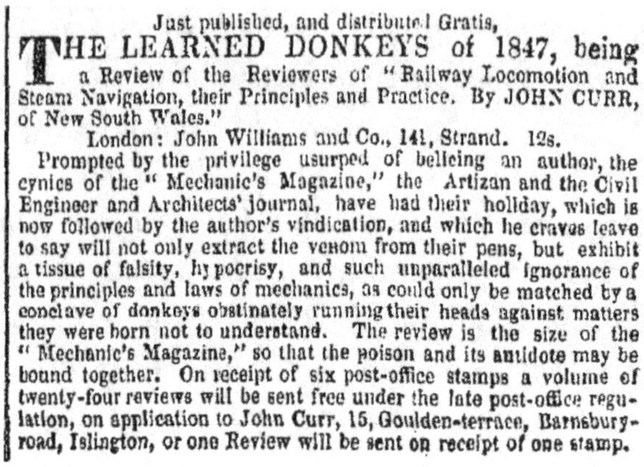

Figure 10.4. Front-page advertisement in the *Morning Chronicle*, Wednesday, 15 March 1848.

> Just published, and distributed Gratis,
> THE LEARNED DONKEYS of 1847, being a Review of the Reviewers of "Railway Locomotion and Steam Navigation, their Principles and Practice. By JOHN CURR, of New South Wales."
> London: John Williams and Co., 141, Strand. 12s.
> Prompted by the privilege usurped of belieing an author, the cynics of the "Mechanic's Magazine," the Artizan and the Civil Engineer and Architects' journal, have had their holiday, which is now followed by the author's vindication, and which he craves leave to say will not only extract the venom from their pens, but exhibit a tissue of falsity, hypocrisy, and such unparalleled ignorance of the principles and laws of mechanics, as could only be matched by a conclave of donkeys obstinately running their heads against matters they were born not to understand. The review is the size of the "Mechanic's Magazine," so that the poison and its antidote may be bound together. On receipt of six post-office stamps a volume of twenty-four reviews will be sent free under the late post-office regulation, on application to John Curr, 15, Goulden-terrace, Barnsbury-road, Islington, or one Review will be sent on receipt of one stamp.

In June 1817 the partnership of Aggs & Curr was dissolved.[473] From this time on Curr concerned himself with tobacco and snuff.

On 14th May 1818 Curr married Martha Melton (1793–1866), by licence, at St Peter Parmentergate, Norwich.[474] According to his nephew, Curr *'married beneath him'*.[475] That Martha was illiterate supports this. Martha's parents, Francis and Mary, who were also illiterate, lived at Great Melton, near Norwich, where Martha was christened. The couple's first child was born in 1818, so Martha was with child at the time of their marriage. Between 1818 and 1835 the couple had nine children.

> *He lived at Norwich where he was an importer of tobacco, enjoyed a considerable income, and kept his carriage and hunters.*[476]

Trade was sufficient that Curr opened two shops to sell his wares, as can be seen from his advertisement in the *Norwich Mercury* shown in Figure 10.5.

Figure 10.5. John Curr III's advertisement in the *Norwich Mercury*, 17 April 1830.

> **JOHN CURR,**
> **TOBACCO & SNUFF**
> **MANUFACTURER,**
> RESPECTFULLY begs to inform his Friends and the Public he has opened Retail Shops in
> **BRIDEWELL ALLEY**
> *And at No. 7, on the Gentleman's Walk*
> For the Sale of TOBACCO, SNUFF, CIGARS, PIPES, and BOXES of every description, which he trusts will be found of the best quality and moderate in price.
> JOHN CURR particularly invites attention to A LARGE STOCK OF GENUINE HAVANNAH CIGARS, which he warrants to be SIX YEARS OLD and of very fine quality, as also to a large assortment of SCOTCH, BRUNSWICK, FRENCH, and HOME made SNUFF BOXES, GENUINE MEERSHAUM PIPES, EAST INDIAN HOOKAHS, &c. &c. which he has purchased from the first market, and offers for sale at reasonable profits.
> *** A Package of Welch high dried Snuff is just received from Lanacomede of particular fine quality and flavour, and which will be found a great treat to all takers of Irish Snuff. (1097)

Curr's father died in 1823 and John Curr III's portion of the estate was some £2,300, a substantial amount at that time. The family lived in some comfort. The Electoral Register records Curr as owning the freehold of a house with land at Thorpe (2 miles east of the city) in 1832 and a freehold estate at St Andrew's Hill, Norwich in 1835.[477] His two oldest sons, John and Charles, were sent to be educated at Stonyhurst College, where they were contemporaries of their cousins Edward, William and Richard, sons of Curr's brother Edward. Edward had been appointed manager of the Van Diemen's Land Company in 1825, which had been established at Circular Head in the north-west of the island. Curr's nephew wrote:

> *In 1833, my father [Edward] being then in England, my uncle John came to Sheffield where my father had taken a house for a few months, and passed some time with us. My father having come from Van Diemen's Land, conversation naturally turned a good deal on colonial ways and doings, the result of which was that my uncle determined to give up his business, and, in the sunny climes of the south, become a grower of the weed of which he had hitherto been an importer. From this step my father endeavoured to dissuade him. Uncle John, however, who had married beneath him, leaving the younger members of his family in England to be educated, sailed with his wife and elder [sic] children for Van Diemen's Land.*[478]

The Currs, together with four children, arrived at Launceston, Van Diemen's Land, in October 1835. Little is known about the family's stay there.[479] From a contemporary newspaper we learn that Curr was a 'respectable merchant'.[480]

According to his nephew,[481] after leaving Van Diemen's Land Curr tried, unsuccessfully, to grow tobacco on a farm near Wollongong. His nephew added that he eventually went to Sydney and again became a tobacco merchant.[482] It appears that he left Launceston for New South Wales in advance of his family. His wife, Martha, and the children arrived at Sydney on 7th July 1841.[483]

His stay in Sydney is marked by letters, on a wide range of topics, to the Sydney newspapers. The only evidence of his tobacco growing is to be found in the first of these:

> Sir,
>
> *In reference to a paragraph in your paper of Tuesday, headed 'Colonial Tobacco,' and purporting to rectify a mistake in last Thursday's* Gazette, *as to whom the first prize was awarded at the late Horticultural Show. I enclose a certificate from Mr. Oliver, (one of the judges), from which you will find it was awarded to myself, and not to Mr. M'Eneroe, for the best specimen of Colonial Tobacco.*
>
> *I also send by the bearer for your inspection, the identical specimen for which the prize was awarded, as under the circumstance of your reporter being secretary to the society, such a mistake, instead of being natural, is the less pardonable.*

> *Not doubting you will rectify the error in your next publication.*
> JOHN CURR, *Sydney, Feb. 13, 1840*.[484]

Beneath Curr's letter appears the following gracious apology:

> [We insert Mr. Curr's letter, and beg to say, we regret that any mis statement [sic] should have appeared in the Gazette. We have read this certificate, awarding first prize to Mr. Curr, signed William Oliver. We may further add, that so far as we are capable of forming an opinion, we should think it difficult, to produce a finer specimen than the one forwarded to us, whoever might be the manufacturer – we are much obliged to Mr. Curr for the same.]

His, supposed, ill treatment at the hands of Messrs. Maudslay, Galloway and Donkin, 'The Learned Donkeys', had been festering since 1814. In an 1842 letter of self-justification to the editors of the *Sydney Morning Herald* he wrote:

> ... it may not be out of place to give a brief account of the first application of two engines in one, dispensing with the necessity of the fly wheel.
>
> The engine was invented and made by the son of the inventor of the rail-road; it was contracted for to be of ten horse power, and was put on board an ugly flat bottomed craft called the Orwell. The owner expected the vessel to go six miles an hour, but the engine could only propel her four and a half. The owners laid the fault on the engine, and the engineer on the craft, when they by the advice of counsel, and in order to obtain the weight of evidence on their side, sent to London, a distance of 110 miles, for three of the most eminent engineers the metropolis contained, whose names were Messrs. Maudesley, of Blackfriars Road, Mr. Galloway, of or near Pimlico, and a Mr. Donkin. These gentlemen after spending six days in examination and trial, and charging ten guineas a day for their labour began to suspect the engine would be found deficient in power, and at the end of four days more they came to the conclusion that the engine was not a ten-horse, but only a five horse-power. So circumstanced the inventor took back his engine, and built a vessel called the Defiance, which it propelled about seven miles an hour, and Mr. Maudesley put a twelve horse engine on board the Orwell, which it could not stir at the rate of three miles an hour.[485]

In August 1843 Curr wrote once more to the editors of the *Sydney Morning Herald*, propounding his calculations on the subject of 'The Aerial Steam Carriage'.[486]

In September his name is mentioned in the *Sydney Morning Herald*:

> *To show the varied nature of the contents of the Sydney periodicals, we subjoin the table of contents* ... The New South Wales Monthly

Magazine, *Vol I, No. 9 ... 'Low steam power instead of sails.' ...*

The article on Low Steam Power is a letter from a Mr. John Curr, combatting certain assertions in an article on the same subject, which appeared in the Colonial Magazine *(London) a few months since. It may be remembered that some remarks appeared in the* Herald *on the same subject, when the article was first re-published in Sydney.*[487]

His next letter, this time on 'Watering the Streets', was published in the *Morning Chronicle* on 11th October.[488]

Hardly allowing his pen to rest he published a pamphlet: *Statistics of Banking in the Colony of New South Wales*,[489] with a short review appearing in the *Weekly Register of Politics*:

A singular amalgamation of shrewd common sense and visionary theory on the subject of the present 'Monetary Confusion.' The main argument is based on a falacy [sic] – which we have already in part exposed: but there is matter in the pamphlet which will repay a perusal.[490]

Another appeared in the *Morning Chronicle*:

'STATISTICS OF BANKING'. – We beg to acknowledge the receipt of this small pamphlet, by Mr. John Curr, containing some very valuable statistical returns of the assets and liabilities of the banks, with their capital and profits, and suggesting a remedy for the present monetary distress of the colony. Mr. Curr has evidently given this all-important subject his most serious attention, and the inferences drawn, and the propositions advanced by him appear to be the result of calm and dispassionate enquiry. We hope to be able to make an extract from his work in our next publication, and in the meantime recommend its general perusal.[491]

Curr must have been heartened by these comments, but they soon turned to ridicule. Firstly, in the *Sydney Morning Herald*:

STATISTICS OF BANKING. – *Under this head there has just been put forth, by a 'John Curr', a pamphlet of sixteen pages. We have read it attentively, and with all the candour and deliberation to which such a discussion is entitled; and we are really sorry to say, we can make neither head nor tail of it. It is a solemn nothing – vox et praeterea nihil!*[492]

And, a few days later, in the *Weekly Register of Politics*:

To Correspondents.

'JOHN CURR.' – We are afraid that in giving this writer credit for a fair share of 'common sense,' our desire to encourage Literature carried us a little too far. A writer on Finance, who cannot discover that money deposited in a Bank is so much circulating medium available

> *to the depositor, must be sadly deficient in that useful quality which we were, we find, too willing to discover in John Curr.*[493]

This provoked another of Curr's letters of self-justification, this time to the editor of the *Australian*:

> *Sir. – the* WEEKLY REGISTER *of Saturday referred to my name in a manner likely to induce a belief that I had corresponded with that paper, which I beg very distinctly to deny.*[494]

Undeterred, he wrote another long letter to the *Australian*, this time a commentary on the 'Report from the Select Committee on Monetary Confusion'.[495]

Almost three years were to pass before Curr's next letter, this time on the subject of railways.[496]

In 1847 Curr sailed for London[497] where his book *Railway Locomotion and Steam Navigation* was published in the same year. Ever busy, while there, he penned an unpublished paper 'On the temperature of steam and its corresponding pressure', which is to be found in the archives of The Royal Society.[498]

In 1850, an Australian newspaper, under the heading 'Important Discovery' refers to a prospective, but probably never published, publication, of Curr's,

> *... announced most energetically and somewhat dramatically, In the following advertisement of a work published in London (1848), by 'John Curr, of New South Wales,' and entitled the 'Physical Properties of steam, in relation to the mobile matter of the Steam Engine.'*[499]

The article reprints the, wordy, advertisement in full, which concludes

> *No payment or any consideration whatever will be required until the work is complete, nor then, if it not be satisfactory; and as a guarantee of the good faith of the inventor,- should he fail to give satisfaction,- either his head, or liberty for life, at the option of any government which may incur the responsibility of employing him, shall be the stipulated penalty.*[500]

Curr returned from London to Sydney, where he arrived on 21st September 1850.[501] His brother, Edward, died at Melbourne on 16th November that year and his nephew noted

> *I saw him in Melbourne the day after my father's death in 1850.*[502]

After this no further publications or letters have been found. It is not known when the Currs removed to Wollongong but Curr and Martha both died at Drummond Cottage, Mount St Thomas, near Wollongong, he on 22nd June 1860, aged 77, and Martha on 23rd October, 1866, aged 75. They were buried in the Old Roman Catholic Cemetery, where is to be found a tombstone bearing a faint inscription of the letters JC, which is thought to be his. A memorial stone, erected years later, gives their details, as well as those of their nine children.[503]

Mary Ann (1786–1868)

Mary Ann Curr was born in 1786 and in 1797 sent to be educated, like her sister before her, at Bar Convent in York. She married, by licence, Louis Amand Beauvoisin (alias Bonvoisin),[504] on 19th July 1821 at Sheffield Parish Church. Her parents had only recently returned from Paris and, in a letter dated 16th August, her father wrote to John Buddle Jnr:

> … *my daughter Mary Ann has changed her name to Beauvoisin and resides in Manchester. She was Married on the Coronation Day, and Mrs Furniss and Harriet went with her to Manchester and spent a few Weeks. Mr Beauvoisin has been many years in Manchester and after spending a few more there, it is their intention to live in France.*[505]

Louis (1785–1850) was almost certainly a French émigré. The baptismal entry for his son, Henricus D'Argenson Beauvoisin, provides a clue to the original family name. A blog in reference to a later Louis says that the family came from Normandy.[506] *Pigot's Directory* for Manchester lists Beauvoisin as a teacher of French.[507]

Mary Ann's brother Joseph served as a priest in Manchester from 1808 until 1822 so perhaps Louis was one of his parishioners and that is how he and Mary Ann met.

The couple had two children: Henry (1823–1882) and Mary Ann (1827–1902). Henry, born at Ecclesall, was baptised by Rev. Richard Rimmer at the Old Chapel, Norfolk Row on 29th March 1823. His godfather, Francis Duchesne, was the priest to whom, in 1822 according to family lore, Henry's grandfather had entrusted £30,000 to invest and had lost the lot.

Mary Ann's father died in 1823 and her share, as for her siblings, of his estate (her husband was a joint executor) amounted to some £2,300. The 1841 census found the family living at Belle Mount, Sheffield, with one servant.

In 1856 an extremely curious notice appeared in the *Liverpool Albion* [Figure 10.6].

After Louis died in 1850 Mary Ann had removed to Southport. Why was Odell seeking her? The explanation may lie in her son Henry's bankruptcy.

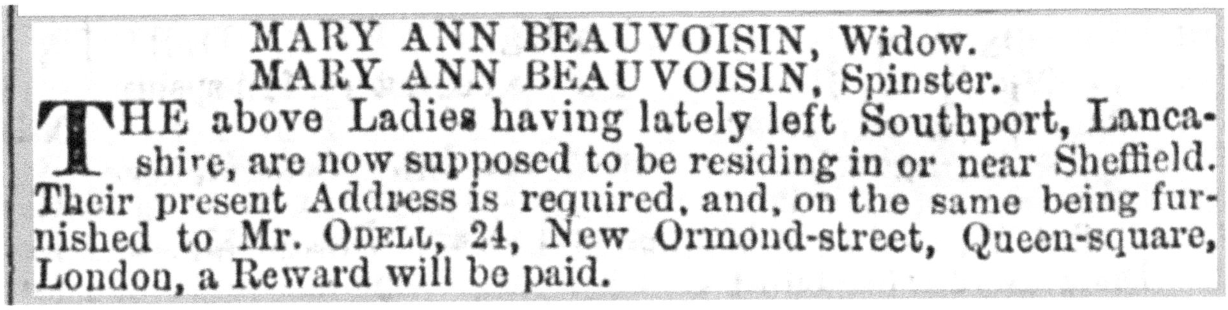

Figure 10.6. Notice in the *Liverpool Albion*, 26 May 1856, seeking information as to the whereabouts of Mary Ann and her daughter.

> MARY ANN BEAUVOISIN, Widow.
> MARY ANN BEAUVOISIN, Spinster.
> THE above Ladies having lately left Southport, Lancashire, are now supposed to be residing in or near Sheffield. Their present Address is required, and, on the same being furnished to Mr. ODELL, 24, New Ormond-street, Queen-square, London, a Reward will be paid.

Henry had over-reached both in business and in furnishing and decorating his home, Belmonte, in Norfolk Rd, Sheffield. The *Sheffield Independent* for 11th August 1855 ran a lengthy description of the bankruptcy hearing against Henry. He owed £2,300, of which £1,750 was to his mother, who had lent him £2,000 in 1853. He had also been lent £400 by his uncle, John Curr III, in December 1853. The Registrar is quoted:

> *It appears that he has been living as if he had a much larger income. He expended of his £2000 capital, at the outset, no less than £1200 in furnishing his house.*

Thomas Odell's occupation was house painter[508] and he had perhaps gone unpaid.

Mary Ann died on 2nd January 1868 at Ypres. Her daughter Mary Ann, educated at Bar Convent, like her mother and other Curr women, had taken the veil and was at St George's Convent, Menin. Her first cousin Mary Anne, daughter of John Curr III, was also a nun there. A grandson of Mary Ann's, Adrian (1852–1906), became a Roman Catholic priest.[509]

Teresa (1790–unknown)

Teresa Curr was born on 1st June 1790 and in 1804 she, too, was sent to be educated at Bar Convent, York.

Of Teresa her nephew wrote: *'I believe another of my father's sisters, called Theresa, also took the veil and died early'.*[510]

A biographical entry for her younger sister Harriet confirms this: *'One of Harriet's sisters became a Trappistine in Stapehill ...'.*[511] Unfortunately, Teresa's name *in religione* is not given. Stapehill Abbey, near Wimborne in Dorset, was a Trappistine Cistercian convent, founded in 1802.

When and where Teresa died is a mystery. In July 1821 she was a witness at the marriage of Mary Ann. Her father died in January 1823 and her name does not appear in his will, so her death was likely between these dates.

What was Teresa like? In a biography of the poet John Holland (1794–1872) of Sheffield Park, the author William Hudson states that Holland wrote a poem on 19th May [1826],[512] 'Theresa's Tree', recalling their childhood. Hudson wrote:

> *This is a beautiful and touching piece of blank verse, containing some very pleasing lines, with tokens of a healthy and charitable largeness of heart. The lady whose name the tree bore was a daughter of Mr. John Curr, of Belle Vue, a mansion which stood at the distance of three fields from Mr. Holland's home, and of which a description is given in* Sheffield Park.[513] *Mr. Curr was an ingenious man, and had the management of the Duke of Norfolk's collieries in the neighbourhood. He was, like his master, a member of the Church of Rome. One of his daughters became an abbess: and Theresa herself became a nun. There was some communication between Mr. Curr's family*

and that of Mr. Holland, who tells of having played, in childhood, without sacrilege, with Theresa's 'curious crucifix of polished jet.' … Mr. Holland's appreciation of the kindliness of several members of the Belle Vue families, probably contributed much to the formation and establishment of those views of Popery which he is known to have held. 'Theresa's Tree' was a thriving willow, which had grown from a twig, received in a religious ceremony at the hand of the officiating priest, and planted by the young lady's own hand in the hope that it would therefore become a more significant and profitable memento of spiritual things than it could have been, if it had been laid in a drawer, After some time she left her home and her friends, to take the veil of a nun, and returned no more; but she could be sung of, in connection with the tree, in the following words:

> *She,*
> *Won by St. Benedict's severest rule,*
> *Vowed, and received the veil. Austere La Trappe*
> *Closed on the virgin victim those stern gates*
> *Which ne'er are opened to the world again,*
> *There changed she her soft raiment for the dress,*
> *The coarse hard habit, of the sisterhood;*
> *She died, was buried, and her grave is – when–?*
> *This thriving tree, once planted by her hand,*
> *Survives, the green memorial of her name,*
> *And seems, as seasons change, to emblem still*
> *The various changes of her transient life;*
> *And long shall its bright buds, its golden palms,*
> *Its silvery leaves, and winter nakedness,*
> *Tell us of her, the spirit of this spot.*[514]

Joseph Richard (1793–1847)

Joseph became a Roman Catholic priest. In 1847, after three priests had succumbed to an outbreak of typhus in Leeds, he volunteered to take their place. On 1st July Joseph, too, fell victim to the disease and was subsequently accorded the status of Martyr of Charity and Servant of God.

In 1829 Joseph's nephews Edward, William and Richard, sons of his brother Edward, were sent from Van Diemen's Land (now Tasmania) to be educated at Stonyhurst College where he visited them. Edward wrote of his uncle:

> *My grandfather's second son Joseph was ten years older than my father and embraced the priesthood. He was well off, an excellent man, published a book connected with the sacraments and duties of catholics entitled 'Familiar Instructions in the Faith and Morality of the Catholic Church adapted to the use of both children and adults'.*

> *The work reached at least three editions. He was a tall man, handsome, stern looking and intellectual. He owned a considerable collection of paintings, was chaplain to Bishop Penswick,[515] and used often to see us at Stonyhurst. He was a very exemplary man, always gave us good advice and money to spend. If I remember right, he died somewhere about 1848 of a disease caught attending the sick.*[516]

Joseph was born in Sheffield on 14th April 1793. In 1802 he was sent to Sedgley Park School[517] and then:

> ... at an early age to the college established out of the ruins of Douay at Crook Hall, Durham ... He was a diligent student, and acquired reputation as a good classical scholar and a sound theologian. He accompanied the college in its removal from Crook to Ushaw in 1808.[518]

He was ordained at Ushaw in 1817 and

> ... appointed assistant to the Rev Rowland Broomhead at Old St Chad's Chapel, in Rook Street, Manchester. There he remained until the opening of the new Chapel of St. Augustine, Granby Row, in 1820, to which he was then transferred.[519]

Joseph delivered Broomhead's funeral oration, which was published and rapidly passed through two editions. Hadfield noted that *'From this time his pen was rarely out of his hand.'*[520]

The Roman Catholic bibliographer Joseph Gillow (1859–1921) included an article about Joseph's life in his literary and biographical dictionary of English Catholics. He commented:

> *The gratuitous attacks on the Catholic Church by the Protestant Bible Association soon engaged Mr. Curr's pen in controversy, followed by other publications, doctrinal and instructive, for the use of his own people.*[521]

Gillow listed the Rev. Joseph Curr's 21 known publications as follows:[522]

1. *A Discourse delivered at St. Augustine's Chapel, Manchester, at the Funeral of the Rev. Rowland Broomhead.* Manchester, 1820, 8vo.; 2nd edit., J. A. Robinson, 1820, 8vo.

2. *A letter to Sir Oswald Mosley, Bart., President of the Manchester and Salford Auxiliary Bible Society.* Manchester, 1821, 8vo., pp. 24.

   > This letter is dated 26 October 1821. Shortly before the erection of St. Augustine's, Granby Row, a large dissenting chapel had been built in the neighbourhood of London Road, at which the Rev. W. Roby was the minister. This chapel is still known as Roby Chapel. These were days when the Bible Society Meetings had become the medium for the spread of Anti-Catholic slander

and bigotry, and a meeting was held in Roby Chapel, in which some Protestant ministers, especially the Rev. Melville Horne, of Salford, and the Rev. W. Roby, distinguished themselves by the virulence and violence of their Anti-Catholic or No-Popery language. A public invitation had been placarded on the walls of the town, asking for a numerous attendance. Mr. Curr was amongst the audience, and in a few days produced his 'Letter to Sir Oswald Mosley', in reply to the assertions of the speakers.

This brought out from one of the speakers, the Rev. Melville Horne, conjointly with the Rev. Nathaniel Gilbert, 'A Letter',[523] which Mr. Curr met with the following rejoinder, entitled:-

3   *Anti-Horne; or, an Address to the Inhabitants of Manchester, in Answer to the Rev. Melville Horne's Warning Voice*. Manchester, 1821, 8vo., pp. 27; dated 28 November; to which the Rev. W. Roby replied in a pamphlet entitled *Protestantism*, dated 6 December 1821, and Mr Curr issued his reply as under:-

4   *Catholicism; or, the Old Rule of Faith vindicated from the Attack of W. Roby*. Manchester, 1821, 8vo., dated 22 December.

The learning, the controversial skill and literary taste exhibited in these rapidly written papers, won for Mr. Curr a character of distinction. His heavy and pressing duties as a priest, and the tendency of Protestant controversialists to exchange argument for physical force, compelled him to relinquish his pen in the controversy with his reply to Mr. Roby's *Protestantism*. The latter, however, published a second tract under the same title, dated Jan. 3, 1822, and a third, on the 22nd of the same month. In reply to the first of these Mr. W. E. Andrews, Editor of the *Orthodox Journal*, issued *A Candid Appeal to the Common Sense of the People of Manchester*, dated Jan. 21, and a Second Appeal, on Feb. 22, 1822, in reply to Mr. Roby's third *Protestantism*.

5   *Particulars of the Conversion of Margaret R- N-*. Manchester, 12mo.

6   *Causes of the Disturbed State of Ireland*. Manchester, 12mo.

7   *An Irish Weaver's Answer to a Publication erroneously styled 'The Catholic'*. Manchester, 1822, 12 mo.

In reply to a series of tracts issued under the name of *The Catholic* by a local Protestant minister, the Rev. W. Gilbert, a weekly publication commencing 24 November 1821, and continuing to 6 April 1822, each number consisting of 4 pp. 8vo., price one halfpenny. It was followed by another four-page tract which seems to have run a course of five weeks, from 13 April to 8 May 1822, entitled *The Catholic Phoenix; or, the Papal Scourge*, price 1d. The publications arising out of the Bible

Society controversy were numerous. *An Irish Labourer* elicited *An English Labourer*; *A Catholic Townsman* 1 January 1822; *A Catholic Youth*; a poem entitled *The Marks of the True Church*, consisting of 44 stanzas in very fair imitation of the ballad *Chevy Chase*. The Rev. W. J. Schofield, of Christ Church, Hulme, was violent Anti-Catholic, and the Rev. R. Strittles, a Bible Christian Minister, came out violent anti-everybody but himself.

Gillow also listed two undated publications:

8 *The Instructor's Assistant.* Manchester, 12mo, which for many years held a prominent place in the Sunday-schools in Manchester and the surrounding district.'

9 *Sermon in Defence of Revealed Religion.* Manchester 8vo.

Gillow explained that:

The boundaries of the three Manchester missions at this time extended far into the country; which very much increased Mr. Curr's labours, and necessitated his keeping a horse.[524]

Hadfield noted:

The great extent of the Manchester missions, added to the other severe labours of missionary duty, at length told on his strength, and necessitated his removal to a somewhat quieter mission at Stockton-on-Tees, where he arrived October 2nd, 1822. There he remained until November 18th, 1826, when he was transferred to Ashton in Makerfield [also known as Ashton-le-Willows], where he wrote:[525]

10 *Familiar Instructions in the Faith and Morality of the Catholic Church; adapted to the Use both of Children and Adults. Compiled from the Works of the most approved Catholic writers.* Printed by J. A. Robinson, Manchester, 1827, 12mo.

Hadfield further noted that this book 'met with considerable popularity'.[526]

Joseph remained at Ashton-le-Willows until 1830[527] and, according to Gillow:

... after a short rest retired to the Monastery of La Trappe in France, to devote some time to further study, recollection and devotion [... and] On his return to England, he took up his residence at St. Cuthbert's College, Ushaw, as a convictor or boarder. Whilst in this retirement he occupied his time with one or other of his spiritual works.[528]

Gillow's list has two more undated works:[529]

11 *Visits to the Blessed Sacrament, and to the Blessed Virgin.*

Translated from the Italian of St. Alphonso Maria de' Liguori.[530] Manchester, 12mo. Frequently reprinted.

12 *The Spiritual Retreat, adapted to the Use of the Clergy serving the English Mission, from the Retreat of Bourdaloue.*[531]

In 1833 Joseph was appointed chaplain to the Claverings at Callaly Castle, Northumberland. Here, according to Hadfield:

> ... *his old controversial spirit manifested itself in several literary encounters with the Reformation Society and the Vicars of Longhorsley and Whittingham, which were published in pamphlet form.*[532]

Gillow lists these 'encounters' as follows:[533]

13 *An Address to all sincere Protestants in Coquetdale and Whittingham Vale, occasioned by the recent Invectives of the Rev. Maurice Farrell, an emissary of the Reformation Society, against the Catholic Faith at Thropton, Longhorsley, and Whittingham.* Newcastle, 1835, 8vo. Dated from Callaly Castle, 29 January.

At Callaly, Mr. Curr was very attentive to the education of the children of the mission, the school having been established towards the end of the last century by the Rev. Thomas Gillow.

14 In 1835 he delivered some Lectures which appeared in the local newspapers, perhaps circulated in pamphlet form, to which the Rev. Robert Green, BA, vicar of Longhorsley, replied with *A Statement of Facts. Observations upon some Lectures delivered by the Rev. Joseph Curr*. Newcastle, 1835, 12mo.

15 *Deception Unmasked: being a Review of a Statement of Facts, &c., by the Rev. R. Green.* Newcastle, 1835, 8vo.

This elicited a rejoinder from Mr. Green, *More Tracts; being a Reply to a Pamphlet published by the Rev. Joseph Curr, entitled Deception Unmasked*. Newcastle, 1835, 12mo.

16 Mr Curr answered, followed by –

17 A third Address, which elicited *To the Inhabitants of Whittingham Parish, in reply to Mr. Curr's third Address*. By the Rev. J. W. Law, Vicar of Whittingham. Newcastle, 1835, 12mo.

18 *The Fox and the Goose; or, a Comico-serio Address to the Good People of Whittingham and all others who have read the Address of the Rev. J. W. Law, of March 27.* Newcastle, 1835, 12mo.

19 Correspondence in the *Sheffield Independent* and *Yorkshire and Derbyshire Advertiser* on the disgraceful attempt to proselytise by the two Protestant Surgeons at Sheffield, published in the *Orthodox Journal*, Nov. 3. 1835, dated from Callaly Castle.

20 In 1836, Mr. Curr published another Letter, which elicited from the Rev. Robert Green, vicar of Longhorsley, *A Letter*

> *to the Protestant Inhabitants of Longhorsley, Rothbury, and Whittingham, in Reply to a Letter from the Rev. Joseph Curr.* Newcastle, 1836, 12mo.

In 1837 Joseph resigned the chaplaincy at Callaly Castle but the exchange continued:

21 In 1840 he wrote two long letters to the *Orthodox Journal*, headed *The Thropton Controversy: to the Rev. Robert Green, Vicar of Longhorsley*, which were probably published in pamphlet form with those of Dr. Corless and others.

Hadfield adds to this that he: '*besides was a frequent correspondent with the* Orthodox Journal, Tablet, Catholic Magazine *and other journals*'.[534]

He was next appointed to the Sheffield mission, staying there until around October 1839, when he took charge of St Alban's, Blackburn, and: '*Whilst there, at the request of the clergy, he presided over the Manchester Conference*'.[535]

In 1842 he went to Whitby, where he remained until 1846. His final post was as chaplain to Bishop Briggs at Fulford. His will, dated 24th June 1847 (a week before he died), states '*late of Fulford, near York*' and leaves bequests to Dr Briggs of two hundred pounds and '*books to the value of thirty pounds*'.

As noted, Joseph volunteered to attend the typhus-stricken people of Leeds. A Leeds newspaper elaborated:

> DEATH OF ANOTHER CATHOLIC PRIEST IN LEEDS. – *Last week was mentioned that the Rev. Joseph Curr, who succeeded the late Rev. Henry Walmsley as head priest at St. Ann's Catholic Church, in this town, only four weeks ago, had been attacked and was then laboring under that dire malady, typhus fever. We regret this week to have to announce his death. Mr. Curr was most highly estimated by his bishop, by the Catholic clergy of this district, and, we believe, by all who had the pleasure of his acquaintance, as a high-minded and devoted clergyman, ready and willing to perform every duty connected with his sacred calling, however (humanly speaking) disagreeable and dangerous the performance of that duty might be.*[536]

Joseph was removed to Huddersfield, and there died 1st July 1847.[537] There is a tiled memorial tablet mounted on the wall of the Mortuary Chapel of St Marie's, erected around 1886, of the clergy who served the Mission [see Figure 7.9].

Joseph's wealth at his death was some £3,000. After some bequests to clergy and charities, he bequeathed 'the sum of nineteen pounds and all my clerical robes and vestments together with my Church plate to the trustees of the Monastery of Saint John the Baptist near Market Weighton in the County of York'; the remainder he left to his family. There is no mention of 'his considerable collection of paintings'.[538]

From his last Will and Testament we learn: *'As a token of my affection'* he left £5 each to Ellen Smith, a niece, and Henry Beauvoisin, a nephew, and his watch to Mary Ann Beauvoisin, another niece. To his brother John he left *'two hundred pounds to be retained by him out of the money he now owes me'*. *'The rest residue and remainder'* he left to *'my brother in law Louis Amand Beauvoisin'*.

Harriet (*in religione* Gertrude) (1795–1868)

Harriet was born in Sheffield on 2nd September 1795 and in 1807 sent to be educated at Bar Convent, York. She *'finished her education in Paris'*.[539]

According to her nephew Edward:

My second aunt, Harriet, meeting I believe with a disappointment in love, took the veil and lived in the convent at York where she became a superioress for about forty five years. She took a cold bath every morning. She bore the reputation of being a very holy and clever woman.[540]

Harriet returned to Bar Convent where she was clothed (admitted to the institute) on 24th January 1825 and professed 23rd January 1827, taking the name *in religione* Gertrude.[541]

… her career as a nun illustrates the versatility and flexibility expected of members before the age of professional training, for she was successively assistant to the School Procuratorix, assistant to the Superior, Dispenser, Mistress of Novices, a member of the foundation in Scarborough, Mistress of Linen and Keeper of Garments.[542]

In 1859 Rev Mother Angela was induced to undertake a foundation in Scarborough. … A small colony of nuns was sent to take possession under the direction of the Rev. Mother's Assistant, Mother Mary Gertrude Curr, who was charged with the duty of organizing the new establishment. A small school for daily pupils was opened, to which was added later a night-school for poor working girls of the congregation. When the little household was fairly settled, Rev. Mother Angela appointed as Superior Mother Mary Alphonsa Ball …[543]

Harriet died on 30th May 1868 and was buried in the conventual cemetery.[544]

Charles (born and died 1797)

There is some doubt about Charles's dates because Father Rimmer amended the entry for his baptism at the Old Chapel, Norfolk Row [Figure 10.7]:[545]

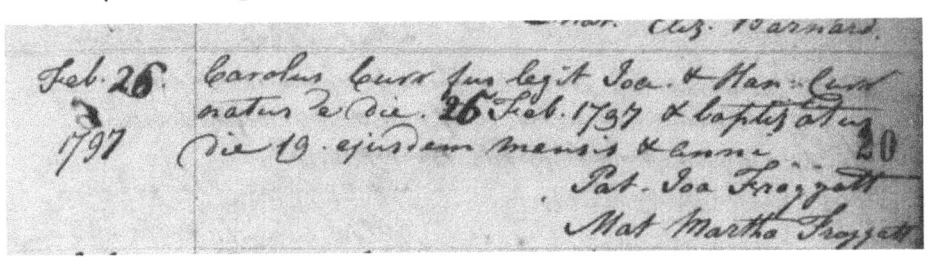

Figure 10.7. Baptism record for Charles.

Poor Charles did not survive long for 'Charles, son of John Curr' was buried at Sheffield Parish Church on 26th October 1797.

Edward Charles (1798–1850)

Edward and his wife Elizabeth had nine sons and six daughters. An account of Edward's life in this book would serve only as a distraction. By any standards his career was quite extraordinary. A contemporary obituary gives an idea of this as well some insight into his character and its flaws:

> THE LATE Mr. EDWARD CURR
>
> *Yesterday, we recorded the death of this gentleman, an event which a lingering illness had prepared his friends to expect. Although Mr. Curr had not for some time appeared prominently before the public, his political history is too closely identified with the colony to be passed without comment, or indeed, without eulogium.*
>
> *Mr. Curr resided for many years in Van Diemen's Land; he was a magistrate and one of the early members of the Legislative council of that colony. Perhaps he may be better known as the Manager of the Circular Head Company, and the title, not yet forgotten, of the "Wizard of the North", serves to indicate the position and influence he acquired. Austere and unflinching, severe and resolute, he made many enemies, and quarrelled with successive governors, but persevered in his own course and adhered to his principles to the hour of his death.*
>
> *Mr. Curr, in Port Phillip, became a more active politician. His previous prejudices, pursuits and experience had addicted him, if the expression be understood, to convict labour and to the last he continued an advocate for transportation. This was his great political crime. In many other respects he was a man fitted to become a leader of the people. It was he who first gave power and substance to the prayers of the Colonists to be separated from the Middle District. To his pen may be traced nearly every argument that strengthened the cause and achieved the victory. Clear in his views and perspicuous in thought and language, he modelled the earliest petitions of the Colonists and gave direction to their intelligence. There is a fine moral to be drawn from the death of one who lived to see the triumph of his labours without participation in the rejoicings which that triumph occasioned.*
>
> *The community looked forward to Mr. Curr as one of the first members of the Legislative Council of Victoria; his loss will be felt, for though comparatively retired, he had employed himself in preparing for the era which has just arrived, and would no doubt have proved himself a practical and efficient politician.*
>
> *The funeral of the deceased gentleman took place yesterday. In the evening, high mass was celebrated at the Church of St. Francis, when the Rev. Gerrald Ward delivered a discourse eulogising Mr. Curr, as a*

member of the Roman Catholic Church, as a parent and as a member of the body politic amongst which he was justly regarded as the "father of Separation".[546]

Juliana (Teresa) (1800–1830)

Juliana was born on 14th August 1800 and in 1814 sent to be educated at Bar Convent.

She married Thomas Ellison of Glossop, a widower, who had succeeded his father as agent to the Duke of Norfolk at Glossop and lived at Glossop Hall. His first marriage, in 1820, had been to Esther Dalton, who died.

On 17th September 1828 Thomas Ellison, of Glossop in the County of Derby, Gentleman:

alleged and made Oath as follows: … 'That he is of the age of thirty four years and upwards, and a Widower, and intends to marry Juliana Curr, of Ashton in Makerfield, in the Parish of Winwick, in the County of Lancaster and Diocese of Chester aged twenty eight years and upwards, and a spinster, and he prayed a Licence to solemnize the said marriage in the Parish Church of Winwick. In which said Parish of Winwick the said Thomas Ellison further made Oath. That the said Juliana Curr hath had her usual Abode for the Space of Fifteen Days now immediately preceding …'[547]

Juliana's brother, the Rev. Joseph Curr, was Roman Catholic priest in Ashton-in-Makerfield from November 1826 until 1830. Juliana and Thomas were married in Winwick parish church, by licence, on 13th October 1828. Sadly, Juliana soon died, as noted in the *Derby Mercury* for 14th April 1830:

On Tuesday the 6th instant, at Glossop, in this county, after three days' illness, in the 28th [sic] year of her age, Mrs Ellison, the wife of Thomas Ellison, Esq agent to his Grace the Duke of Norfolk.'

Juliana's burial is recorded in the name Juliana Teresa Ellison, aged 30, on 13th April 1830, at Glossop.[548]

Thomas was married for a third time on 4th June 1832:

On Monday, the 4th inst, at Manchester, Thomas Ellison Esq of Glossop Hall (agent to the Duke of Norfolk) married Mary, eldest daughter of George Gibson Esq of the former place, cotton merchant.[549]

APPENDICES

Appendix 1
A life preserved... 1788

INSTANCE OF
A Life preserved after falling into a Coal-Pit
From the *Gentleman's Magazine*, 5 March 1788

John Boys, a collier, employed on the coal works belonging to the hon. the late Lady Windsor, and the late Mr. Alderman Simpson, of Newcastle upon Tyne, at Lanchester Common, in that neighbourhood, going to his work very early one morning in 1763, and according to custom, on his turn to descend the shaft, in waiting to take the ascending hook, in order to his making a loop to introduce his thigh for that purpose, the pit, casting up very strongly a thick dense vapour, deceived him in the attempt of laying hold thereof, and, by his throwing his centre of gravity, unsupported, too far over the mouth of the shaft, he unfortunately fell to the bottom; a depth of 42 fathoms, or 84 yards.

Immediately on his falling, a cart was sent for, to convey the body home, as no person had ever been known to survive such an accident to such a depth; but, to the great surprise of the other colliers, on his being sent to-bank, or drawn out of the pit, in a corf, and after having recovered in some degree from the violence of the fall, he was found, on examination, neither to have a broken or dislocated bone or joint, nor any external wounds, or even marks of contusion; yet the delicate compages of the human frame had received such a shock and derangement, from the momentum of his striking the bottom, that he was never able afterwards to walk without the assistance of two sticks.

He was a pretty jolly man at the time of the accident, of about 12 st. weight; and survived about twenty years, getting his livelihood by cobbling old shoes, not being able to work any more in the coal pit.

Many people have attributed this very remarkable escape to the resistance he met with in falling from the force of the strong up-cast current of air in the pit, having retarded the acceleration of his descent: but

I think that reason of little consequence; it ought rather to be attributed to his having fallen perpendicularly, and without having been dashed and reverberated from side to side in the shaft (as generally happens when anything is dropped down a pit), and from his having struck the bottom in the most favourable position for the preservation of his head, &c., and the consequent saving of his life.

It is very remarkable, that he broke the strong chain on the rope at the bottom of the pit, consisting of links, made of round iron, near three quarters of an inch in diameter. On his being asked concerning his sensations during the fall, he said he descended very smoothly; but, as his descent was confined only to a few seconds, it cannot be supposed that he could, during so short a space of time, employ the power of perception in any considerable degree.

JOHN BUDDLE

Bushblade's Colliery, March 5, 1788

M. A. Richardson, *The Borderer's table book; or, gatherings of the local history and romance of the English and Scottish border*, in eight volumes. Vol. VII: *Legendary Division*. Newcastle-upon-Tyne, printed for the author, 1846: 'Instance of a life preserved after falling into a Coal-Pit from the *Gentleman's Magazine*', pp. 410–11. The article is signed JOHN BUDDLE. Bushblade's Colliery, March 5, 1788.

Appendix 2
The New Chapel; history of financial dealings

Extract (verbatim) from Hadfield 1889: *A history of S. Marie's Mission and Church, Norfolk Row, Sheffield*, Pawson & Brailsford, Sheffield, pp. 33–37.

The deed of conveyance to Messrs. Curr, Smilter and Furniss is dated the 10th day of June, 1814, and is made between Thomas Wybergh, of Clifton Hall, in the County of Westmoreland, barrister-at-law, and Vincent Henry Eyre, late of Sheffield, in the County of York, but then of High Field, near Chesterfield, in the County of Derby, Esquire, of the first part; the Most Noble Charles, Duke of Norfolk, Hereditary Earl Marshall of England, of the second part; and the said John Curr, John Smilter and John Furniss of the third part. The property thereby conveyed is described as follows:

> *All that piece or parcel of ground, situate and being in Sheffield aforesaid, extending all the length of a street called Norfolk Row, and bounded by that street on or towards the south, by the street called Far Gate on or towards the west, by the street called Norfolk Street on or towards the east, and by property belonging to Mr Abraham Hawley, on or towards the north, and containing 3,680 superficial square yards or thereabouts, be the same more or less. And all that capital messuage or mansion house, situate and standing thereon, and which for a great many years past has been occupied by the land agent for the time being of the said Duke and his predecessors, with offices, outbuildings, and appurtenances to the same belonging. And also all that other messuage or dwelling-house thereon erected, and which had formerly been occupied by Messrs Walkers, Eyre and Stanley, as a bank, but which had then been lately occupied by Miss Steel, with its appurtenances. And which last-mentioned messuage or dwelling-house and 99 superficial square yards of land adjoining, and including the site of the said dwelling-house, were by indenture dated the 2nd April, 1794, demised to Vincent Eyre, Esquire, deceased, from the 25th March, then last past, for 99 years, at the yearly rent of £1 5s. And also all that other building thereon erected, and occupied as a Catholic Chapel; and the stable and outbuildings thereon nearer to Norfolk Street. And all other erections and buildings upon the same land and every part thereon erected. Also another piece of land on the south of Norfolk Row, mentioned to have been demised by an Indenture of Lease dated 21st October, 1790.*

In 1816 Messrs Curr, Smilter and Furniss being indebted to Messrs Walkers, Eyre and Stanley the bankers, in the sum of £2,800 for money borrowed to pay the purchase money for the land above mentioned, Mr Curr paid out of his own pocket £2,000 to Messrs Walkers, Eyre and Stanley, and the property was conveyed to Mr. John Iredale [a Sheffield Catholic], of Sheffield, currier, upon trust, to raise £2,000 and the interest thereon. In 1821 some portions of the property purchased from the Duke of Norfolk's trustees in 1814 having been sold, the remaining portion was conveyed by Messrs Curr and Furniss (Mr Smilter being then dead), and by Mr John Iredale, the son and heir-at-law of Mr. John Iredale, mentioned above, to Mr Benjamin Schofield, upon certain trusts, one of which was for sale, and to pay out of the proceeds the sum of £1,710 17s., the balance of the £2,000 due to him to Mr John Curr. Mr John Iredale, the father, died on the 30th day of June, intestate as to trust estates, and John Iredale (the son) was his eldest son and heir-in-law. Mr John Smilter died on the 15th day of February, 1817. Mr John Curr died on the 26th day of January, 1823, having made his will, dated the 16th day of the same month of January, whereby he appointed John Curr,[550] of the City of Norwich, tobacconist; Joseph Curr,[551] of Liverpool, Clerk in Holy Orders; and Louis Amand Beauvoisin[552] of Sheffield, and afterwards of Manchester, gentleman, his executors. In 1823, out of the £1,710 17s. owing to John Curr's estate, there remained only £1,200 principal and £154 for law charges and expenses due. In 1824, John Iredale, the son, on the request of John Curr's executors, conveyed to Benjamin Schofield all the property that had not been sold. In the deed of conveyance it is described as follows:

> *All those the tenements, pieces or parcels of ground, buildings, and hereditaments hereinafter particularly described, being such of the messuages, buildings, pieces or parcels of land and hereditaments comprised in and conveyed by the above mentioned indentures of the 30th and 31st days of August, 1816, as have not been since the date of the same last mentioned indentures sold and conveyed; that is to say: All that piece or parcel of land situate, lying and being in Sheffield, in the County of York, and fronting to a street there called Norfolk Row, and containing 1,146 superficial square yards or thereabouts, be the same more or less, bounded by the said street called Norfolk Row on or towards the south, by other parts of the land and premises comprised in the said indentures of the 30th and 31st days of August, 1816, respectively, sold and conveyed to Messrs Robert Frederick Wilkinson and Michael Ellison, on the east; by premises late of Abraham Hawley, deceased, on the north; and by other part of the land and premises comprised in the said indentures of the 30th and 31st days of August, 1816, sold and conveyed to Messrs Nathaniel and German Wheatcroft, on the west; together with the buildings now erected on the said piece or parcel of ground.*
>
> *And also all that other piece or parcel of land, situate in Norfolk Row aforesaid, bounded northwards by the said street called Norfolk Row; west by a common passage between the same and land now*

or late part of the settled estates of the late Duke of Norfolk, several years ago demised to Benjamin Broomhead and Sarah Broomhead; south, by a road or passage, about six feet wide, running alongside the yard belonging to the Presbyterian Chapel; and east by ground other part of the settled estates demised to the said Benjamin Broomhead, and containing in the whole, including one half of the said common passage, as far as it runs along the said piece or parcel of land or ground, 415 superficial square yards or thereabouts, be the same more or less; together with all that messuage or dwelling-house standing on the last mentioned piece or parcel of land, now in the occupation of the Rev. Richard Rimmer.

By a deed dated 25th day of March, 1824, and made between Benjamin Schofield, of Sheffield, in the County of York, auctioneer, of the first part; John Bernard Furniss, of Sheffield, aforesaid, merchant, of the second part; and Thomas Smith, of the City of Durham, Doctor of Divinity, Thomas Penswick. of Liverpool, in the County Palatine of Lancaster, clerk, Benedict Rayment, of the City of York, clerk, and Richard Rimmer, of Sheffield aforesaid, clerk, of the third part, Mr Benjamin Schofield, by the direction of Mr John Bernard Furniss, conveyed the property vested in him to Dr Thomas Smith and the Reverends Thomas Penswick, Benedict Rayment, and Richard Rimmer. This deed was enrolled in Chancery, and appears to be the first trust deed in connection with the Church property in Norfolk Row. The balance of the mortgage debt due to Mr Curr's executors does not seem to have been paid off at this date (1824). On the 6th November, 1826, Mr Curr's executors assigned their mortgage debt to the Rev William Croskell, of the City of Durham, clerk.

On the 7th November, 1826, William Croskell signed a declaration of trust, from which it appears that the mortgage debt, £1,354, vested in him as trustee, was, in fact, the proper moneys of Thomas Smith, Thomas Penswick, and Benedict Rayment, in the following proportions: £1,000 of the said Thomas Smith; £200 of the said Thomas Penswick; and £154 of the said Benedict Rayment.

Thomas Smith died, leaving the said Thomas Penswick and William Croskell and the Rev Richard Thompson his executors, and the £1,000 due to Thomas Smith's estate was paid to his executors, William Croskell and Richard Thompson, signed the receipt for the same on the deed of 7th November 1826.

Thomas Penswick died, and by his will appointed the Right Reverend Dr Briggs, the Rev William Croskell, and the Rev Richard Thompson his executors, and they signed the receipt for £200 due to Thomas Penswick's estate. This receipt is indorsed on the deed of 7th November, 1826, but is not dated.

Benedict Rayment was the survivor, and by a deed dated 17th January, 1837, the property was conveyed to him, freed and discharged from any mortgage. This deed states that prior to the 17th January, 1837, various sums, amounting altogether to £610 17s., had been paid at various times to William Croskell, and that on that date, £743 3s. was paid to him by the said Benedict Rayment, and for this amount a receipt is indorsed and signed by William Croskell.

On the 18th January, 1837, the day after the property was freed from mortgage, it was conveyed by the Rev Benedict Rayment to trustees, these being: the Right Reverend Thomas Youens, of Ushaw College, in the County of Durham, Doctor of Divinity; the Rev Joseph Curr, of the City of York, clerk; the Rev Henry Walmsley of Leeds, in the said County of York, clerk; the Rev Charles Pratt, of Leeds, aforesaid, clerk; the Rev Robert Thompson, of the same place, clerk; the Rev John Furniss,[553] of Doncaster, in the same County, clerk; and the Rev Randolph Frith, of Manchester, in the County Palatine of Lancaster, clerk.

Part of the property purchased in 1814, was, in 1820, sold by Messrs. Curr and Furniss, by the direction of Mr. John Iredale, to Messrs. Nathaniel and German Wheatcroft. The quantity of land conveyed was 672 yards. It had a frontage to Norfolk Row of 24 yards. The price paid by Messrs. Wheatcroft was £352 16s. On the 26th August, 1846, the property was repurchased from Messrs. Wheatcroft for £1,400, and was conveyed to the Right Reverend John Briggs, of Fulford House, in the County of York, Doctor of Divinity, Roman Catholic Bishop of Trachis, and Vicar Apostolic of the Northern District; the Rev. Charles Pratt, of Sheffield, clerk; the Rev. Thomas Billington, of the City of York, clerk; the Rev. Robert Hogarth, of Marton, in the said County of York, clerk; and the Rev. James Platt, of Bishop Thornton, in the said County of York, clerk.

Appendix 3
Full prospectus for Curr's
The Coal Viewer

The full prospectus is reproduced here. From *The Sun, Advertisements and Notices* as it appeared on Friday 8th November 1793 (no. 347). From: *Seventeenth and Eighteenth Century Burney Newspapers Collection* (https://www.gale.com/intl/c/17th-and-18th-century-burney-newspapers-collection).

PROPOSALS FOR PRINTING BY SUBSCRIPTION,
THE COAL VIEWER and ENGINE
BUILDER's PRACTICAL COMPANION.
Containing

1. A full Explanation (with accurate Drawings and Directions to Workmen) of an improved Method of conveying Coals or Ores from the Face of the Works to the Shafts of Collieries or Mines in Corves or Trams, on Cast Iron Rail Roads of new Constructions; whereby a burthen of 3½ Tons or upwards is rendered a moderate Draught for an ordinary Horse on a deal Level; and which improvement may be applied with advantage above ground for conveying Coals, Limestone, Iron, Ironstone, &c. &c. in many situations in the Kingdom.

2. A new and easy method of Hurrying, Putting, or Conveying of Coals in Collieries where the rise of the Seam or Bed exceeds three inches per yard, whereby the full Corf descending on the Barrow Way or Rail Way, by its gravity draws up the returning empty one to a proper place to be taken to the face of the Workings. Also of making these full and empty Corves open Trap Doors without the expence of Boys to attend them.

3. Plans, with Practical Directions to Masons, Blacksmiths, and Carpenters, for the erection of Fire Engines of every size, from 24 to 72 inches diameter of Cylinder; with the Scantlings, Lengths, and Contents of Timber, and the quantities and weights of every other component Material, all regulated and adjusted to the size of the Cylinder.

4. Estimates of the Expences of erecting Fire Engines of the separate sizes of 30, 40, 50, 60, and 70 inches diameter of Cylinder, showing the Weights and Quantities of every article of importance contained in each Engine, which Estimates and Directions are not only applicable to Collieries, but to Works and Manufactories of all kinds where Fire Engines are introduced.

5. Several Estimates of the Expences of Winning and Opening New Collieries of different Depths.

6. An extensive Set of the most useful Tables for expediting the Business of Colliery Erections, and promoting the Dispatch of Estimates and Valuations of Stock. Among which are the following, viz.

Table of the Counterpoising Powers of Cylinders from 10 to 100 inches diameter, with sundry Depths of Shafts, and Diameter of Bores, &c. &c.

Table of the Weights of Cylinders, Pistons, Pumps, Buckets, and Clack Trees, Winbores, &c. of different Dimensions, and proportionate Thickness of Metal.

Table of the Number and Dimensions of Boiler Plates and Weights thereof, suitable for Boilers, from Eight to Sixteen Feet and a Half Diameter, increasing by Six Inches, and according to improved Plans, with Sections of them annexed to the Work.

Table of the Weights of Forged and Rolled Iron of many different Scantlings, Flat, Round, and Square.

Table of Weights of Ropes of the most useful Sizes for the Business of Collieries.

Table of the Quantity of Coal contained in a Statute Acre, in Beds or Seams from 18 Inches to 8 Feet Thick, in the most useful Denominations of Weights, Measures, and Stacks used in the Kingdom.

This Work will be furnished with several Plates, describing the Author's Inventions of Rail Roads, Corves, and other Improvements therein treated of, as well as of the Fire Engine Houses, Plans, and Sections of the Castings and Beams, &c. &c. and will be comprised in One small Volume Octavo, for the convenience of the Pocket.

By JOHN CURR, of Sheffield, in Yorkshire,

Colliery Agent to his Grace the Duke of Norfolk; and Patentee of a New Method of Drawing Coals, Ores, &c. by Conductors in Shafts, which prevent the Breakage of Coals, enables a great Quantity to be drawn at One Shaft, and lessens the wear and tear of Corves, &c. &c. as also Patentee of a New Method of applying two or more Ropes to the Drawing of Coals, instead of one, which affords considerable savings.

N.B. Every Improvement and Invention herein pointed out may be seen in actual practice at the Collieries on Attercliffe Common, near Sheffield, and at Sheffield aforesaid, where the curious have a general invitation from the Author to view and examine the same, and which are now adopting in several other parts of the Kingdom.

CONDITIONS

I. This Work will be elegantly printed in Octavo, and the Plates wrought of on French Paper.

II. The Price to Subscribers to be Two Guineas in Boards, to Non-Subscribers Two and a Half, to be paid for on delivery of the Book.

III. As soon as a sufficient number of Subscribers offer, the Work will be put to the Press.

Subscribers are desired to signify their intentions as soon as convenient, by a Line addressed to the Author at Sheffield aforesaid.

Appendix 4
Coalbrookdale Report: Curr's letter of 25th May 1793

Coalbrookdale Report: letter from John Curr to Richard Dearman, 25th May 1793. From: Reprinted in Bland 1930–31, pp. 127–29. Cited in Trinder 1973, p. 133, note 77. Original letter is now in the Coalbrookdale Museum.

Sheffield.

Sir,   May 25th, 1793.

On my travels from the Dale I have made it my study to hit upon the best plans of getting the crates out of the Boats into the Pitts, disposing of them on the Roads at the Jinneys; in which I have had the good fortune to please myself and doubt not of its meeting your approbation and Mr. Reynolds. I however thought it prudent to give you short sketches of the separate subjects before I proceed to draw regular plans.

1st the Taking off of the Crates and mode of hanging. The Plan here proposed is to move the 2 Inner Cranes about 4 feet to make a Platt form or Cage (shewn by dotted lines at the Ends of the Crates to contain 2 Crates) with Carriages under them, and this cage is suspended by a double rope at each end of it and each Pair of Ropes have their separate Pulleys, which will be about $7\frac{1}{2}$ feet as under Center and Center. As soon as the Crates run to the Pit Top the Crane (c) will be ready to lift them into the Boat in the Navigation (B), and the Crane (d) will have a pair of empty Crates ready to sett upon

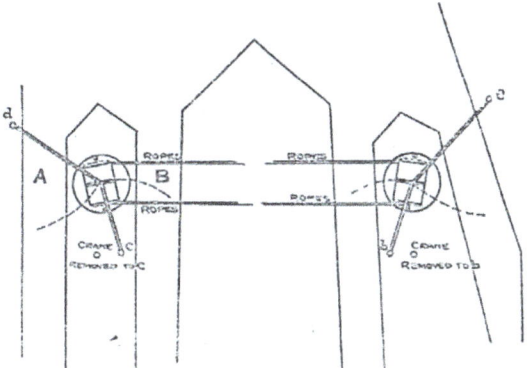

FIG. 10.   MODE OF DISCHARGING CANAL BOATS.

the Carriages. The Conductors here are out of the way of the Crates; as are also the 2 pr of Ropes which suspend the Cage and goes over the Pulleys.

The Cranes must have a swivel to them to admit the Crates turning into a position to suit the Boats. By this means we can apply 4 Ropes to each pair of Crates.

The mode of disposing of the Crates at the Jinneys.

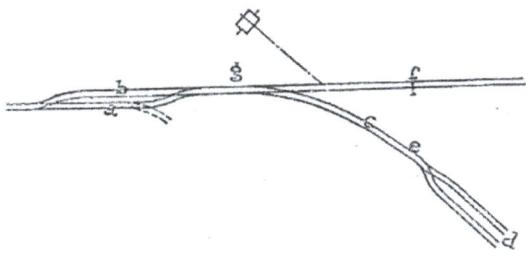

FIG. 11.   ARRANGEMENT OF ROADS AT SURFACE.

When the Horse brings his load of Crates from the "Pass Bye" upwards to the "Pass Bye" (*a*) he there leaves them. By this time we are to suppose there are 6 Crates standing upon the Road (*c*) which the Horse is hooked too and he draws them to the "Pass Bye" (*b*). He then hooks again to his former load at (*a*) and setts down on the road (*c*) from whence they are took by the Jinneys. By this means the road must be level from (*e*) to the main Road at (*f*), it measures 6 or 7 Yards and from (*f*) to (*g*) is 17 or 18 yards.

(*d*) shews the descending Jinney Road. This plan being different from what I propos'd when at the dale which was to go off at the dotted Lines near (*a*) it makes this Alteration in the "Pass Byes" (Viz. that they must each begin and end 11 or 12 Yards nearer the Upper end of the Road than where I staked them out.). We can now easy alter the "Pass Bye" Places 11 or 12 Yards but can't alter the intended place for the Jinneys.

The mode of disposing of the Crates in the Pit Bottoms.

FIG. 12. ARRANGEMENT OF ROADS AT PIT BOTTOM.

The Horse will bring the Crates to the "Pass Bye" (*a*); When the Cage has set down with two full Crates on the Pit Bottom then 2 Men will each take a Crate and sett them upon the Road (*b*). They will then take each a Crate from the Road (*c*) and place them upon the Cage at the Pit Bottom. They then Bid them go at Pit Top and Walks through to the other Shaft while the Crates are ascending and goes through the same operations with the other Shaft. The Road being level from the "Pass Byes" to the Pits, 2 Men will do any Quantity of business that can possibly be wanted. I think they will have no guiding of Crates nor hooking or unhooking, nor shall we upon this plan want hooks and links put upon each corner of the Crate as I before proposed, Nor will there be one brick to move in the Tunnel or any other openings made.

I should be glad that Mr. Reynolds & you might pay a little attention to the Plans I have here proposed, which after mature deliberation am in hopes will be found practicable and easy, and if you approve of them, I will attend particularly to the different points and get proper Moddles made to suit them.

I am, Sir,

Your obd. & hbl. Sert.

John Curr.

Appendix 5
Newcomen-type steam engines in the Attercliffe–Orgreave coalfield

The text and table [Figure A.1, overleaf] are from John Hunter, Early pumping engines in the Attercliffe–Darnall–Orgreave coalfield and the role of John Curr. *Transactions of the 2nd International Early Engines Conference*, Vol. 1, IEEC & ISSES, 2021, pp. 73–74.

John Hunter in the 'Conclusions' of his paper notes, *inter alia*, that:

> *The surface sites of these early collieries have mostly been lost beneath industrial, housing and highway development in subsequent centuries. The location of the shafts at High Hazles Colliery is still evident, as is part of the colliery tramway. The site of the Greenland engine pit may also have survived. The most obvious surviving structure, however, is the oldest one, i.e. the medieval Attercliffe Weir.*

Additionally, he adds:

> *The following table provides a summary list of the Newcomen-type steam engines which were erected to drain the Barnsley Coal workings in the Attercliffe–Orgreave coalfield, together with the limited technical details that have been discovered to date:*

Engine name & location	Dates	Built / operated by	Manufactured by	Purpose	Technical details / comments
Attercliffe (village)	1747-<1763	Jonathan Smith	Unknown	Pumping	Unknown
Darnall Colliery #1	1757-1795	Jonathan Smith?	Unknown	Pumping	Single external boiler
Attercliffe Common Colliery #1	1787->1810?	John Curr	Walker & Co? with Hartop, Binks, & Booth	Pumping	Twin external boilers, 50 inch cylinder, 7ft. 5in. stroke
Attercliffe Common Colliery #2	1788-1800?	John Curr	Binks, Booth & Hartop	Pumping	Twin external boilers, 46½ inch cylinder, 6ft. 10in. stroke
Darnall Colliery #2	1788?-1795	Unknown	Unknown	Pumping	Unknown
Attercliffe Common #3 (later, Greenland #2)	1789-1825	John Curr	Binks, Booth & Hartop	Pumping	Twin external boilers, 61 inch cylinder, 8ft 6in. stroke, 63hp. Moved to Greenland in 1809
Dore House Colliery	1792/3-1802?	John Bargh?	Unknown	Pumping / winding	50 inch cylinder
High Hazles Colliery	1796/7->1811	Unknown	Unknown	Pumping / winding	50 inch cylinder, single external boiler
Greenland #1 (Attercliffe #4)	1795->1819	John Curr	Unknown	Pumping	50 inch cylinder, twin external boilers
Dore House Colliery	1796->1820	Unknown	Unknown	Winding	18-inch cylinder, with flywheel (moved to High Hazles)
Attercliffe Common Colliery	1796-?	John Curr	John Curr	Winding	27 inch cylinder
Flat Pasture Pit	Late 1790s->1820	Unknown	Unknown	Winding?	
Finch Well Pit	Late 1790s->1820	Unknown	Unknown	Winding?	

The following additional early engines, located outside the coalfield and not previously identified, were discovered as a result of the research undertaken to prepare this article:

Engine name & location	Dates	Built / operated by	Manufactured by	Purpose	Technical details / comments
Killamarsh Colliery #1	Pre-1780	Unknown	Unknown	Pumping	Offered for sale by George Curr
Killamarsh Colliery #2	Pre-1780	Unknown	Unknown	Pumping	Ditto
Park Furnace #1	1784-?	Binks, Booth & Hartop	Hartop, Binks, & Booth	Pumping	45 inch cylinder,
Park Furnace #1	1785-?	Binks, Booth & Hartop	Hartop, Binks, & Booth	Pumping	29 inch cylinder,
Ballifield Colliery	1780s->1799	Unknown	Unknown	Pumping	

Figure A.1. Table of Newcomen-type engines. From: Hunter 2021, p. 74.

Appendix 6
New Punches or Props

'NEW PUNCHES OR PROPS', as advertised in the *Derby Mercury* for Thursday, 3 June 1802:

>NEW PUNCHES OR PROPS
>for
>Coal Mines, &c.
>
>The general scarcity and advancing price of the Wood used for the purpose of supporting the Roofs of Collieries and other Mines, having become a very serious grievance, particularly to the Proprietors of Coal Mines, a mode of relief, securing the certainty of supply, and assuring a very considerable saving in the actual expence, is now recommended to their attention; viz. the adoption of PROPS or PUNCHES of CAST IRON instead of Wood, or any other material; for which his Majesty's LETTERS PATENT have just been obtained. …
>
>Cast Iron Punches, the length of 4f. 6in. and 56lb. wrought each, have been in use upwards of Seven Months, at the Adelphi-Colliery, Duckmanton, near Chesterfield; during which time it is in proof that not One Punch has broken in its place, or been lost by the breaking in of the Roof: so that they have answered the intended purpose, to the great profit of the Proprietors, and also to the entire satisfaction of the Workmen, who have by this means no Punches to cut or exchange, and who can now go thro' the Business with more comfort and safety, than they did with Wood Punches and Blocks.
>
>It appears from the above experiment that the expence of the purchase and carriage of Wood Punches used in one year, or less, will clear all expences attending the establishment of Cast Iron ones; which must operate as a strong inducement to their being immediately adopted, especially as a trial may be made in any Colliery on as small a scale as may be judged expedient without incommoding the general plan of working.
>
>The fullest information may be obtained by personal inspection at the Adelphi-Colliery above mentioned, where Mr. JOHN CHARLTON, Agent to the Colliery and Patentee, will attend any person making application.
>
>Enquiries may also be made. Or orders given at the Iron-Works of Messrs. SMITH and Co. Chesterfield; or at the Foundry of Mr. JOHN CURR, Coal-Viewer and Engineer, Sheffield Park, and Patentee of the New invented Flat Rope, for drawing Coals, &c. out of Mines. May 25, 1802.[554]

Appendix 7
Rope making

This text is quoted verbatim from pages 73–74 of H. W. Dickinson's 'A condensed history of rope-making', *Transactions of the Newcomen Society*, **23** (1942–43), pp. 71–91 and 4 unnumbered pages of plates XVII–XX (with permission from the Newcomen society).

Hemp Fibre Rope-Making

SPINNING THE YARN. In 1393 we have a representation of the first stage in rope making – that of spinning the yarn – taken from *Mandelscheses Portrat Buch* in Nuremberg. This shows no advance on previous methods. The 'tackle board' does not appear before about 1500. So little difference from what was practised for the next five hundred years in Europe is shown that this may serve as a text for a fairly full description of the art during the period indicated; supplemented by personal observation of a ropewalk still working as in the old way.

In the first place the fibre of hemp *(Cannabis sativa)* had been found by long experience to be the most suitable fibre then known for cordage. After harvesting the hemp is prepared by 'retting' i.e., prolonged soaking in water to promote fermentation and convert the mucilage of the fibre into soluble substances, then drying, breaking and hackling to remove the tow and lay the fibres, as far as possible, parallel to one another. The fibre is smooth but the staple is long – 3 ft. ordinarily and in some varieties of hemp three times as much – hence the necessary amount of twist to be put in the yarn is not great.

The spinner wraps the bundle of prepared fibre round his waist [...] with the ends brought to the front; he pulls out the proper number of fibres, twists them between thumb and forefinger and attaches this part to the hook of a 'whirl' rotated by multiplying belting from a wheel driven by an assistant. The twist is right-handed, and this probably arises from the natural tendency to twist in that direction seeing that the right hand is employed, or it may be a legacy from the remote past. Pulling out more fibres, he walks backwards along the 'rope-ground' or 'rope-walk', which may be as much as a quarter of a mile long to accommodate what subsequently became the British standard length of rope, viz., 120 fathoms. The feed of the fibre and the motion of the whirl are kept in correlation, the art of the spinner being to feed the fibres with the right hand while he smoothes them to a sliver with a piece of coarse cloth or flannel held in the left [...].

Arrived at the end of the rope-walk a second spinner repeats what the first has done. The yarn is taken from the whirl, given to an assistant who attaches it to a reel and winds it on, the first spinner advancing *pari passu* keeping the yarn from untwisting, which it has a tendency to do. The yarn of the second spinner is joined to that of the first, reeled up and so on.

WARPING THE YARNS. The next process is the important one of warping or stretching by main force the yarns supported on the posts at intervals along the rope-walk, until each yarn sustains the same pull.

An intermediate stage, when such is needed, is that of tarring the yarn.
[…]

LAYING THE ROPE. The concluding part of rope-making consists in 'laying' i.e. combining yarns into strands and strands into hawsers and cables, the principles of stretching the component parts and reversing the 'hand' of the twist at each stage being essential.

Laying is effected by bunching together the ends of the components and twisting them on a single hook while the components at their other ends are separately attached to contiguous hooks, separately rotated in a framing; care is taken that the components receive at these ends as much twist (or a trifle more known as 'forehard') as is taken out of them by the single hook.

Uniformity in twisting the components together is ensured by a 'lay-top' – a conical plug of wood with three lengthwise semi-circular grooves – squeezed along as the closing proceeds.

Appendix 8
Descriptions of three of Curr's patents

The texts below are from William Chapman's *A treatise on the progressive endeavours to improve the manufacture and duration of cordage*. London: printed for W. H. Wyatt, 1808, pp. 37–39.

Patent no. 2270

November 17th, 1798. Mr. John Curr, of Sheffield, obtained a Patent for 'a method of forming and making a flat rope, intended to be used in drawing coals and other minerals and water out of mines of any kind'. His specification says 'the said flat ropes may be formed by connecting two or more small ropes sideways together, by sewing or stitching, lapping, or interlacing with thread or small ropes, etc.' He then describes his machine for sticking them together.

It was found necessary to make the constituent ropes alternately of a right hand and left hand twist, to keep the flat rope in a quiescent state; and, in this form, composed of four small shroud-laid ropes, they have now been used some years, in the midland counties, for the drawing of coals. These ropes wind up in a spiral upon themselves.
Chapman 1808, XI

Patent no. 2914

In March 1806, Mr John Curr, of Sheffield, Took out a Patent for 'a method different from any that has hitherto been invented or known, of spinning hemp for making ropes or cordage'.

His method of spinning consists in regulating the number of twists in the yarn, to the length moved by the spinner, so that they may elongate equally on being untwisted when made into a strand. To effect this desirable end, he has a barrel connected with the spinning-wheel, and consequently with the whirls by which the yarns are twisted. A cord is wound up on this barrel, which, when the spinners are going to set off, is attached to any one of them, and the rest are to keep pace with him. His speed is of course regulated by the unwinding of the cord from the barrel. He observes, that much of the strength of the strand of a rope depends on the equal elongation of the yarns, upon their being untwisted by the counter-twist of the strand. Mr. Curr's method will tend to produce that equality; but much must also depend on the threads being all spun of an equal and uniform thickness.
Chapman 1808, XXIV.

Patent no. 2960

In August 1806, Mr. John Curr obtained a Patent for a 'method of laying or twisting the yarns that compose a rope, by which method the yarns of a rope have a better and more equal bearing than they have in the common way.'
This Patent, like the one preceding for spinning yarns, is designed to give the uniformity of twist in the rope: for this purpose, in laying the strands into a rope, he describes a method of regulating the motion of the roper's top; and also to give, a regular motion to a perforated implement, which is to perform the office of Mr. Balfour's top-minor. (Vide invention No. V.) A simple and accurate method of regulating the motion of the top is known to all ropers: viz. they, mark the strands with chalk, where they cross any of the stake-heads, at a distance from the strand-hooks at the head of the ropery. The motion of the top is then given by the after-turn or twisting of the rope behind it, and the foreman regulates its progress by the angle of the spiral, which he designs the strands to form round their common axis. If the chalked part of the strands advance towards the top, the men at the strand hooks are directed to heave faster, because they have not put so much twist into the strands, as has been taken out of them by the twisting of the rope. On the other hand, if the chalked part recede from the top, it follows that more twist has been put into the strands than taken out by the rope, and the men at the hooks are directed to heave slower.

The twist put in at the strands, and taken out by the rope during the process of laying, is not, as has generally been conceived, alike as to quantity: but the requisite turns of the strand are fewer than those of the rope, the disparity increasing with the obliquity and following determinate laws, so that when the degree of obliquity of the twist is predetermined, and correspondent motions given to the rope, and to the strands, according to the invention No. X, where the whole process is carried on in a rope-walk, by machinery connected by an endless rope, and moved by a steam-engine, the man attendant at the top has, similarly to the common practice, the chalked parts of the strands for his guide. If they recede from him, he lets the top move a little quicker, and if they advance towards him he holds back. Thus by both the last described means there is the utmost certainty of the twists of the strands remaining precisely the same in the rope, as they were before conjoined. It may, however, be altered afterwards, and without disadvantage, provided all the comparative twists be previously arranged accordingly. A knowledge of the causes and effects acquired by the experience arising from breaking, on a machine which will weigh their strength, a sufficient number of ropes formed under different properties of relative twists throughout the whole process, must form the best basis for making cordage to the greatest perfection; and, as it depends on some regular combination of twists to produce that desired end, it follows that machinery duly arranged and connected, and capable of being set to a minute diversity, must have a decided advantage over any unconnected operations by hand.
Chapman 1808, XXV.

Appendix 9
Flat rope: Curr's pamphlet addressed to Tyneside colliery owners

Seeking orders for his flat ropes from north-east England colliery owners, Curr had a pamphlet printed (by Akenheads, printers, Newcastle-upon-Tyne, 5th June 1804) as an open letter for circulation promoting his flat ropes. From: TCR NEIMME/Wat/1/26/1.

> Newcastle on Tyne, June 5th, 1804.
>
> Sir,
>
> As ropes are become expensive articles in collieries, I beg leave to hand you a fair and impartial statement of my flat ropes introduced by Mr Sober Watkin, at Painsher Colliery, on the river Wear, which was set to work on the 30th day of January, 1803. It will be necessary to remark, that the round ropes introduced on the west side of the shaft generally wore out nearly two on the east side thereof, and it is now very evident that the same effect will be produced in my flat ropes. On the 12th of December, 1803, the flat rope was taken off on the east side, when it had done nearly the work of seven of the common round ropes, and the first rope laid on the west side is yet at work; and it is highly probable will work out two flat ones on the other side, as it is likely to work two or three months yet to come, or, nearly 1½ years from its first laying on. It will be proper to state also, that a saving of at least seven shillings per week in the article of grease has been made, which the round ropes required to make them wind more pleasantly on the rope roll; that the shaft coal is greatly abated, and, of course, by the flat rope not twisting round, the corves and pit sides are very much preserved, and the pitmen like them better than round ropes, conveying them so steadily, and more particularly in not being liable to the dangerous circumstance of GLUEING.
>
> It will be also necessary to remark, that Mr Watkin and Co. having laid on a pair of my ropes, on a pit at FATFIELD COLLIERY with a very narrow shaft, (and consequently very destructive to ropes of any description when the motion is so very quick,) one of these flat ropes, by some accident or cause, broke, after having done the work of more than two common round ropes; but this rope, by the introduction of my splice which has been used for some months past at Painsher Colliery, is in a state of doing a great deal of work, on a proper shaft of less depth, and Mr Watkin proposes laying them on again as soon as he has a pit ready for work, which he is now preparing. A similar accident has happened to a round rope on the same shaft I'm told very lately.
>
> Mr Fenwick has also laid on a pair of my flat ropes on the E Pit, at Lambton Colliery, and has now used them for six calendar months; the depth of the shaft is 93 fathoms; the rope roll, formerly used with the round rope was 8 feet in diameter, and this pit had been previously wrought nearly two years, with the same machine and engine (which was the same case at Painsher Colliery, above stated) and having introduced my round faced pullies and nitches on the rope wheel, the flat ropes now in use, and which are just 1-3d heavier than the common round ropes, have worn out 5 pairs of them; and are, at this time, in a good working condition, and from their present appearance, it is highly probable, that they will go a considerable time longer; It is also found, that the same quantity of work is done in much less time in the pit with the flat than the round ropes; and the flat ropes not twisting do not produce near so much shaft coals as before, from the corves ascending more steadily in the shaft.

The above assertions will bear the test of strict inquiry, and I must now beg leave to make some general reflections, which experience has dictated; and have to say, that where the rope wheels of your machines are already small (from 5 to 6 or 7 feet diameter) that flat ropes may be immediately introduced at a very small expence, enabling more coals to be drawn in the same time, and embracing the sundry articles of saving above-mentioned; and your counterbalances in this case may be thrown aside, which subject you to repairs, and occasionally stop the work, as the flat ropes, lapping on themselves on proper rope barrels, never fail to make a counterbalance; and such machines as have now large rope barrels, and in their present state are not capable of embracing every accommodation above set forth, I have no doubt the owners of such machines will soon see their advantage in reducing them, and increasing the nut wheels when opportunity offers, in order to embrace the full accommodations afforded by the flat rope, and in particular by throwing aside their counterbalances.

I am, your obedient and humble Servant,

JOHN CURR.

⁂ Address at Belle-Vue House, near Sheffield; and may be met with at the Three Indian Kings, Quay-side, Newcastle, on Saturday the 16th of June Inst.

Ropes may be supplied in two or three months from the time of ordering.

Copy of a letter received from Mr Anstice, of Made[ley, ...] day of June, 1804.

Sir,

Agreeable with your request, I transmit you an impa[rtial account of the] merits of your flat rope, according to our experience:—We have a[pplied them] working at Madely Wood Colliery, near Shiffnal, Shropshire, upon a sha[ft ...] yards deep, where sundry pairs of the common round ropes had been applied, p[rior] to the introduction of your flat rope; and by making a small alteration in our [rope] wheel to form nitches, and applying the broad round faced pullies, we have [found] that the present pair of flat ropes has already done the work of 5½ pairs of [round] ropes, which were 6¼ and 6½ inches round, and the flat ropes are about 1-[?] or 1-40th heavier; the quantity drawn in one band is from 14 to 18 cwt. [and] the quantity drawn up in 12 hours from 70 to 80 tons. With this work, a pa[ir] of round ropes, of the size abovementioned, was destroyed in about 13 weeks, tho[ugh] our engine was placed about 40 yards from the pit, to prevent the side wear of the round ropes on the barrel, which is an inconvenience by no means necessary when the flat ropes were introduced, yet our engine has not been removed.

There is certainly, also, a considerable saving in what you call the shaft coal, as well as the wear of the shaft itself, which, in the use of the flat rope, is very inconsiderable, and in the use of the round rope, in many instances, a considerable loss, to which, we may add a saving in the article of grease, the cost of which on a pair of new round ropes during the first month may be near £3.

We may add, had there not been some obvious conveniencies as well as advantages in the use of your flat rope, our colliers would not give (as they decidedly do) the preference to them, compared either with chains or round ropes.

Not doubting, from the present appearance of these ropes, that more extensive experience will enable us to add further and stronger testimony to the advantages your flat ropes certainly possess, and with every wish for their extensive application for the public good, and your emolument,

We are, Sir,

Your obedient humble Servants,

WILLIAM ANSTICE, Successor to the late W. Reynolds, Esq.
WILLIAM SMITH, Agent to the Colliery.

P. S. We have for a few weeks past had a pair of your flat ropes at Blest's Hill, where they promise as well as those already tried; and we expect to apply another pair in our colliery before long.

W. A.

Appendix 10
Letter from John Buddle to
John Curr II, 23 February 1806

A four-page letter from J Buddle Jnr to John Curr Snr, dated 23rd February 1806, provides an insight into the performance of flat rope. From: NEIMME archive: TCR NEIMME/Bud/15/134–137. This letter is transcribed below

134

Walls End Colliery Feb'ry 23, 1806

Dear Sir

I have inspected the flat Ropes at the Different Collieries on the Wear, and refer you to the following statement viz. Washington 2 Pairs of Ropes on the D Pit with Double Corf.

From	Nov'r 8th 1804 to & with	Dec'r 31st	1154 - 13
	Dec'r 31st 1804 --------	Dec'r 31st 1805	8590 - 14
	Dec'r 31- 1805 --------	Jan'y 10th 1806	159 - 2
		Total	9904 - 9

When the above Ropes were taken of[f] on the 10th Jan'ry Mr Humble sent to Walls-end and got a pair of the last sent Ropes to replace them, and on examining them last Monday I perceive on[e] of them failing in 2 or 3 different places, and the other appears sprained. I conceive the reason is their being too weak for the double Corf, and must therefore beg of you to send a pair; the proper strength <u>without delay</u>, as I'm sure the present ones will soon be destroyed.

Lambton. The last new ropes sent to this Colliery are not yet set to work.

The performance of the others are as below viz. E Pit. Ropes laid on Dec'r 5 1803. Sept 4th 1804. The North Rope taken of & drew   2017 - 0
Sep 7   The South Do.   2042 - 0
                                                              Total -- 4059 - 0

Sep 4: 1804 Laid on the 2nd North Rope, which
            was taken off worn out Feb 26 1805 drew   –   1155
Sep 7: 1804 Laid on the 2nd So [South] Rope, was taken
            off worn out May 30th 1805 drew –   –   1879
                                                     Total   3034

135

Feb'ry 26th 1805 Laid on the 3rd No [?North] Rope which was taken
            off worn out Nov'r 6 1805, and drew   1718
May 30th 1805 Laid on the 3rd So [South] Rope, is still on
            and drew to & with Nov'r 29 1805

when the Pit was laid off and
has not worked since   1020

[The units used above are for tons & hundredweight of drawn coal used as a measure of work performed by each rope (Charles Dixon).]

Pinsher. Mr Laverick informs me that you have an account of all the Ropes except the last pair which were laid on at the new Pit or E Pit Aug't 8 1805 and have drawn to & with Feb'ry 12th 1806 4210. One of those Ropes the Easternmost broke about a fortnight ago and was spliced[,] it bears evident marks of decay and has become unlaid in several places towards it[s] lower end and where it is secured by Capping. The other Rope is in very good trim and looks as if it would do as much more work as it has already done and why so great a difference between them I do not know as on examination the external appearance are perfectly similar – it arises most likely in the different quality of the hemp which compose the two Ropes or in some difference in the mode of one manufacturing of which you perhaps may be aware. When the E. Rope broke it gave no previous notice nor bore any appearance

136

of decay but snapped at once with the full Corf in an oblique direction which at first sight I thought followed the direction of the perforation made by the lacing needle but on close examination found it had much more obliquity. It strikes me from the laying of this Rope having given way that some material defect has attended its Manufacture as I do not perceive any cause either in the Rope wheel Pulley or in the Shaft to injure that more than the other. -- The stopping of Water in our sinking Pit at W. End has so completely employed the Carpenters since my last so that no opportunity has occurred for laying on the flatt Ropes but I hope we are now nearly through our difficulties. -- On asking Mr Taylor respecting the Timber left at Percy main he informed me that Crowther got it. I therefore applied to Crowther who states the Acct as follows viz-

Dr: Mr Curr in Acct with Ph: Crowther [ ]

1803	£ s d	1803	
June 18th To 12 Cold Chisels	0 - 18 - 0	By 55. 20 ft 11 In. 3/4 Deals @ 3/8	
Attendance at Rainton Coll'y several times with Mr Curr on altering the Mach'e from round to flat Ropes			£ 10 - 3 - 11
	1 - 1 - 0	Signed for Percy main	
To Bal'ce due Mr Curr	8 - 4 - 11	John Robinson	
	£ 10 - 3 - 11		

137

An Acc't of Rail Road sold viz.
    Walls end in 1805

		Cwt Qr Lb	
Jan'y 1	To 4 Switch Plates	5 - 2 - 0 @ 16/-	£ 4 - 8 - 0
Ap: 11	To 2 Siding [ditto]	1 - 1 - 0 @ 16/-	1 - 0 - 0
Sep: 3	To 180 4 feet 8 In. Plates	67 - 2 - 0 @ 15/-	50 -12 - 6
Oct'r 4	To 160 [ditto] [ditto]	57 - 0 - 0 @ 15/-	42 - 15 - 0
			£ 98 - 15 - 6

    Sold to Elswick viz.
1805

Ap:	To 126 4 feet 8 In. Plates	47 - 1- 0 @ 15/-	£ 35 - 8 - 9
Sep:	To 40 [ditto] [ditto]	15 - 0 - 0 @ 15/-	11 - 5 - 0
			£ 46 -13 - 9

I shall endeavour to get you some Money for the above in the Course of next Month which is the usual Time of Paying Tradesmen. – You will please to give me directions as to the Settlement of Crowther and Surtees & Co Accts and with best respects to all the good folks at Bellevue.
        I remain
           Dear Sir,
                Yours truly
                  Jno Buddle Junr

Jno Curr Esq

Appendix 11
Rope – advertisements

The full texts of some of Curr's advertisements are presented below.

The Hull Advertiser, 5th December 1807

Cables applied to the Windlasses and Capstans of Ships and Vessels of all descriptions, under His Majesty's Royal Letters Patent.

JOHN CURR, of Belle Vue House, near Sheffield, having invented a method of applying Cables, whether flat or round, upon The Windlasses and Capstans of Ships, and having had a flat Cable in use for 17 months, on a Vessel of 100 Tons Burthen, called the Grand Trunk, trading from Gainsboro' to Hull, belonging to the Gainsboro' Boat Company, recommends it to the particular attention of the Owners of vessels.

The flat Cable, which should be composed of four patent round Ropes, is not required to be as long as a round Cable to secure the Ship at Anchor, as on account of its flatness more friction is occasioned in the water; and as a small Rope is considerably stronger in proportion to its size than any large one, the flat Cable is not required of so large dimensions as the round one, and being thin and pliable bends round the windlass, and passes through the Hause Hole, without having any tendency to break or strain any of the yarns, as is evidently the case in large round ropes. The flat Cable being both smaller and shorter than a round one, is less expensive at the first cost, and as it cannot untwist is preserved from being penetrated by the water, and the *surging* of the Cable round the Windlass being done away the Patentee is of opinion, that the flat Cable will wear as long as 8 or 10 round ones. The flat Cable is all wound upon the Windlass, or Capstan, and a Break or friction Lever applied upon it, which is used in bringing a vessel to anchor, and which prevents the necessity of *Folding on*, and also of conveying away or removing the Cable from place to place, which is a great saving of the Rope as well as time and Labour; and as the purchase of the Windlass is increased by a simple means as much as 2 and 4 to 1, the Anchor is lifted in much less time, and the Cable being always in its place, a large Vessel may be brought to anchor almost instantaneously by one or two men only, which in cases of emergency and danger is of considerable importance, and without doubt will often prevent the loss of Ship and Cargo, and perhaps of the Crew. In Large vessels where Capstans are used, the *Messengers* and *Nippers* will not be required. The means of drying the Cable, when taken out of the water, is simple and effective, and only requires the Windlass to be turned a few turns backward, which separates the coils of rope, and admits the air, and the Cable being covered by a piece of canvass, when coiled upon the Windlass, is preserved from the weather. The Windlass being always occupied by the flat Cable, a small Windlass is commodiously introduced, which is used for any purpose that the common Windlass is occasionally wanted for. After 17 months trial of the flat Cable, during 16 months of which the Captain has never used the round one, it appears to be scarcely injured. The Patentee has received the following Letter from MR. BRIGHTMORE:

Trent Point Wharf, Gainsboro', Nov. 6, 1807

SIR,- I have asked Captain Herratt respecting the flat Cable; and he prefers the flat to the round Cable, as commonly applied, for the following reasons:-
1st. Because he can bring his Ship to Anchor in one third of the time that is usual with round ones.
2nd. Because he can bring her to Anchor by the Break, with the power of one man instead of three.
3rd. Because the flat Cable is always ready upon the Windlass, and needs no coiling away.
4th. Because, not being liable to friction, it will be much more durable. So far Captain Herratt's Report.
It is evident, with regard to wear, it is much less liable to friction, therefore will be more durable; and in a late examination, I have found the Cable where it has been most exposed to the water is quite fresh in the inside, so that, without exception, it appears to have every advantage (upon your plan) over the round one worked in the old way; and Herratt and his Son, and the Men he has usually employed, are I believe of this opinion.

I am Sir, your obedient humble Servant,
W. BRIGHTMORE.
For Gainsboro' Boat Company and self.

When the above declaration, made by a respectable and disinterested party, and arising from actual experience, shall have been duly weighed and attended to by the Owners of Vessels, and recollecting that either the flat or round Cable may be applied, the Inventor conceives that the prejudice of Sailors to their old system of using the Cable will be removed; and he is now ready to make engagements with the Owners of Vessels, to furnish them with flat Cables, made lighter and shorter than round ones, at 14d. per lb. (at the present price of hemp,) warranted to do the work of (?) round ones, as they are usually applied. And in case they do not wear as long as three, in the opinion of two impartial men, he will engage to supply them with others, free of expence, till the work of three be done, if the destruction does not arise from wilful injury, or from casualty.

The expence of applying the flat Cable will be from 17 to 50 (?) 60 Guineas, and in very large ships the expence may be greater.

JOHN CURR is also the Inventor of a Patent Round Rope, (of which the Patent Flat Rope is composed,) which is nearly twice as strong as common round Rope, as appears from the following experiment, publicly made at Gainsboro', October 13th, 1807, which Invention is now used at the Rope

Works of Messrs. HALL and Co. Barton-upon-Humber; Messrs. WILLIAM BOURN and SON, Gainsboro', and Messrs. WILLIAM TONGE and SON, Stockwith, near Gainsboro'.

"We, the undersigned, were present at the following experiment made under J.CURR's Round Rope Patents, at Messrs. W. BOURN and SON's Ropery, in order to prove the strength of his Patent Round Rope, compared with Ropes made in the common way, and do hereby attest, that the Rope was made 4 ½ inches circumference, contained 51 threads per strand, was shroud laid, and, on a stretch of 17 feet in length, took 9 Tons, 6 Cwt. 3 Qrs. to break it down. Witness our hands,

GEORGE MILNES,	JOHN MAW
WILLIAM FRETWELL	WILLIAM BURTON
ROBERT PEATFIELD	JOHN FRETWELL
JOSEPH RHODES	W. BOURN
H. DEWDNEY	W. BARLOW

The Patentee's Invention also extends to the application of Round Ropes on the Windlass and Capstans, in a way something similar to the flat ones; but as the round Cable on account of its size and length, occupies so much of the Windlass, and not being so well adapted to bear a strain as the flat one, and more subject to friction, he recommends the flat Cable in preference to a round one.

Specimens of flat Cable, for Ships of 800 or 900 tons burthen, may be seen at Mr. La MARCHE's, Merchant, Hull.

Letters addressed to the Patentee will be duly attended to.

November 13, 1807. ....

The Patentee, with a view of getting his invention into general use as soon as possible, will supply the Owners of Ships, with flat Cables, and apparatus for using them, (if ordered any time within four months of this date) and will not expect to be paid for these till two years hence, when, if they are not approved of, he will allow them to be returned without making any charge for the use of them.

Applications may be personally made to the Patentee, at Mr. La MARCHE's, Merchant, Hull, in the beginning of December next.

Messrs. BRIGHTMORE and CO. have applied a Flat Crane Rope about 16 months, which has now done the work of two Round ones, and the Patentee is of the opinion it will wear six or seven years longer.

The *Chester Chronicle*, Friday, 9th November 1810

Patent Flat Ropes and Cordage

THE Flat Rope invented by John Curr, of Sheffield, has been applied for upwards 12 years in great variety of concerns, to the drawing of minerals, cranes, &c. and its almost incredible superiority over the common ones indisputably proved. The Patentee, however, is constrained to admit, that in some instances the Flat Rope has failed considerably short of its usual durability arising from mismanagement in the manufacture of the round ropes that compose it, from which he considers himself obliged in justice to himself and the public, to establish a manufactory of the round as well as the flat ropes, for which he has erected a covered ropery near that town. His Patent improvements in the round, conduces greatly to the superiority of the flat rope. A round patent rope 4 ½ inches circumference, was tried at the East India House, in London, and exceeded in strength any rope they had in record, taking upwards of 10 tons 18 cwt. to break it; his patent whale line, 2 ½ inches circumference, took upwards of 3 tons 4 cwt. to break it; and one of his flat ropes, applied as cable on one of Messrs. Brightmore & Co.'s vessels, of Gainsboro', the only one that has been used by the Captain for nearly 4 ½ years (although he has always a round one ready for use) to all appearance will work several years longer.

The inventor having prepared his works for the manufacture of flat ropes and round cordage, and lines of every description, begs leave to inform the public, that he henceforth engages every flat rope, of his manufacture, if equal in bulk to a round one, from 7 ½ inches to 6 ½ inches circumference, shall do as much as the average of six common round ones; and his flat rope, equal in bulk to a round one from 6 ½ to 6 inches circumference, shall do as much as the average of 3 common round ones; and his flat rope, equal in bulk to a round one from 3 ¼ to 3 ½ circumference, shall do as much as the average of 4 common round ones, casualties and wilful injury excepted: and if in any instance his flat rope shall be found deficient to the above, and the said rope shall have been remitted for within six months of delivery (the limits of his credit) he hereby engages to make an abatement proportionate to the said deficiency, and its extra charge over the round ones. The inventor, however, flatters himself his flat ropes will in general far exceed the above statement, for which no extra charge will be made; that his patent round rope of 6 or 7 inches circumference will do the work of two common ones, and the public may also rely on his round ropes, hauling lines, whale lines, &c. whether patent or common, being of a superior quality. For further particulars, address the Patentee, at Belle Vue House, near Sheffield.

18 October 1810.

The *Leeds Mercury* Saturday 21st December 1811

Patent Flat Ropes

IMPROVED, and Guaranteed in the Wear under increased Penalty by the Inventor, JOHN CURR, of Belle Vue, near Sheffield. In the present very high Price of Hemp, (which advanced £20 per ton in two Months,) it is most important to the Proprietors of Coal Works to know, that the Inventor has greatly improved the Flat Ropes by improving the Round ones that compose them, and engages, under a printed Declaration, on the Bill of Parcels, that each Rope he shall hereafter furnish shall perform as follows:- Ropes composed of 4 Round ones, weighing from 5½ to 6 lb. per Yard (equal to from 6½ to 7½ Inches Circumference of common Round Rope capable of lifting from 19 to 26 Cwt. per Pull, including Basket or Hudge, Chains, &c. attached to it,) shall do the Work of 7 common Round Ropes; or, in other Words, where the Depth of the Pit does not exceed 70 Fathoms, the Corf, Hudge, Chains, &c. does not exceed one-fifth of the Weight of the Goods lifted, the Quantity of Coals or other Minerals and Water drawn does not exceed two hundred and forty Tons per Week, that the Pair of Ropes of this Size so working shall last for three and a half Years. If more Coals, &c. be drawn weekly, or the empty Corf, &c. be heavier, the Ropes of course are supposed to be worn out in proportionably less Time.

    Ropes weighing from 4½ to 5 lb. per Yard, (or from 6¼ to 6½ Circum. Round Rope, lifting from 14 to 16 Cwt.) as much as Common Round ones.

    Ropes from 4 to 4¼ lb. (5⅞ to 6⅛ Circum. Round, lifting 9 to 13 Cwt.) 5 Round ones.

    Ropes from 3¼ to 3½ lb. (5⅛ to 5⅜ round, lifting 4 to 5 cwt.) 4 Round ones.

In order to ascertain the Wear of common Round Ropes, the Colliery Books shall be referred to, and the 4, 5, 6, or 7 Round Ropes above specified, last worn out on that Shaft, and Quantity drawn shall be the Standard for the Flat Rope's Performance. On new Shafts, where no Reference can be had, the nearest similar Situation of Depth, &c. shall be the Standard. If the Flat Rope be ordered of proper Strength to lift its Berthen, and shall not perform the Work as above specified (wilful Injury and Casualties excepted) the Patentee engages to return 1½ d. per Pound for every common Round Rope's Work they shall fall short of performing, or 2d. per lb. on the large Size, for every Year they shall fall short of working three Years and a Half, as above specified, provided he be informed of the Failure within ten Days of the Time the Rope is rendered unfit for Use, with a Statement of what the common Round Ropes have done in the same Shaft; and that a Remittance be made for Ropes sent from Sheffield, six Months after the Date on Bill of Parcels by Bills on London at two Months.

    *** As Hemp is expected to be much higher this Winter, the Patentee recommends early Applications.

For further Particulars address the Patentee.

December 1, 1811.

Royal Cornwall Gazette, 3 December 1814

Flat Cordage for Drawing Minerals, &c. out of Mines, by the original inventor and patentee, JOHN CURR, of Sheffield

Where these flat Ropes have been established for nearly two years, and are properly applied in the mines of Creaver, Oatfield, Wheal Abraham, and Wheal Fortune, in this County. The Inventor has the satisfaction to observe, that they have done the work of from four to six times as much as common round ropes of equal weight per yard, both in perpendicular and underlay shafts: or in other words have worked night and day, for 8 to 12 months, on the steam whins. The Inventor recommends from 10 to 20 yards of chain, at the least, at the bottom-end of each flat rope, nor can the full advantage be obtained from flat ropes, without the application of them, on the steam whins or horse gins: at the same time, beginning to lap on a small diameter of rope barrel, (about 2 feet) the counter-balance, or more properly speaking the equalizing power, will be more effectively obtained. By the operation of the flat rope lapping on itself, it affords a great saving in coals and horse-labour, and is the most simple and least expensive counter-balance that can possibly be applied.

This flat cordage being now in general use in many mines, in Cornwall, the Inventor proposes to keep a few ropes at PENRYN, to supply in cases of emergency: requesting, notwithstanding, that in general, the orders may be transmitted to SHEFFIELD, where Ropes of all lengths, up to 510 yards long, and any required strength, may be had, on transmitting the weight which it will be required to raise, on giving an order six months before they are wanted.

Of the utility of flat cordage, and the saving in labour &c. any one may be satisfied, on application to Captain Andrew Vivian, and John Gumlry, or the Agents employed under them: also on application to Mr. Woolf, Engineer, of Pool, near Redruth: or Mr. Peter Rogers, of Penryn, who will supply any one with flat cordage, from the Repository at Panryn, to whom orders may be given, or to the inventor, at SHEFFIELD.

N. B. Mr. CURR, has obtained a Patent for applying his flat ropes as Cables to Ships, on proper windlasses,[555] on which the whole cable is laid, and has now one on board the Grand Turk, a ship of 100 tons, John Herett, master, belonging to Mr. Brightmore of Gainsboro', and trades from Thence to Hull, which has been used for 8½ years, and is yet in condition: besides it does not require above two-thirds of the hands to weigh the anchor, and one man can, by means of a brake, bring the vessel to anchor in half the time it could be done with the common Cable. A brig of 230 tons, belonging to Ayr, in Scotland, rode out a very heavy storm in the Downs, with a single flat Cable, whilst many other vessels at the same time parted their Anchors, which were using the common round Cables.

The Inspectors at Lloyd's (on account of insurances,) have seen and approved of this invention.

Appendix 12
Auction notice in *The Iris*

From *The Iris; or, The Sheffield Advertiser*, October 24th, 1820

TO BE SOLD BY AUCTION
By Mr. T. N. BARDWELL

Without Reserve, on Wednesday, Oct 25th, 1820, upon the premises of Mr. Curr's Foundry, in Sheffield Park. Sale to commence at Ten o'clock in the forenoon.

Lots
1. Quantity of wood.
2. Do.   do.
3. Do.   do. planks.
4. Do.   do.
5. Fifteen rope reels.
6. Large wheel.
7. Wood-work for a crane.
8. Two levers belonging to lacing machine.
9. Large reel.
10. Seventeen reels.
11. Eight do.
12. Wood tops.
13. Barrel rollers
14. Quantity of wood.
15. Do.   do.
16. Lacing machine frame.
17. Windows.
18. Desk and frame.
19. Large pulley or wheel.
20. Step ladder.
21. Two blocks with hackle teeth.
22. Large tub.
23. Pair large iron blocks.
24. Do.   do.
25. Five Wood blocks.
26. Scales and beams.
27. Long new thick rope.
28. Iron.
29. Grindstone and frame.
30. Cast metal barrel for crane.
31. Five tin barrels.
32. Eleven metal wheels.
33. Iron.

34. Iron and locks, &c.
35. Metal weights, &c.
36. Stool, hammer, &c.
37. Crane wheel and chain.
38. Tar and tub.
39. Garden spade.
40. Quantity of stone.
41. Metal plate.
42. Three pieces of metal.
43. Large stone.
44. Grindstone.
45. Large piece of metal.
46. Carriage to make ropes.
47. 14 metal rope stretchers.
48. Gasometer, 9 feet long, 7 feet wide, 6 feet deep.
49. Tarring machine.
50. Small crane with rope.
51. Twelve pieces of metal.
52. Quantity of wood.
53. Carriage for rope making.
54. About 30,000 common bricks.
55. Stone.
56. Metal pump, 7 yards.

GLOSSARY

This glossary has been collated from various sources, as noted in the text, with further material provided by John Hunter.

Basset

An outcrop; the edge of a stratum.

Coal: 'hard' & 'small'

Coal classification is complex. For instance, according to W. S. Gresley: *'There is no standard size distinguishing coal from slack'*.[556]

Hard[s], or Barnsley, coal is the highest rank of coal, slow-burning and with a high heat output. Hards is a low-ash, low-sulphur, low-volatile coal. Of high value, it was used in *cutlers' hearths*.

Small[s] coal, or slack, is what the name suggests: coal broken into small pieces. Slack was of low value and was known as *engine coal* as it could not be used in furnaces, nor was it favoured for parlour fires. Indeed, according to Galloway, *'small coal in this era was of such little account as to be largely burnt in order to be got rid of '*.[557]

Barnsley coal seams comprised four constituents. The bottom third was *bottom softs*; the middle was *hards*; *top softs* were above the hards, and *bags* were next to the roof. The seam worked in the Sheffield area was known as the Silkstone, which was generally 5 ft thick, including the dirt band. At Nunnery Colliery it was 5 ft 7 in. thick, including 10 inches of dirt. Barnsley and Silkstone softs were favoured for domestic fires. Hards and softs were found together in the same coal seam, but as separate bands.

Figure G.1. Iron corf plate castings at Hemingfield Colliery (Elsecar). Photo © John Hunter.

Corf

A container for coal, originally a basket (Latin *corbis*, German *Korb*), afterwards an iron receptacle. Corf is the term used by Curr for a wheeled truck on rails. In South Yorkshire corfs were made of wood, strengthened by iron bands and mounted onto a wrought-iron frame to which the axles were attached [Figure G.1].

Engine

Curr employed the 'common' or atmospheric steam engine. Developed originally by Thomas Newcomen in 1712, it was mostly operated in reciprocal motion, and, crucially, the steam was condensed inside the cylinder. Such engines were simple and inefficient, but cheap to operate given a cheap supply of coal. Curr used them mainly for pumping water (the see-saw motion of the beam was well suited to the operation of pumps) but sometimes for winding coal. In the latter instance it is probable that he modified the engine to produce additional rotary action by means of a crank and flywheel fitted next to the pumps. Despite the flywheel, the winding will have been both slow and jerky.

The 'common engine' effectively became obsolete in 1800 on the expiry of James Watt's patent for the separate condenser. Before 1800 attempts to meddle with Watt's patent were usually met with immediate legal challenge. After 1800 it became easier for other engine builders to design and make new (condensing) rotary steam engines without fear of litigation.

The expiry of Watt's patent coincided with Curr's dismissal by the Duke of Norfolk. As noted, Curr's foundry supplied engines. Those built after 1800 remained of the 'common' variety.

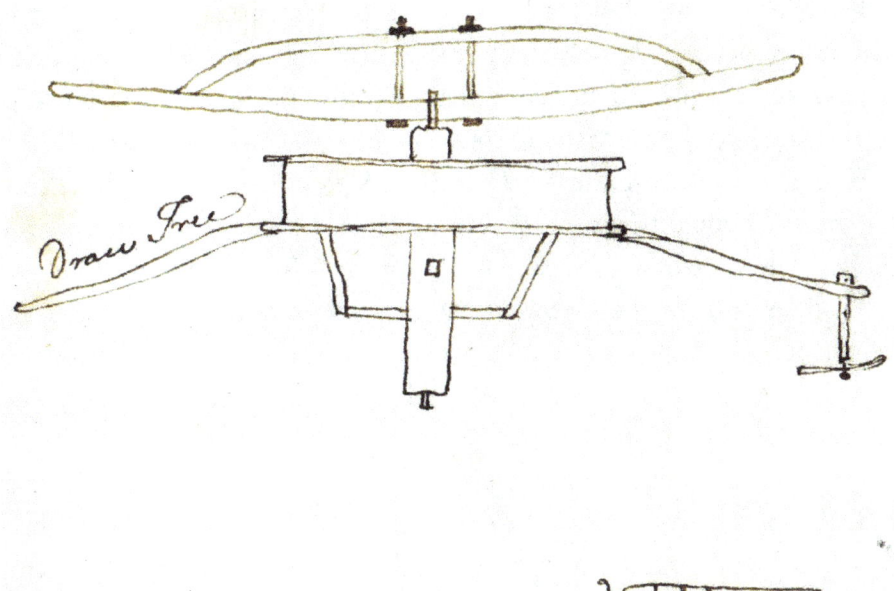

Figure G.2. Horse-gin. This sketch is from J. Buddle Snr's Sheffield colliery notebook, 1773. From: TCR NEIMME/Bud/68.

Gin or horse-gin

A horse-driven winding engine [see Figure G.2]. Small steam engines used for winding were known as 'whimseys'.

Hewer

The man who digs the coal.

Hurrier (haulier)

A boy or man who goes with a pony or horse in a pit to attend the trams.

Hurry

To haul, pull or push trams of coal in a mine.

Inby

See Rolley.

Iron: cast, wrought iron and bar

Cast iron is made from pig iron, which is the product of melting iron ore in a blast furnace. It tends to be brittle.[558] It will fracture if subject to torsional or bending stress, but it is ideally suited to many purposes if used correctly. Importantly, cast iron (of the correct grade – there are several types) can be used to make castings in intricate shapes, but the metal cannot be forged or forge-welded.

Wrought iron is resistant to fracturing. Initially wrought iron was much more expensive than cast iron due to its lengthy production process. In Curr's time it was made from pig iron by decarburisation (burning out the carbon content from 4% to 0.1%).

In 1783 Henry Cort patented improvements in the puddling process involved in making wrought iron. Cort's process used coke, instead of charcoal, as fuel, which further lowered the cost of manufacture. However, the process remained expensive until more innovations were introduced by the Merthyr Tydfil ironmasters, Crawlay and Humfray, in the early 1790s.

Wrought iron can be forged, forge-welded, and rolled into bars, rods and sheets. Wrought-iron rails were to replace those made of cast iron because they were easier to make in long lengths in a rolling machine and were lighter and tougher. However, wrought iron had its drawbacks, for instance it had a tendency to de-laminate. In the 1860s wrought-iron rails were replaced in turn by Bessemer steel rails, made by forging cast steel ingots through rolling mills.

Bar iron is similar to wrought iron (almost pure iron) – but not produced in a blast furnace from molten metal. It was imported as bars from Scandinavia and Russia, where it had been made by charcoal smelting below the melting point of iron. Sheffield imported large quantities of this metal product – it was the raw material for making 'cementation' and 'crucible' steel (because it had none of the impurities present in local iron ores). If Curr's waggon-way employed iron wear-strips on the wooden rails, he may have used bar iron, possibly hammered down to a thinner gauge.

Methods of extraction

Two methods, with variants, were used during the period in question: the *longwall* and *bord and pillar*. In South Yorkshire bord and pillar was known as *short work*.

Lewis described the first of these:

The longwall method ... allowed all the coal to be dug, but nearly always resulted in subsidence. The working face was pushed forwards in a continuous line, and the 'goaf', the area from which the coal had been dug, was filled up with rubbish behind the miners as they advanced. Passages or gates were left through the goaf to carry the coal out.[559]

Flinn described the second, which had the disadvantage that much of the coal could not be extracted:

... By this method – the 'pillar and stall', 'bord & pillar', or 'short and narrow' method – coal was taken from passages – the bords or stalls – between pillars many yards square. Those at George Humble's Houghton colliery in County Durham won in 1749–50 were sixteen and a half feet thick and in lengths from sixty to 120 feet. ... This system was universally used in the North-East and in many other coalfields. Pillars were either square or rectangular. Pillars of this size allowed less than fifty per cent of the coal to be removed. Slightly smaller pillars, such as those left in the Barnsley district of south Yorkshire in the early eighteenth century, permitted up to two-thirds of the coal to be removed.[560]

Fortunately, a contemporary description of the methods employed at Sheffield Park colliery was provided by Hatchett:

> *In the working places when they have removed a certain portion for 20 yards or more of the face of the seam of coals, they support the open part behind them with Punches or Props of timber which as they advance they remove and the incumbent strata gradually fall in and fill up the space. This can only be done when the Roof is brittle and not too strong (the stratum of coal dips about three and a ½ inches in a yard). When otherwise they can only cut away 6 or 8 yards of coal and then leave 8 yards as a support alternatively as at Newcastle.*[561]

Outby

See Rolley.

Patents (English patents 1617–1852)

Before October 1852, details of granted English patents were simply recorded (enrolled) in the Patent Rolls at the end of a long and costly application process. No numbering system was in place. The details recorded in the Rolls usually included the name, rank and address of the patentee, the title of the invention, a formal recitation of the terms of the monopoly (patent) granted and the date of grant. From the 1730s onwards patentees were also obliged to file a specification which described their invention in technical detail.

During the 1850s the British patent system was subject to major reforms. As part of the reforms, English patents granted between 1617 and 30 September 1852 were identified from the Rolls, sorted into chronological order and then numbered in a single continuous sequence. The sequence runs from GB1 of 1617 to GB14359 of September 1852.

From October 1852 British patents covered the whole of the United Kingdom and were printed and published each week in pamphlet form. The text of the patent document identifies the patentee and provides a technical description of the invention. If a technical specification was found, then just the details of the grant recorded in the Patent Rolls were transcribed. The transcribed text of each patent and any drawings were then printed and published as a numbered pamphlet. The patents were printed in London by George E. Eyre and William Spottiswoode and published at The Great Seal Patent Office.

John Curr's patents, in pamphlet form, were acquired by the author disbound. Each (about 27 x 18 cm) has the original printed front blue wrapper but not the back, as well as a plate, if called for and of varying size, attached. They were printed in 1855–56, and most are lithographs.

Put

To convey coal from the working face to the tramway.

Putter (haulier)

A man or boy who conveys coal from the working place to the tramway.

Putting ponies

Ponies 10 or 11 hands high used in mines for hauling.

Rolley

Used only on Tyneside, where a 'rolley' was an underground flanged-wheel waggon; 'outby' was the area of a coal-mine between shaft and transfer-point of corves from tram to trolley; and, 'inby', the area of a mine between face and transfer-point of corves from tram to trolley.[562]

Sough

A drain.

Stack (coal)

To build up coals into heaps on the surface.

Staith

Staging where corves are unloaded.

Win

To sink a shaft or drive a drift to a workable seam of coal, etc.

Winning

A new mine opening. The term had a wider context, especially in Curr's time in Sheffield. The working limits of a colliery were defined by the area of the royalty (i.e. the property boundary of the coal being leased from its owner) and also by the lowest elevation from which water was drained. Attempting to work the coal seam below the water level was difficult, because the water had to be pumped (or baled) out. New winnings were expensive. They involved the expenditure of capital in sinking shafts and, if necessary, the installation of steam engines to pump water.

The collieries in Curr's charge in the Silkstone seam at Sheffield Park were drained by gravity by a very long sough. Because of the big hill at Sheffield Park, a large area of coal could be worked without drainage because it was above sough level. The Silkstone sough probably drained into the River Sheaf. The Parkgate sough drained into the River Don. The area under Sheffield Park is not more faulted than elsewhere – in fact it is less so. Soughs were expensive to maintain and to keep open. Men were employed to clear out the sediment, and to keep the roof propped up (such work was known as *feying*). Long soughs required air shafts at intervals for ventilation. The next 'winnings' after Sheffield Park Colliery (below its sough level) were, in succession, Ponds, then Crooks Croft, then Nunnery Colliery, all of which required shafts and pumps.

Working practice

The main tools of the collier consisted of a pick, shovel, hammer and wedges. The first operation in working the coal was to undercut or hole the face at the bottom to a depth of 2 feet 6 inches to 3 feet 6 inches with a pick, although vertical grooves were sometimes made. It was during this stage that considerable quantity of slack was produced that could only be used in the Newcomen engine. The coal was brought down in large pieces by wedges hammered into the coal face. In the Orrell coalfield the wedges used measured 12 inches long by 2 ½ inches wide by 1 inch thick in the middle, tapering to one end with the striking end octagonal in shape. Shortly before the opening of the Attercliffe Colliery, records show payment for 5000 wedges at 1d. per score.[563]

Hewers brought down the coal for fillers to load the corves that were hurried to the shaft bottom by boys leading horses. Hangers-on fastened the corf to the winding rope to be wound up the shaft by a gin boy supervising the horse-drawn whin gin, later replaced by engine 'tenters'. Trap boys worked the ventilation doors, 'wood getters' cut the punchwood and masons lined the shafts, horsegates and soughs. Other workers drove the headings, banksmen supervised the surface, including the coal stackers, carpenters, blacksmiths, fire-pan and engine 'tenters'.[564]

John Buddle Snr, in his 1773 notebook, illustrated several implements, noted in Figures G.3 and G.4.

Figures G.3 & G.4. Pick & Mell (left) and Wheelbarrow (right). These sketches are taken from J. Buddle Snr's Sheffield colliery notebook, 1773. From: TCR NEIMME/Bud/68.

BIBLIOGRAPHY

Agricola 1556
Agricola, Georgius. *De Re Metallica*. 1556

Ashton 1948
Ashton, T. S. *The Industrial Revolution*. London: Oxford University Press, 1948

Ashton 1963
Ashton, T. S. *Iron & steel in the Industrial Revolution*. Manchester, 1963

Ashton and Sykes 1929
Ashton, T. S. and Sykes, Joseph. *The Coal Industry of the Eighteenth Century*. Manchester University Press, 1929

Baxter 1966
Baxter, Bertram. *Stone Blocks and Iron Rails*. Newton Abbott: David & Charles, 1966

Blackwell 1828
Blackwell, Henry. *The Sheffield Directory and Guide*. Sheffield: John Blackwell, Iris Office, 1828

Bland 1930–31
Bland, Fred. John Curr, originator of iron tram roads. *Transactions of the Newcomen Society*, **11** (1930–31), pp. 121–30

Blue Books 1842 (see *Children's Employment Commission* 1842)

Bracegirdle 1973
Bracegirdle, Brian. *The Archaeology of the Industrial Revolution*. Fairleigh Dickinson University Press, 1973

Buddle Snr 1785
Buddle, John Snr. Observations on the Raising of Hedges, in a letter to the author, included in: Robert Callender, *A Practical Essay on the Raising and Management of the Crataegus Oxycantha; Or, Common White Thorn For Hedges …*. Newcastle upon Tyne: printed by L. Dinsdale, 1785, pp. 27–41

Buddle 1788
Buddle, John. A life preserved after falling into a coal-pit, from the *Gentleman's Magazine*, 5 March 1788 (included in Richardson 1846)

Catholic Encyclopedia
Herbermann, Charles G. (ed.) & others. *The Catholic Encyclopedia*. New York: Robert Appleton Co., 1907–14

Chadwick and Knights 1988
Chadwick, Roy and Knights, Martin C. *The Story of Tunnels*. London: Andre Deutsch, 1988

Chapman 1808
Chapman, William. *A treatise on the progressive endeavours to improve the manufacture and duration of cordage*. London: printed for W. H. Wyatt, 1808

Children's Employment Commission 1842
Royal Commission of Inquiry into Children's Employment in Mines and Manufactories. *First Report of the Commissioners: Mines*. Presented to both Houses of Parliament by Command of Her Majesty. Published as part of the British Parliamentary Papers. London: printed by William Clowes and Sons, Stamford Street, For Her Majesty's Stationery Office

Children's Employment Commission 1842, Appendix 1
Royal Commission of Inquiry into Children's Employment in Mines and Manufactories: Children's Employment Commission, *First Report of the Commissioners, Mines, with appendices. Part I: Reports and Evidence from Sub-Commissioners*. Published as part of British Parliamentary Papers. London: printed by William Clowes for HMSO, 1842

Clark 1943
Clark, Ronald H. East Anglia's first steamboats. *Model Engineer*, 29 January 1943

Coleridge 1887
Coleridge, H. J. (ed.). *St. Mary's Convent, Micklegate Bar, York, 1686–1887*. London: Burns & Oates, 1887

Condition and Treatment of the Children Employed 1842
The Condition and Treatment of the Children Employed in the Mines and Collieries of the United Kingdom; Carefully compiled from the appendix to the first report of the [Children's Employment] commissioners appointed to inquire into this subject, with Copious Extracts from the Evidence, and Illustrative Engravings. Privately compiled and published in London, by William Strange in 1842, with a Preface [signed] W. C., May 13th, 1842.

Curr, E. M. [EMC Memoranda]
Curr, Edward Micklethwaite. *Memoranda concerning our family, 1877*. Melbourne: State Library of Victoria, La Trobe Library, MS 8998, unpaginated

Curr [III] 1847, *Learned Donkeys*
Curr, John. *The Learned Donkeys*. London: John Williams & Co., 1847 (copy held by the State Library of New South Wales)

Curr [III] 1847, *Railway Locomotion*
Curr, John [III]. *Railway Locomotion and Steam Navigation: Their principles and practice*. London: John Williams & Co., 1847

Curr 1797 (1970), *The Coal Viewer*
Curr, John. *The Coal Viewer, and Engine Builder's Practical Companion*. Sheffield: printed for the author by John Northall, 1797; 2nd edn: Frank Cass & Co., 1970, with an Introduction by Charles E. Lee

Dawson 2001
Dawson, Charles. PS *Orwell*, 1813–14. *Mariner's Mirror*, vol. 87, May 2001

Dendy Marshall 1938 (1971)
Dendy Marshall, C. F. *A History of British Railways down to the year 1830*. Oxford University Press, 1938 (reprinted in 1971)

Dickinson 1942–43
Dickinson, H. W. A condensed history of rope-making, *Transactions of the Newcomen Society*, **23** (1942–43), pp. 71–91, and 4 unnumbered pages of plates, XVII–XX.

Dunn 1844
Dunn, Matthias. *An Historical, Geological, and Descriptive View of the Coal Trade of the North of England*. Newcastle-upon-Tyne, 1844

Dunn 1855
Dunn, Matthias. Scientific and practical mining college proposed to be erected in Newcastle-on-Tyne. *Mining Journal*, vol. 25, issue 1048, 22 September 1855 (pp. 602–3)

EMC *Memoranda* (see Curr, E. M.)

Emerson 1754
Emerson, William. *The Principles of Mechanics*. First published 1754

Farey 1811–17 (1989)
Farey, John. *General View of the Agriculture and Minerals of Derbyshire*. Printed by B. M'Millan, Bow Street, Covent-Garden, in 3 vols, 1811–17. Peak District Mines Historical Society reprint, 1989.

Farey 1827 (1971)
Farey, John. *A Treatise on the Steam Engine, Historical, Practical, and Descriptive*. Volume 1. London: printed for Longman, Rees, Orme, Brown and Green, 1827. Reprint: Newton Abbot: David & Charles, 1971.

Fawcett 1911
Fawcett, J. W. *A history of the parish of Dipton*. Durham, 1911

Flavell 1996
Flavell, Neville. *The Economic Development of Sheffield and the Growth of the Town c. 1740–c. 1820*, PhD thesis, University of Sheffield, February 1996.

Flinn 1984
Flinn, Michael W. *The History of the British Coal Industry; Volume 2, 1700–1830: the Industrial Revolution*. Oxford: Clarendon Press, 1984

Fordyce 1855
Fordyce, W. *The History and Antiquities of the County Palatine of Durham*. Newcastle, 1855, vol. 1

Galloway 1882, *History*
Galloway, R. L. *A History of Coal Mining in Great Britain*. London: Macmillan and Co., 1882

Galloway 1898, *Annals*
Galloway, R. L. *Annals of Coal Mining and the Coal Trade*, vol. I. London: Colliery Guardian, 1898

Gatty 1873
Gatty, Alfred. *Sheffield Past and Present*. 1873

Gillow 1885–1902
Gillow, Joseph. *A literary and biographical history, or bibliographical dictionary, of the English Catholics from the breach with Rome in 1534, to the present time*. Published in 5 volumes. London: Burns & Oates, 1885–1902. Joseph's entry is on pp. 608–12

Gooch 1989
Gooch, Leopold. *From Jacobite to radical: the Catholics of north east England, 1688–1850*. Thesis, University of Durham, 1989

Gresley 1883
Gresley, William S. *A glossary of terms used in coal mining*. London: E. & F. Spon, 1883

Haddock 2015
Haddock, K. *British Opencast Coal, a photographic history, 1942–1985*. Ipswich: Old Pond Publishing, 2015

Hadfield 1889
Hadfield, Charles. *A History of S. Marie's Mission and Church, Norfolk Row, Sheffield*. Sheffield: Pawson & Brailsford, 1889

Hall 1858
Hall, T. Y. Discussion at meeting on 7 October 1858, recorded in *Transactions of the North of England Institute of Mining and Mechanical Engineers [NEIMME]*, vol. 7 (1858)

Harrison 1956
Harrison, David. *Along Hadrian's Wall*. London: Cassell and Co. Ltd, 1956

Hatchett Diary (Raistrick 1967)
Raistrick A. (ed.). *The Hatchett Diary: A tour through the Counties of England and Scotland in 1796 visiting mines and manufactories*. D. B. Bradford Barton, 1967

Heesom 2004 (*ODNB*)

Heesom, A. J. John Buddle (1773–1843), mining engineer. *Oxford Dictionary of National Biography*. Oxford University Press, September 2004

Hiskey 1978

Hiskey, Christine E. *John Buddle (1773–1843), agent and entrepreneur in the North-East coal trade*. M. Litt. thesis, University of Durham, 1978

Holland 1820

Holland, John. *Sheffield Park: A descriptive poem*. Originally published by J. Montgomery, Sheffield, in 1820 (reprinted 1859).

Holland 1826

Holland, John. *Theresa's Tree*, 1826

Holland 1837

Holland, John. *The tour of the Don, a series of extempore sketches [chiefly by Holland] made during a pedestrian ramble along the Banks of the river and its Tributaries*, 2 vols. London, 1837

Hopkinson 1958

Hopkinson, Geoffrey Gill. *The Development of Lead Mining and of the Coal Industry in North Derbyshire and South Yorkshire 1700–1850*. PhD thesis, University of Sheffield, 1958

Hudson 1874

Hudson, William. *Life of John Holland of Sheffield Park: from numerous letters and other documents furnished by his nephew and executor, John Holland Brammall*. London: Longmans, Green & Co., 1874

Hunter 2021

Hunter, John. Early pumping engines in the Attercliffe–Darnall–Orgreave coalfield and the role of John Curr. *Transactions of the 2nd International Early Engines Conference*, Vol. 1, IEEC & ISSES, 2021, pp. 54–76

Jacob 2014

Jacob, Margaret. *The First Knowledge Economy: human capital and the European Economy, 1750–1850*. Cambridge University Press, 2014

Journal of the Franklin Institute

Journal of the Franklin Institute of the State of Pennsylvania, series 2, vols 13–14, 1834.

Kirkus (Sister Gregory) 2001

Kirkus, Sister Gregory. *An IBVM [Institute of the Blessed Virgin Mary] Biographical Dictionary of the English Members and Benefactors (1667–2000)*. London: Catholic Record Society, 2001

Leader 1901

Leader, R. E. *Sheffield in the Eighteenth Century*. Sheffield, 1901

Lee 1960–61

Lee, Charles E. Some railway facts and fallacies. *Transactions of the Newcomen Society*, 33, 1960–61, pp. 1–16

Lewis 1974

Lewis, M. J. T. *Early Wooden Railways*. Routledge & Kegan Paul, 1974 (first hardback edn 1970)

Lewis 1992, map

Lewis, Michael. Map: An historical atlas of County Durham, produced for the Durham County Local History Society, 1992

Liffen 2008

Liffen, John. The discovery of a manuscript version dated 1795 of John Curr's *The Coal Viewer* (1797). Railway and Canal Historical Society, Tramroad Group. Occasional Paper **184** (January 2008)

Lock 2015 (*ODNB*)

Lock, Alexander. John Curr (1756–1823), coal viewer and engineer. *Oxford Dictionary of National Biography*. Oxford University Press, September 2015

Lock 2016

Lock, Alexander. *Catholicism, identity and politics in the Age of Enlightenment: the life and career of Sir Thomas Gascoigne, 1745–1810*. Boydell Press, 2016

McConnell 2004 (*ODNB*)

McConnell, Anita. William Chapman. *Oxford Dictionary of National Biography*. Oxford University Press, 2004

MacKie 1801–50

MacKie, Charles. *Norfolk Annals, a chronological record of remarkable events in the nineteenth century*, vol. 1, 1801–50

Malster 1971

Malster, R. *Wherries and Waterways &c*. Lavenham, 1971.

Mateaux 1885

Mateaux, Clara L. *George and Robert Stephenson*. Cassell, 1885

Medlicott 1982

Medlicott, Ian R. *The Landed Interest and the Development of the South Yorkshire Coalfield 1750 to 1830*. Unpublished MPhil thesis. Milton Keynes, UK: The Open University, 1982

Medlicott 1983

Medlicott, Ian R. John Curr and the development of the Sheffield collieries, 1781–1805. *Transactions of the Hunter Archaeological Society*, **12** (1983)

Medlicott 1999

Medlicott, Ian R. John Curr, 1756–1823, mining engineer and viewer. In: Melvyn Jones (ed.), *Aspects of Sheffield, volume 2: Discovering Local History*. Barnsley: Wharncliffe Publishing Limited, 1999, pp. 63–78

Miscellanea IV: 1907 & VI: 1909
Miscellanea. London: Catholic Record Society, IV: 1907; VI: 1909

Miscellaneous Correspondence
Miscellaneous Correspondence in Prose and Verse, containing a variety of subjects ..., London: Benjamin Martin Publisher, printed by W. Owen, vol. II (1757 & 1758, printed 1759); vol. IV (1762)

Mott 1969–70
Mott, R. A. Tramroads of the eighteenth century and their originator: John Curr. *Transactions of the Newcomen Society*, **42** (1969–70), pp. 1–23

Odom 1926
Odom, [Canon] William. *Hallamshire worthies: characteristics and work of notable Sheffield men & women*. Sheffield: J. W. Northend, 1926

Oldroyd 2007
Oldroyd, David. *Estates, enterprise and investment at the dawn of the industrial revolution: estate management and accounting in the North-East of England, c. 1700–1780*. Burlington, VT: Ashgate, Aldershot, 2007

Partington 1822
Partington, Charles F. *An historical and descriptive account of the steam engine*. London: The Architectural Library, 1822

Phillips 1844
Phillips, John (F.R.S., F.G.S.). *Memoirs of William Smith, L.L.D*. London: John Murray, 1844

Raistrick 1967 (see *Hatchett Diary*)

Richardson 1846
Richardson, M. A. *The Borderer's table book; or, gatherings of the local history and romance of the English and Scottish border*, in eight volumes. Vol. VII: *Legendary Division*. Newcastle-upon-Tyne, printed for the author, 1846

Rieuwerts 1981
Rieuwerts, J. H. *A Technological History of the Drainage of the Derbyshire Lead Mines*. PhD thesis, University of Leicester, 1981

Robinson 1995
Robinson, John M. *The Dukes of Norfolk*. Phillimore, 1995

Sargent 2015
Sargent, Edward. 'A history of rope making'. Docklands History Group, 2015

Skempton (ed.) 2002
Skempton, A. W. (ed.). *A Biographical Dictionary of Civil Engineers in Great Britain and Ireland*, vol. 1 (1500–1830). Thomas Telford, 2002

Smiles 1857

Smiles, Samuel. *The Life of George Stephenson, railway engineer*. London: John Murray, 1857

Tattersfield 2011

Tattersfield, Nigel. *Thomas Bewick: the complete illustrative work*. British Library, 2011

Taylor 1859

Taylor, T. J. On the progressive application of machinery to mining purposes. *Proceedings of the Institute of Mechanical Engineers*, vol. 10, issue 1, June 1859

The Coal Viewer (see Curr 1797 (1970))

Thurston 1895

Thurston, Robert H. *A History of the Growth of the Steam Engine*. London: Kegan Paul, Trench, Trübner and Co., 1895

Tomlinson 1915

Tomlinson, W. W. *The North Eastern Railway; its rise and development*. Newcastle: A. Reid & Co.; London: Longman, Green & Co., 1915

Trinder 1973

Trinder, Barrie. *The Industrial Revolution in Shropshire*. Phillimore, 1973

Trinder 1974

Trinder, Barrie. *The Darbys of Coalbrookdale*. Phillimore, 1974 (several editions available)

Welford 1895

Welford, Richard. *Men of Mark 'twixt Tyne and Tweed*, vol. I. London/Newcastle-upon-Tyne: Walter Scott Ltd, 1895

Wilson 1843

Wilson, Thomas. *The Pitman's Pay & Other Poems*. Gateshead: William Douglas, 1843

Wilson 2022

Wilson, A. N. Our weird 'unis' are increasingly pointless. *The Times*, 13 August 2022

Woolhouse 1832

Woolhouse, Joseph. *A description of the town of Sheffield*. 1832 (no further details available)

Newspapers and periodicals

Advocate, Tasmania, 25 October 1926
Edward Curr 'Father of the Coast' by H. R. Haslock

Aris's *Birmingham Gazette*, 13 June 1803
Patent flat ropes

Bath Chronicle, 23 November 1815
A royal visit

British Palladium
The British Palladium, or Annual Miscellany of Literature and Science. Mathematical problem-solving. London: printed for D. Steel, 1771–73

The Builder, April 1843

Chester Chronicle, 9 November 1810
Patent flat ropes and cordage

Cornwall Chronicle (Launceston), 24 October 1835
Oct. 19 – Arrived the barque *Ann*, 339 tons, John Virtue, Commander, from London. Passengers Mr. and Mrs. Curr and 4 children … with a general cargo assigned to … John Curr …

Derby Mercury
Sale of Norwood-Nook colliery – apply George Curr, 3 March 1780
New punches or props, 3 June 1802
A colliery to be either let or sold, 13 April 1809
Death of Juliana, 14 April 1830

The Farmer's Friend and Freeman's Journal, York, 27 November 1852
Aislaby Hall to be let

General Magazine
The General Magazine of Arts and Sciences. London: printed for W. Owen, 1764

Gentleman and Lady's Palladium
The Gentleman and Lady's Palladium. Printed for J. Scott, 1760, 1761

Gentleman's Diary
The Gentleman's Diary or, The Mathematical Repository; an Almanack. London: printed for the Company of Stationers, 1760, 1761, 1763, 1764, 1768, 1771, 1772 (Mathematical problem-solving)

Gentleman's Magazine
Buddle, John. A life preserved after falling into a coal-pit, 5 March 1788
Country News: The violent thunder-storm…, 1 July 1810
Obituary of Buddle Jnr, January 1844

Hull Advertiser, 5 December 1807
Cables applied to windlasses and capstans of ships

The Iris, or, The Sheffield Advertiser, 14 March 1820
John Curr II, Belle Vue house advertised for sale

Leeds Intelligencer
Curr was advertising for subscriptions for his book, 2 December 1793
Horrible death of (old) Mrs Curr, 6 July 1801
Carpet machinery to be sold, 25 February 1805
Steam engine to be sold, 8 May 1819

Leeds Mercury, 21 December 1811
Patent flat ropes

Liverpool Albion, 26 May 1856
The two Mary Beauvoisins missing

London Gazette, 4 October 1817
Notice of partnership dissolved

Mechanics' Magazine, Museum, Register, Journal and Gazette
2 & 30 October 1847

The Monthly Review, vol. 29, August 1799
London: printed by A. Strahan

Morning Chronicle, 15 March 1848
Advertisement by John Curr for his pamphlet, *The Learned Donkeys*

Newcastle Courant, 6 January 1798
'This day was published The Coal Viewer ... & etc.'

Norfolk Chronicle
Aggs & Curr advert, 5 January 1811
Report on *Courier* explosion, 5 April 1817

Norwich Mercury, 17 April 1830
John Curr [III] Tobacco & Snuff manufacturer (advertisement)

Royal Cornwall Gazette, 3 December 1814
Flat cordage

Sheffield Daily Telegraph
New trustees for John Curr [II]'s will, 10 November 1873
City stadium threatens distinguished house (Belle Vue), 20 May 1939

Sheffield Independent
Penniston obituary (claim to be inventor of flat rope), 27 September 1828
Henry Beauvoisin's bankruptcy, 11 August 1855

Sheffield Register, Yorkshire, Derbyshire, & Nottinghamshire Universal Advertiser
Weekly sailings to Hull, 30 May 1789
Now let to George Curr, 4 June 1790

The Town and Country Magazine, or, Universal Repository of Knowledge, Instruction and Entertainment. London: printed for A. Hamilton, vol. II, September 1770

Transactions of the Midland Institute of Mining, Civil & Mechanical Engineers, vol. XL, 1887–89

The Universal Magazine of Knowledge & Pleasure, vol. 55, London: published by John Hinton, December 1774

York Herald, 14 June 1851
Death of Hannah Curr

Yorkshire Archaeological Journal
Yorkshire Archaeological Journal, 1987, vol. 59

Archives

Some Endnotes cite archival material for which the author has not been able to find further information.

Sheffield [City] Archives (SA)

Papers & correspondence (the Arundel Castle Manuscripts [ACM] relating to Charles Howard, 11th Duke of Norfolk). These were presented to Sheffield City Library in 1959.

ACM/S/179
Estate accounts 20 December 1777: *'Received of Mr John Curr the Remainder and in full for the overend both of the Wood and Manor Pits due Xmas 1775 and 1776 £134 1s. 6d.'*

ACM/S/185
Estate accounts 2 December 1778; refers to Curr viewing and examining several collieries for which he was paid £4 5s. 2d.

ACM/S/185/1–10: 1–3: Vincent Eyre's accounts 1753–60; 4–10: Henry Howard's accounts 1761–69.

ACM/S/198
Sheffield Park colliery bill books, 1781–87.

ACM/S/199
Sheffield Park colliery bill books, 1789–98.

ACM/S/200
Account book, 1789–1801.

ACM/S/201
Accounts paid by John Curr for expenditure at Attercliffe colliery, 1786–89.

ACM/S/202
Attercliffe Colliery disbursement accounts, including wages, 1789–98.

ACM/S/205
Vincent Eyre's account for the joint collieries of the Duke of Norfolk, 1801–5.

ACM/S/214
J. Curr, Letter of John Curr to the Duke of Norfolk: *Report of Inventions introduced by John Curr in the Sheffield and Attercliffe collieries; and reasons why collieries of late years have been unproductive*, 23 October 1801. A draft of the Duke's reply is included in this archive.

ACM/S/215
Unsigned (by John Buddle Snr). *Report on a Colliery in Sheffield Park belonging to the Duke of Norfolk*, 1773.

ACM/S/216
Report by John Curr on Mr. Swallow's colliery at Parkin Wood, Ecclesfield leased from the Earl of Surrey, 1779.

ACM/S/217
Curr, J. *Report on Sheffield pits in the tenure of Messrs Townsend and Furniss*, signed John Curr, 27 August 1779.

ACM/S/218
Curr, J. *State of Sheffield Colliery from the improvements in 1774 to this time*, unsigned but in Curr's hand, (25th?) March 1780.

ACM/S/219
Curr, J. *Estimate on Sheffield colliery*, unsigned but in Curr's hand, 6 November 1780.

ACM/S/221
Curr's report of 1784 (Medlicott 1982, p. 233, footnote 21).

ACM/S/222
Estimate on the Manor Colliery; John Curr, 1785.

ACM/S/223
J. Buddle, *Report of John Buddle to the Duke of Norfolk: Report relative to Sheffield Park and Attercliffe Common collieries*, 7 April 1787.

ACM/S/224
J. Buddle & John Stephenson, *Report of John Buddle and John Stephenson to Vincent Eyre, Esq., on Attercliffe Common Colliery*, 24 April 1789.

ACM/S/225
Statements by Curr relating to Attercliffe colliery, 1795.

ACM/S/226
Sheffield and Handsworth Collieries; Wm. Stobart, 1817 & 1820.

ACM/S/231
Accounts of coal and ironstone got by …, 1805–16.

ACM/S/232
A valuation of the seams of coal …, 31 August 1840.

ACM/S/274
Duke of Norfolk v. Staniforth, at York assizes, 1803–4.

ACM/S/474/5
Question whether the Duke as Lord of the Manor can erect five engines and buildings on [Attercliffe] Common. For the purpose of extracting coal from beds under the Common, 1785.

SD: Sheffield Deeds

ACM/SD/3, 4, 9, 10
Lease amended for renewal to Eyre's executors and copies of the new lease, May 1805.

ACM/SD/18
Renewal of lease of the seams of coal …, 31 August 1840.

ACM/SD/520
Indenture made the 22nd Day of November 1803 BETWEEN Charles Howard Duke of Norfolk … and John Curr of Belle Vue house in

Sheffield Park, Engineer … from the 29th Day of September now last past …

ACM/SD/521

Indenture made the 22nd Day of November 1803 … let unto the said John Curr ALL those four Closes or parcels of land near the Mansion house of the said John Curr in Sheffield Park …

ACM/SD/666

Vincent Eyre's lease. Lease to Vincent Eyre … of all the seam called the Attercliffe and Darnall seam lying under Attercliffe Common and under Sheffield Park, with all works and engines; for 21 years, for one full equal fourth share of the profits, in consideration of costs and charges in opening the seams. [Alongside this next sentence 'See SD 4' is handwritten] Also Eyre's declaration concerning the lease, citing articles that the profits of the Duke's moiety were to be for repaying the money advanced by Eyre for opening the new colliery. 1st September, 1789.

ACM/SD/870/59

99-year lease, dated 1 January 1781, from the Duke of Norfolk's estate, to Richard Wilson, book-keeper, for a parcel of ground in Pond Street, Sheffield.

Other

ACM/MAPS/5/69

'A plan of forty eight Tenements in the Park erected by John Curr, 1790'

FC/P/SheS/1196S

'Sheffield: Sycamore Street. A plan of the late John Curr's freehold property, 1823'

FC/P/SheS/1408S

'Sheffield: Young Street. A plan of the late John Curr's freehold property, 1823'

FC/P/SheS/800L

1823 plan of John Curr's leasehold property at Sheffield Park

The circular and letters below came to light recently, too late to be mentioned in the book:

WWM/Stw P7(iii)-80

Curr's printed circular promoting flat rope, together with covering letter to Charles Bowns, auditor for Lord Fitzwilliam, *Concerning I know it is in my power to render you a service in the Article of Ropes for drawing yr. Coals at Lord Fitzwilliams Colliery I have taken the liberty of sending you this Circular*, 1811

WWM/Stw P7(iii)-85

Bowns's letter to Joshua Biram, steward for Lord Fitzwilliam, enclosing Curr's circular *but without any wish to recommend his Ropes to be tried*, 1811

Durham County Record Office (DRO)
National Coal Board Records, first deposit: Papers of John Buddle
References DRO NCB1/JB

Buddle/Curr correspondence

345: Curr to Buddle, 7 February 1800
re flat rope.

346: Curr to Buddle, 17 February 1800
re flat rope, rails and improved corf wheels.

347: Curr to Buddle, 11 March 1800
re flat ropes, Fenwick's trial & Chapman's patent.

348: Buddle to Curr (copy), 25 March 1800
re flat rope (Chapman) rail plates & hay.

349: Buddle to Wm. Bourn, 27 March 1800
re remittance for flat ropes

350: Wm. Bourn to Buddle, 19 May 1800
re request to settle account.

351: Curr to Buddle, 21 April 1800
re 'I have found all your Durham freeholders …', gives names and whether they will vote for Major
Russell rather than Mr. Taylor, e.g. 'Mr. Gee says he has made a promise to Mr. Taylor but says he will consult some of his friends whom I found he had rather offended by voting lately for Mr. Taylor …'.

352: Curr to Buddle, 12 May 1800
re flat rope, Fenwick's experiment doubted.

353: Buddle to Curr, 3 June 1800
re rails and flat rope, Fenwick

354: Curr to Buddle, 6 June 1800
re castings, improvement in [corf] wheels & flat rope, Chapman & Fenwick's statement doubted.

355: Buddle to Curr, 16 June 1800
re flat rope, Chapman

356: Curr to Buddle, 24 June 1800
re flat rope and rail roads

357: Curr to Buddle, 16 July 1800
re price of hay, shipping costs, quotation for engine and flat rope.

358: Curr to Buddle, 13 August 1800
re flat rope and shipping costs.
'I informed you in my last I was in hopes of making a perfect flatt Rope or in other Words a flat Rope that has little or no inclination to turn, in which I now have the confidence to say I have succeeded to my wishes...'
Addendum: Answered 19 August – Ordered 500 yards of Rail Roads.

359: Curr to Buddle, 30 August 1800

re rail roads and flat rope.

'I think there is little doubt now of a machine for lacing being wanted at N. Castle and think the lower part of the Quay side would be favourable especially as I propose to have a warehouse for my Rail Roads which my Agent would be able to look after also.'

360: Curr to Buddle, September 1800

re rail roads, *'improvement in the waggon'* and ropes for Hebburn.

361: Curr to Buddle, 30 September 1800

re ropes & railroads arrived at Gainsborough.

362: Curr to Buddle, 4 October 1800

re shipping troubles, possible new route (Tinsley – Newcastle), order for ropes and rails.

363: Curr to Buddle 17 October 1800

re flat rope. Addendum – order from Wm. Russell for rails.

364: Curr to Buddle, 30 October 1800

re travel & other arrangements (re ropes for Hebburn).

365: Curr to Buddle, 29 December 1800

re trial of shaft conductors at Wallsend & flat rope at Hebburn.

366: John Curr [Jnr, III] to Buddle, 26 May 1803

re Percy Main Engine.

367: John Curr [Jnr, III] to Buddle, 16 June 1803

re not much of an apology

368: Curr to Buddle, 12 October 1815

re travel arrangements

369: Curr to Buddle, 16 August 1821

re return from Paris, concerned about *'not having received my interest from last year'* from Messrs. Neesham & Co., news of family.

Further letters

370: John Curr [Jnr, III] to Buddle, 19 July 1814

re Tyne Steam Packet & giving details about his steamboat at Norwich.

371: John Curr [Jnr, III] to Buddle, 8 August 1814

re law suit over his boat & asking for further detail on Tyne Steam Packet.

372: Unsigned & unaddressed, 5 August 1814

re considerable particulars relating to the Tyne Steam packet (in JBJ's hand?).

373: 'Proportions of Engine (Tyne Steam Packet)', dated 25 July 1814 and initialled "JBJ". (This note is written on a fragment of paper.)

Other

117: Includes two items:

Wm. Bourn, 1 May 1820

printed circular 'To the Proprietors of Coal & other Mines'. [Author's

note: Bourn had purchased Curr's flat rope machinery]
Wm. Bourn to Buddle, 20 June 1821
re touting for business.

National Maritime Museum
The Caird Library and Archive:
ADM 354/233/98, 6 September 1808
(ADM 354 consists of bound out-letters of the Navy Board to the Admiralty held in the Manuscripts Section of The Caird Library).

North of England Institute of Mining and Mechanical Engineers (NEIMME)
Buddle–Atkinson collection
References: The Common Room (TCR) NEIMME…

Volumes numbered Bud/65–75 comprise a set of notebooks mainly of John Buddle Senior containing rough colliery memoranda and also copy agreements, valuations and accounts.

Bell/3/327–346
Very rare pamphlet entitled *An account of an improved method of drawing coals and extracting ores &c. from mines*, by JOHN CURR, printed in Newcastle 1789 (bound in with others).
Title page (1), text (10), 2 plates: Fig. 1, signed J Buddle Junr., &c. W (1) & Figs. 1–5 & 7–20, signed J Buddle Junr. &c. (1), other pages blank.
Letters patent, specification and plans of an improved method for drawing coal by John Curr. 1789.
Author's note: This is indeed very rare. It is not recorded in ESTC; JISC (Jisc Library Hub Discover) records two copies at Leeds, to which WorldCat adds one at Yale.

Bud/13/110 & 113
110: Printed invitation to tender for machine at Pensher Colliery, August 1805.
113: John Curr III's proposal, September 1805.

Bud/15/…
To John Curr Snr:
6 & 7: 15 September 1802
re Curr to take a long tour (Ireland), arrival of Mr Wade's engine, flat & round ropes at Cowpen & Hebburn, amount of railroad disposed of & carriage charges.
8 & 9: 28 September 1802
re payment for ropes & cast iron goods, and the substitution of flat ropes for round.

62, 62a, 62b & 63: 12 February 1804

re asking for information about Wyken colliery, near Coventry.

65 & 65a: 27 February 1804

Asking for Curr's advice on a colliery in South Wales.

70, 70a & 71: 27 March 1804

re Wyken colliery.

74, 74a, & 74b: 11 April 1804

re Wyken colliery & Mr. Wade's account.

82, 83 & 84: 4 November 1804

re Wyken colliery.

134–137: 23 February 1806

re flat ropes at Washington, Lambton & Penshaw.

224–225: 23 August 1807

re plates for Elswick Colliery, ropes for Kenton Colliery and the rope trade.

To John Curr [Jnr, III]:

17–19: 24 May 1803

re Percy Main engine

21 & 22: 12 June 1803

re Percy Main engine, rail road & flat ropes

26a: 17 July 1803

Re Percy Main engine failure.

47: 27 January 1804

re 50 to 60 fathoms of 8 inch pipe for Wallsend Colliery

67: 7 March 1804

re a tender for a winding engine at Durham & Mr. Peareth's engine

Bud/53/…

These letters all relate to the supply of flat ropes for the inclined plane at Benwell Colliery:

33, 33a & 33b: 10 March 1811

J Curr Snr to J Buddle Jnr

34: 27 April 1811

P. Crowther to J Buddle

34a & 34b: 5 May 1811

J Buddle to P Crowther

35 & 35a: 8 April 1811

J Curr Snr to J Buddle Jnr (of which the first half also relates to the inclined plane at Benwell; the 2nd half is about the Wyken trial at Westminster)

Bud/65–67: Bushblades colliery notebooks
NRO3410/Bud/65: 1 May 1768 to 22 January 1770
NRO3410/Bud/66: January 1770 to December 1773
NRO3410/Bud/67: Christmas 1773

Bud/68

Sheffield Colliery Note Book [John Buddle Sen.] visit for Duke of Norfolk Oct 1773 & April 1776 (including Expence of a Journey to Sheffield to do business for his Grace the Duke of Norfolk).

Bud/70

Memorandum by Vin. Ayre [sic] for Report to Duke of Norfolk on Attercliffe Colliery, and effects on Darnell [sic] & Sheffield Park Collieries – 17 April 1789. (Including Vincent Eyre's request to Messrs. Buddle & Stephenson.)

Bud/137/5/25

'*Plan for drawing Coals by Conductor in a Basket or Hazell Corf to prevent them dashing against the Pit side supposing a single corf drawn at a time, by J. Curr*', n.d., unsigned, <u>hand-drawn original</u>. (Fair copy & in a different hand to ZD/50/1.)

Wat/1/26/1, 1804–1845 (includes four documents in total)
'Mr Curr's [printed] letter introducing his Flatt ropes to the different colliery owners', Newcastle on Tyne, 5th June 1804.

ZD/50/1, 1a & 1b
1: (Cover) '*Mr Curr's Plan of Drawing Coals by Conductor in the N Castle Collieries*', <u>initialled</u>.
1a & 1b: Title as Bud/137/5/25, n.d. & unsigned, <u>hand-drawn & written original</u>.
Plan and sections of a device for drawing coal by conductors to prevent the corves from being damaged on the sides of the pit, scale 11/3: 1", patented 1788, with explanation and proposals to erect the device for a trial, by Mr. Curr.

Further material

The items listed here are extra material, of interest but not used directly in this book.

Bruton 2015
Bruton, Roger N. *The Shropshire enlightenment: a regional study of enlightenment activity in the late eighteenth and early nineteenth century.* PhD thesis, University of Birmingham, 2015, especially pp. 45 & 48–51

Parkes 1818

Parkes, Samuel, *The Chemical Catechism*, 8th edn. Collins & Co, 1818; pp. 489–91. '*Coak [sic] Ovens. At the Duke of Norfolk's colliery, near Sheffield, several of these ovens are built on the side of a hill, occupying spaces formed within the bank. … I am indebted to Mr. Curr, steward to His Grace the Duke of Norfolk, for these particulars, who very politely attended me through the works in the year 1802, and assisted me in taking the necessary drawings, admeasurements, &c. This description will be better understood by referring to a copper-plate engraving of these ovens in the second volume of my* Chemical Essays …'

Parkes 1823

Parkes, Samuel. *Chemical Essays*, vol. I, 2nd edn. London: Baldwin, Cradock & Joy, 1823; pp. xxxii–iii (description of the plate), pp. 363 & 430. *Plate XIII, Apparatus for Making Coke. Fig. 1 represents the ground plan of one of the coke-ovens belonging to His Grace the Duke of Norfolk, at Sheffield in Yorkshire, communicated by Mr. Curr, steward to His Grace.* Author's note: This is the only contemporary illustration of any part of the Norfolk collieries at Sheffield discovered.

Pratt 1912

Pratt, Edwin A. *History of inland transport and communications in England*. London, 1912

Sheffield Star, c. 1963

The hauntings of Belle Vue house etc., c. 1963. Author's note: It would have been nice to have found this article, but the *Sheffield Star* is not included in the British Newspaper Archive.

John Curr

Curr family

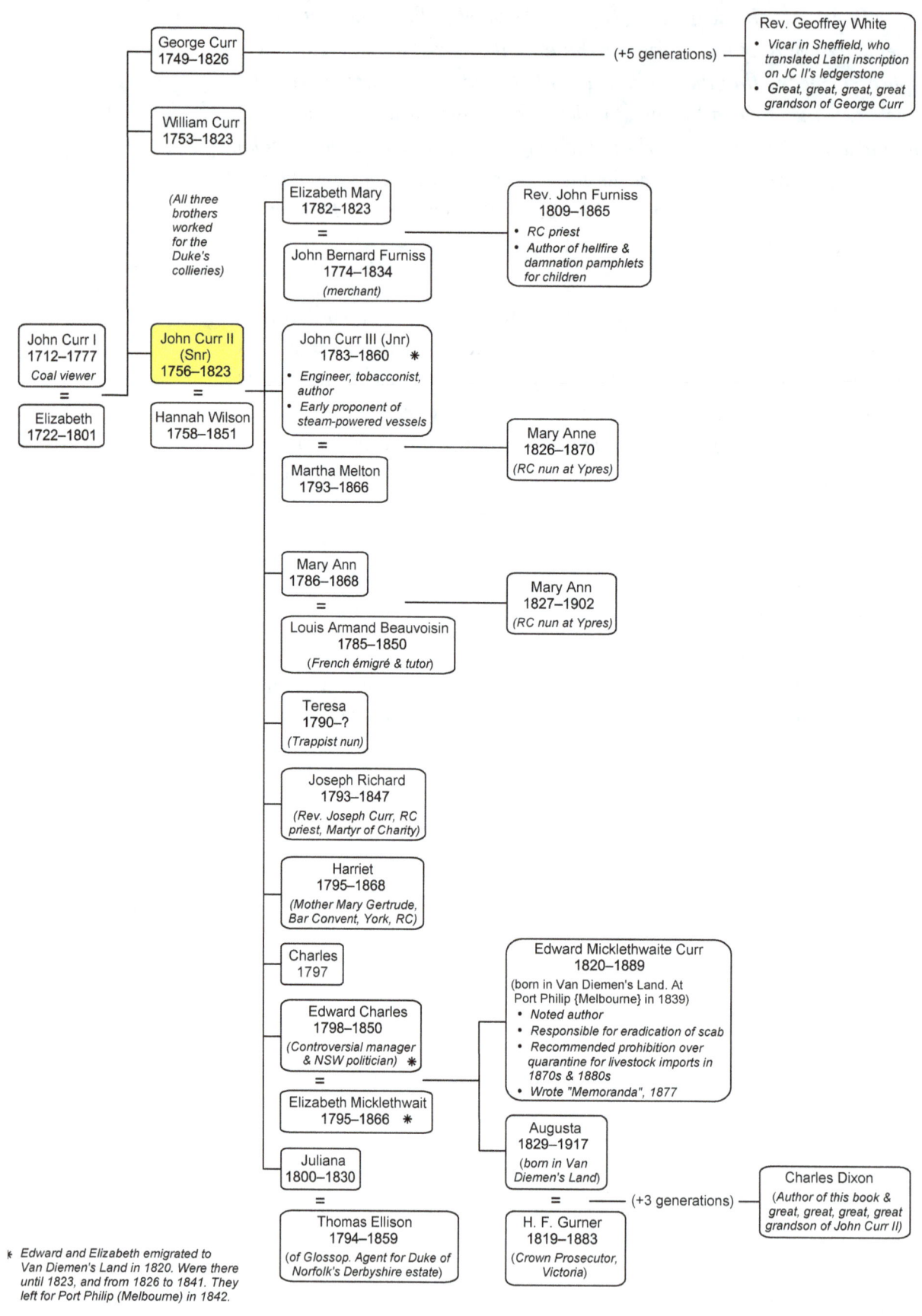

Working relationships

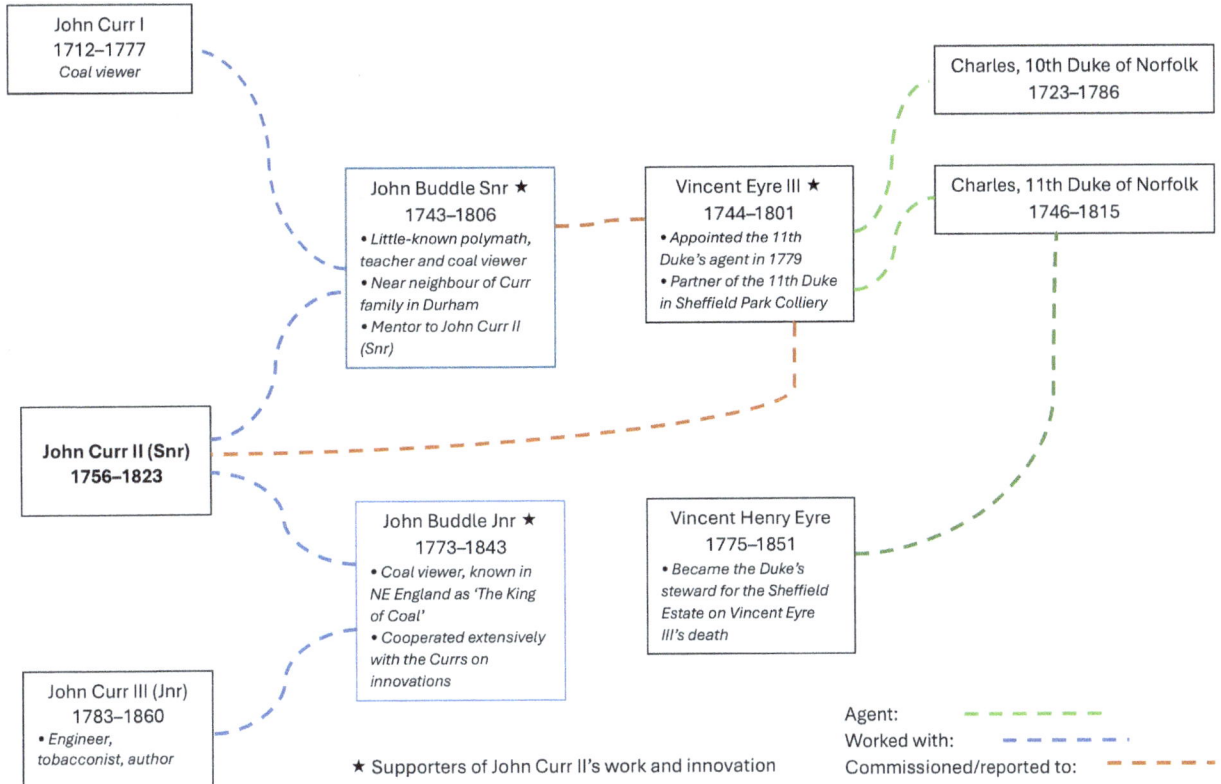

ENDNOTES

For the most part, a short title format has been used, with full details provided in the relevant bibliography section.

Preface

1. 'Johnsoniana', *The European Magazine*, vol. 7 (January 1785), 51–55, p. 54.
2. [Samuel Johnson], 'Preface' to Richard Rolt, *A New Dictionary of Trade and Commerce, Compiled from the Information of the Most Eminent Merchants, and from the Works of the Best Writers on Commercial Subjects, in all Languages* (London, 1756), sig. A1r.

1. Introduction

3. Bland 1930–31, 'John Curr, originator of iron tram roads', pp. 121–30. See also Lock 2015 (*ODNB*), 'Curr, John (1756–1823)'.
4. Dunn 1855, 'Scientific and practical mining college proposed...', pp. 602–3.
5. Galloway 1882, *History of Coal Mining ...*, p. 182.
6. Ashton and Sykes 1929, *The Coal Industry of the Eighteenth Century*, p. 64.
7. Hopkinson 1958, *The Development of Lead Mining and of the Coal Industry...*
8. Charles E. Lee (1901–1983), transport historian and past president of the Newcomen Society.
9. From the dust-wrapper of Curr's *The Coal Viewer*, 2nd edition.
10. Charles E. Lee, quoted in Mott (1969–70), 'Tramroads of the eighteenth century and their originator: John Curr', p. 19.
11. Trinder 1973, *The Industrial Revolution in Shropshire*, p. 181.
12. Medlicott 1999, 'John Curr, 1756–1823, mining engineer and viewer', pp. 63–78.

2. Family

13. Fawcett 1911, *A history of the parish of Dipton*.
14. There is an undocumented entry on *Ancestry* for one Thomas Curry, 1682–1745, born and died in Chester-le-Street, who married Jane Brown (1682–?).
15. Hall 1858, Discussion at meeting on 7 October 1858, recorded in *Transactions of the NEIMME*, p. 6.
16. Paraphrased from Gooch 1989, From *Jacobite to Radical ...* (thesis).
17. Charles Lee in his introduction to the reprint of Curr's *The Coal Viewer*, Alexander Lock in *ODNB* (2015) and Ian Medlicott (1999, p. 73) all refer to John Curr I as coal viewer at Bushblades. None provide a source.
18. Burial in the Anglican churchyard; not to be confused with the Catholic All Saints, also in Lanchester, which opened in 1926.
19. Northumberland & Durham Burials, transcribed by Northumberland & Durham

20 *Sheffield Register*, Now let to George Curr, 4 June 1790.
21 *Leeds Intelligencer*, 6 July 1801.
22 *Gentleman's Magazine*, Country News: 1 July 1810, p. 80.
23 Between 1754 and 1837 marriage was legal only when it took place in a Church of England parish church. A marriage licence could be obtained by Catholics, such as John Curr, from the Church of England for a fee, together with a sworn declaration that there were no legal impediments to the marriage. Such a licence waived the necessity for the usual banns to be read.
24 Sheffield Archives (SA) ACM/SD/870/59.
25 Will of Richard Wilson of Sheffield Yorkshire 30 April 1723, PROB 11/1670/32, The National Archives. Available at: https://discovery.nationalarchives.gov.uk/details/r/D150850.
26 EMC *Memoranda*.
27 Adult ages were rounded down to the nearest 5.
28 Hannah's burial card can be seen in *Miscellanea IV*, 1907, p. 366.
29 Prerogative & Exchequer Courts of York Probate Index, 1688–1858. The Probate Act books were for recording a brief summary of each will proved. Usually just a line or two.
30 1851 census: A. Macartney, aged 56, Catholic Priest, born Ireland – lived at Egton Bridge [6 miles south of Whitby]. Father Macartney was executor.
31 Inheritance tax was first introduced in 1796 and by 1857 was applied to any estate with a value of over £20.
32 Durham County Record Office (DRO) NCB1/JB/369.

3. Education

33 Oldroyd 2007, *Estates, enterprise and investment at the dawn of the Industrial Revolution...*, 2007, pp. 79–80.
34 Oldroyd 2007, pp. 82 & 109.
35 Hiskey 1978, *John Buddle (1773–1843), agent and entrepreneur in the North-East coal trade* (thesis), pp. 13 & 17.
36 Flinn 1984, *The History of the British Coal Industry; Volume 2, 1700–1830*, p. 61.
37 This cross-examination exchange and letter appear on p. 1 of Hiskey 1978.
38 Wilson 2022, Our weird 'unis' are increasingly pointless. *The Times*, 13 August.
39 Jacob 2014, *The First Knowledge Economy...*, p. 47. Jacob continues (p. 78): 'Overseers ... knew enough about profit and measurement to present the owners with economic costs and choices. The relationship between good engineering practices and profit had been spelled out in print only in 1744, in Desaguliers' *A course of experimental philosophy*. To lower expenses, the owners of coal mines, he admonished "need to find a philosopher to come and find a means to bring down the end of a beam [of a water pump] without men or horses." His experimental course became the textbook of early industrial development ... the great engineers of the eighteenth century (who thought of themselves as natural philosophers) ...'
40 The date and time of his birth can be worked out from the question published in *The Gentleman's Diary, or the Mathematical Repository; An Almanack*, 1763. New Mathematical Questions proposed to be answered in the next Year's *Diary*.
(1) Quest. 242, by *Mr. John Buddle. Required the Year, Month, Day, and Hour*

when I was born, from the following Equations*, where x represents the Year; y the Months from the Beginning of the Year; z the Days of the Month; and v the Hours after Noon? ... [*the equations], p. 43, and *The Gentleman's Diary*, 1764: Answers to the Questions in the last Year's Diary, p. 24, (1) Quest. 242 answered. *Consequently, he was born July 13, 1743, at 7 o'Clock in the Evening*, p. 24.

41 'Buddles' is found, along with the more usual 'Buddle' in *The Gentleman's Diary* for 1760, 1761 & 1768.

42 Hiskey 1978, p. 4.

43 *Miscellaneous Correspondence, Containing a Variety of Subjects, Vol. II, For the Year 1757 and 1758*, printed 1759 (a journal). Mathematical Questions Answered. Question 203, answered by Mr. John Buddle at Mr. Robinson's School, at Biddick, p. 941, December 1758.

44 *The Gentleman's Diary, or the Mathematical Repository; An Almanack*, 1760. Answers to the Questions in the last Year's Diary, (1) Quest. 186, answered by Mr. John Buddle. p. 20: (2) Quest. 187, p. 21, Messrs. Buddles [sic] ... favoured us with good Solutions to the above, p. 22, p. 24: (4) Quest. 189. Answered by Mr. John Buddles [sic], of Chester-le-Street. The same answered by Mr. Tho. Robinson, p. 24, (5) Quest. 190, p. 25, Messrs. Buddles [sic] ... favoured us ..., p. 26.

45 *The Gentleman and Lady's Palladium*, by the author of the *Royal Astronomer and Navigator*. Printed for J. Scott, 1760. NEW QUESTIONS. XIII. Question 174. By Mr. John Buddle, at Chester le Street, near Durham, p. 40.

A MASTER of a Ship being at SEA, on May 19th 1759, in Latitude 40 20' N. took the Sun's equal Altitudes, observing the Times of those Observations, by his Watch, at 9h 12m, in the Forenoon, and 2h 59m in the Afternoon: *Required*, from thence, the time by his *Watch*, on that Day, when the Sun passed the Meridian, or was at his *greatest* Altitude?

46 These are examples of answers given by Buddle in *The Gentleman and Lady's Palladium* of 1760:

p. 53: II, Q 147, was answered by Buddle (among others).

p. 59: IX, Q 154, as above.

p. 60: X, Q 155, Mr. John Buddle answered the same very concisely.

p. 60: XI, Q 156, answered by Mr. John Buddle of ...

p. 67: XV, Q 160, The same answered by J Buddle.

47 *Gentleman and Lady's Palladium*, 1760, p. 42.

48 *Gentleman and Lady's Palladium*, 1761, p. 16, VII. Quere 122. By Mr. John Buddle of Chester-lee-Street [sic].

49 *Gentleman and Lady's Palladium*, 1761, p. 21: VIII. REBUS by Mr. *John Buddle*

To the name of a Tree that is tall and upright,

Which is frequently seen, and which three Letters write;

If you add what 'tis sold for, the Name 'twill complete

Of an amiable Damsel, both rich and discreet.

50 *Gentleman and Lady's Palladium*, 1761, p. 17: IV, Question 186. By Mr. *Buddle*. A Traveller calling at a Gentleman's House, drank out of a *Silver Tankard*, full of Strong Beer, in Form of conical *Frustum*, till he just saw the Edge of the Bottom, and drank four Ninths of the (?) beer; the Height of the *Tankard* was 6, and Bottom diameter (illegible) inches: Required the *Tankard's* Top-Diameter?

51 *Gentleman and Lady's Palladium*, 1761, VIII, pp. 38, 42 & 47: Q169, XI, Q172 & XVI, Q177 were solved by, among others, John Buddle.

52 *Miscellaneous Correspondence*, vol. IV, 1762, pp. 863–64, 879 & 883.

53 Hiskey 1978, chapter 1, p. 4 & endnote 13: University of Durham, Department of Paleography, schoolmasters' licences.

There was no religious requirement to prevent a Unitarian from obtaining a licence in the eighteenth century.

54 Welford 1895, *Men of Mark 'twixt Tyne and Tweed*, p. 425.

55 Galloway 1898, *Annals of coal mining...*, p. 308, with Galloway citing p. 3 of the *Newcastle Weekly Chronicle* of 24 July 1873. From that article we find: 'We have heard that he may have originally been a pitman, but whether this may have been the case or not, he was particularly well versed in the practice as well as the theory of coal-mining.'

56 Charles Hutton (1737–1823), mathematician, was born in Newcastle-upon-Tyne into a coal-mining family.

William Emerson, 1701–1782, mathematician, born at Hurworth, near Darlington. He later lived on his estate at Castle Gate, near Eastgate in Weardale. 'Emerson maintained a remarkable output of textbooks. In 1735 he married Elizabeth, the daughter of the Reverend Dr John Johnson, the Rector of Hurworth. The rector had offered a £500 dowry with his daughter's hand, but he disapproved of her choice. He treated his scruffy son-in-law with contempt and refused to pay, so Emerson loaded all his wife's clothes in a barrow and wheeled it round to the parsonage, saying he refused "to be beholden to such a fellow for a single rag". He also vowed to prove the rector wrong.' From: https://mathshistory.st-andrews.ac.uk/Biographies/Emerson/ (citing Chris Lloyd, Cloddish beer drinker who played the numbers game, *The Northern Echo*, 25 January 2010).

Author's note: I was born in the Old Parsonage at Hurworth-on-Tees.

57 *The Builder*, April 1843, p. 77.

58 *The General Magazine of Arts and Sciences*, 1764. Plate, *A Representation of a COAL WAGGON, by John Buddle, 1764*, p. 284, and text, *Description* of a Coal-Waggon, pp. 285–86.

59 Chester-le-Street Parish Registers, vol. 5, 8 May 1768.

60 DRO, EP/La4, 14 April 1777 (cited in Hiskey 1978, p. 146, n. 14).

61 *British Palladium...*, 1771: Answers to all the Enigmas in last Year's *PALLADIUM* (p. 24), Mr. *John Buddle*, of *Kyo, Durham*, answered the *Prize-Enigma*, by *Slack's Puffs*, at his *Printing-Office, Newcastle*, which he happened to see (p. 25).

Buddle Snr's name appears several times in this issue: ANSWERS to the QUESTIONS in last Year's *PALLADIUM* ... Consequently the Lady's *Name*, who has *no Equal*, is found to be COLE, *required*. ... - Mr. *John Buddle*, of Kyo, near *Lanchester*, in the County of *Durham*, (a former ingenious Correspondent) determined the *Lady's* Name, as above, by a short Deduction; ... (p. 30). II. Q402: ...Mr. *John Buddle*, of *Kyo*, in the County of *Durham*, determined the Tower's height as above, by a short and elegant Process: ... (p. 31). VI. Q406, answered by Mr. *John Buddle*, of *Kyo*, Durham (p. 34). PRIZE-QUESTION answered by Mr. George Coughlan ... Mr. *Buddle*, of Kyo, near *Lancaster* [sic], determines by a short Process, the Height of the lowest Tower=49.169712, and of the highest, 53.169712 Yards (p. 50).

62 *Gentleman's Magazine*, January 1844, obituary of Buddle Jnr, 'He was not only a great lover of books, but a great reader of them ...', p. 100.

Salkeld's sale catalogue, 1880, item 499, Essay upon Pronunciation and Gesture founded upon the Rules of the Ancients, 1750, 12 mo., old calf, 2s 6d. Bookplate of John Buddle [whether father or son is not stated].

63 Newcastle: Printed and sold by T. Slack; and by G. Robinson, Pater-noster-Row. 8vo, viii + 54 pp., 1778. Buddle's Preface (p. ii) notes Kyo, *near* Lanchester, 18 June 1778.

64 Matthias Dunn (1788–1869), *History of the Viewers*, MS, n.d., based on conversations in 1811–12 with Samuel Haggerstone. TCR NEIMME/East 3a: *A*

minute from the books of the late W. Brown of Throckley gives the following list of engines at work drawing water in 1769 – Bushblades, 2 engines, diameter of cylinders 42, 52.

65 Tattersfield 2011, *Thomas Bewick…*, p. 207.
66 Buddle Snr 1785, Observations on the Raising of Hedges… In: Robert Callender, *A Practical Essay on the Raising and Management…*, pp. 27–41.
67 Buddle Snr's letter 'Instance of a life preserved after falling into a Coal-Pit from the *Gentleman's Magazine*' is signed JOHN BUDDLE. Bushblade's Colliery, March 5, 1788. It is included in Richardson 1846 (*The Borderer's table book*, vol. VII), pp. 410–11.
68 TCR NEIMME/Bell/3/327.
69 Welford 1895, p. 425.
70 Jacob 2014, p. 80.
71 A thorough search by Jen Hillyard of the NEIMME collection has failed to find the document referred to by Hiskey in her 1978 thesis, chapter 1, p. 4: … *his inscription in the front of an old book of boring and sinking notes suggests that he was taking an interest in such matters as early as 1765*. Endnote 15: p. 303, NEIMME, shelf 46, vol. 15, Bushblades Colliery Memoranda; vol. 27, *Notes on Boring and Sinking*, bears the inscription *'Mr Jno. Buddle, 1765.'*
72 TCR NEIMME/Bud/65–67: 1 May 1768 to 22 January 1770; January 1770 to December 1773 & Xmas 1773.
73 For instance, Galloway 1898, *Annals*, p. 309: '… *it was probably immediately afterwards* [his publication in 1778 of a new edition of The Marquis of Worcester's Century of Inventions; see note 63 above] *that Mr. Buddle abandoned the "delightful task" for the more arduous and more remunerative profession of colliery viewer*'; and Welford 1895, p. 425: '*His first appointment in that capacity* [colliery manager] *was at Greenside, near Ryton …*'.
74 Flinn 1984, p. 61: '*The prestige of the north-east viewers as the leading experts of the day meant that they received commissions from beyond their own coalfield.*'
75 This theory is explored in greater detail in the chapter 'Early Career'.
76 SA ACM/S/215.
77 SA ACM/S/218.
78 Flinn 1984, pp. 100–1 and his note: NEI[MME] Bell 14/3/396–400.
79 Fordyce 1855, *History and Antiquities…*, p. 181.
80 Hiskey 1978, p. 5. From the present author, citing Flinn 1984, p. 60: In this capacity he was largely concerned with advising on the terms of the leases.
81 Galloway 1882, *History*, p. 129.
82 Ashton 1948, *The Industrial Revolution*, p. 64.
83 Harrison 1956, *Along Hadrian's Wall*, p. 36.
84 *Newcastle Journal*, 11 November 1843, cited in Hiskey 1978 (p. 5, n. 23). This article 'pointed out … [that] an insinuation in a London paper that Buddle had risen from being "a mere pit lad" was erroneous'. For Buddle, see Heesom 2004 (*ODNB*), 'Buddle, John (1773–1843)'.
85 From Hiskey 1978 (p. 5, n. 22): *Morning Post*, 25 June 1842 (quoting letter from Buddle to Londonderry, 14 June 1842); NCB1/JB/2556, *Minutes of Evidence taken before the House of Commons Committee on the S. Durham Railway Bill*, p. 25, 6 June 1836.
86 *The Town and Country Magazine*, vol. II, September 1770: '*Answers to the mathematical questions proposed in the eighth number …*', p. 491. '*Mr. B. Hart, of Bristol, Mr. John Curr, of Bushblades, in Durham … answered all the questions*', p. 492. Other names are listed as having answered from one to three of the questions.

87   *The Gentleman's Diary; or, The Mathematical Repository; an Almanack*, London, printed by W. Bowyer and J. Nichols for the Company of Stationers, 1771, pp. 35–46. Curr answered questions 347–352 & 354–356. 'The PRIZE Quest. 356, answered by … Messrs. … John Curr …sent very good Solutions to this Quest., …'

88   *Gentleman's Diary*, 1772, pp. 31–46.

89   *British Palladium…*, 1772, pp. 35–37. Curr answered questions 423–433, 435–436, 439, 442, & 445–446. He received special mentions as follows:

      III. Question 425: *Mr. John Cur [sic], of Bushblades, observes, that the number of Posts, and also of Rails, making a Load, should have been given.*

      IX. Question 431: *Mr. John Curr, of Bushblades, Newcastle, answered it by drawing a Triangle round the Square, trigonometrically, demonstratively, and by Calculation.*

      XI. Question 433: *Mr. John Curr, of Bushblades, Newcastle, determined, by an elegant Process, $x=17.908609$, and $y=.844952$.*

      XVII. Question 439: *Mr. John Curr, of Bushblades, Newcastle, gives 2567.732, &c. Miles the Height required, from considering the Pole nearer the Earth.*

      Thomas Robinson of Biddick, Buddle Snr's old teacher, is listed as answering seventeen of the questions, as well as being the proposer of Question 443.

      There is an interesting comment under question 446, which Curr answered: '*This sort of question (more laborious than curious) is to exercise the young Algebraist and Geometrician.*'

4. Character

90   Author's note: E. M. Curr sailed for England from Van Diemen's Land in 1829, so never knew his grandfather who died in 1823. Edward and two of his brothers were educated at Stonyhurst College, a school of Roman Catholic foundation.

91   EMC *Memoranda*.

5. Wealth

92   Medlicott 1983, John Curr and the development of the Sheffield collieries…, p. 54: '… *and between 1792 and 1801 it supplied goods to the value of £14,069* …'.

93   SA: ACM/MAPS/5/69, FC/P/SheS/1196S and FC/P/SheS/1408S.

94   From an article in the *Sheffield Daily Telegraph*, 10 November 1873.

95   EMC *Memoranda*.

6. Belle Vue House

96   ACM/SD/520.

97   ACM/SD/521.

98   EMC *Memoranda*.

99   ACM/SD/520.

100   ACM/SD/521.

101   'City stadium threatens distinguished house', *Sheffield Daily Telegraph*, 20 May 1939. It is thought that the *Sheffield Star* printed a piece 'The Hauntings of Belle Vue House' around 1963.

102   Holland 1820, *Sheffield Park…*; the stanzas quoted are taken from the new edition of 1859.

103   'City stadium …', *Sheffield Daily Telegraph*, 20 May 1939.

7. Worship

104 Woolhouse 1832, A description of the town of Sheffield: *'In going up Fargate there was houses built on both sides. The Lords House stood a little on the North side of the present Norfolk Row. A very elegant old House, it was inclosed by a Wall in a half Circle and Palisaded. The present Duke of Norfolk was born in this house. This I expect is why it was called the Lords house, he being l. Of the Manor.'* [The house was probably built in 1712.]

105 EMC *Memoranda*.

106 Hadfield 1889, *A History of S. Marie's Mission and Church...*, p. 31:

'In 1791 the most important Act of Relief, passed in favour of the Catholics, deservedly styled the "Toleration Act" received the Royal assent. By its provisions Catholics were exempted from prosecution as recusants, for not resorting to the Established Church, for being a Papist or reputed Papist, hearing or saying Mass, being a priest or performing any religious rite. Places of Catholic worship were permitted if certified at the Quarter Sessions ...'

The position and prospects of Catholics had now immensely improved ...'

107 Personal communication: The Rev. Geoffrey White, a descendant of John Curr II's older brother, George.

108 Odom 1926, *Hallamshire Worthies...*, pp. 58–59. Odom continues:

'An obituary in the *Iris*, probably written by the late John Holland, says: "Of Mr. Rimmer's private character and benevolence according to his means, it is difficult to speak so as to do justice to his memory. Whatever opinions were entertained respecting the principles of that Church of which the deceased was a minister, individuals of different denominations in Sheffield all concurred in bearing testimony to the excellencies of the kind-hearted, humane, and charitable man whose death is recorded. In holding his own religious opinions, he stood at the very furthest remove from bigotry; and as he was deservedly beloved, so he will be sincerely lamented, not more by Catholics themselves than by those Protestants who knew his worth. In relieving distress, Mr. Rimmer constantly illustrated that excellent grace of charity which has often been so largely exemplified by many holy men in the Roman Catholic Church; but he did not confine himself to the members of his own community, it being a maxim with him, which he often repeated, that it is immaterial what religion persons profess, if they stand in need of help, it is then our duty, as Christians, to relieve them."'

109 Author's note: Three generations of Fairbanks, all surveyors, are notable, in particular, for their maps of the Sheffield area. They were employed extensively by the Duke of Norfolk's Sheffield estate office.

110 Hadfield 1889, pp. 39–41.

111 Hadfield 1889, p. 33:

'Upon this land there then stood a messuage or mansion house, which for many years past had been occupied by the land agent for the Duke of Norfolk; another messuage or dwelling-house which had been occupied by Messrs Walkers, Eyre and Stanley, the bankers; another building occupied as a Catholic Chapel, a stable and outbuildings. On the south side of Norfolk Row there was another piece of land which was purchased at the same time. The quantity of this last plot is not stated in the deed of conveyance, but in subsequent deeds it is stated to contain 415 square yards.'

112 Author's note: From EMC *Memoranda*: 'My grandfather was buried at Sheffield, and when a boy I often saw on the wall of the Sheffield church a marble tablet bearing a commemorative description of him.'

113 Hadfield 1889, p. 37

114 Blackwell 1828, *The Sheffield Directory and Guide*, p. 6.
115 From Hadfield 1889, p. 68: 'The plan given is a *fac simile* of one prepared at the time ... as a careful record of certain existing graves.'
116 Hadfield 1889, p. 68.
117 Mott 1969–70, p. 22.

8. Early career

118 SA ACM/S/179.
119 Hunter 2021, Early pumping engines in the Attercliffe–Darnall–Orgreave coalfield and the role of John Curr.
120 SA ACM/S/215. [The printed catalogue of the Arundel MSS has a hand-written archivist's note 'S215, Sheffield Park Colliery. Probably by J. Curr (2 copies)'.]
121 SA ACM/S/217 (1779); SA ACM/S/224 (1789) and see also pp. 69, 98 & 99.
122 TCR NEIMME/Bud/68: Sheffield Colliery Note Book Oct 1773 & April 1776.
123 Mott 1969–70, p. 16: 'DISCUSSION: ... a joint contribution from Dr. M. J. T. Lewis and W. N. Slatcher ... that it seemed probable that Curr went to Sheffield in or about 1773 ...'.
124 SA ACM/S/217.
125 SA ACM/S/215.
126 *The Universal Magazine of Knowledge and Pleasure*, vol. 55, December 1774, p. 330.
127 'The waggon-way started in James Mellor's Broad Oak field and ran almost parallel with City Road to a coalyard at the bottom of Park Hill, where Duke Street and South Street joined Broad Street.' (from Medlicott 1982, *The Landed Interest and the Development of the South Yorkshire Coalfield 1750 to 1830*, p. 258, note 55).
128 Curr [III] 1847, *Railway Locomotion*, pp. 179–80.
129 EMC *Memoranda*.
130 Edward Curr, 'Father of the Coast', by H. R. Haslock. *Advocate*, Tasmania, 25 October 1926.
131 Smiles 1857, *The Life of George Stephenson, railway engineer*.
132 Mateaux 1885, *George and Robert Stephenson*.
133 SA ACM/S/217.
134 Leader 1901, *Sheffield in the Eighteenth century*, p. 341 (cited in Lewis 1974).
135 SA ACM/S/198 (cited in Lewis 1974).
136 Lewis 1974, *Early Wooden Railways*, p. 132.
137 Robinson 1995, *The Dukes of Norfolk*.
138 Henry's son, Bernard Edward, was to succeed as 12th duke.
139 The 11th duke renounced his Catholicism in order to enter Parliament, but remained a staunch supporter of Catholic emancipation.
140 SA ACM/S/179 (cited in Medlicott 1982, pp. 272–73).
141 SA ACM/S/185 (cited in Medlicott 1982, pp. 272–73).
142 Medlicott 1999, p. 64 (no source given).
143 SA ACM/S/217.
144 SA ACM/S/218.
145 SA ACM/S/216.
146 SA ACM/S/219.
147 SA ACM/S/217. Interest was payable at £10 per £100 (10%) on £3,200 and £5 per £100 (5%) on £2,000.
148 Leader 1901, p. 85: '... in 1774, at the pit hill were: Hard 3s. 4d.; hard and small, 2s. 8d., small 2s. per load of eight corves.' Mott 1969–70: 'Since "hards" brought,

at the stage 3s. 4d. compared with 2s. 2d. for house coal', quoting from Curr's report of 1779 (SA ACM/S/217).

149 SA ACM/S/215. *'The colliery at that time [1773] was worked at a depth of 50 fathoms.'*

150 SA ACM/S/217.

151 Further comment from John Hunter (Hunter 2021, p. 61, note 62): For example, George was a witness at Curr's wedding in 1781, and also at the baptism of some of his children.

152 Hunter 2021, p. 61.

153 Medlicott 1982, p. 273, note 28.

154 Author's note: Literally, *'English coal road'*. The term applied strictly to waggons running on wooden track in early seventeenth-century German mines.

155 Lee 1960–61, Some railway facts and fallacies, p. 7.

156 Vitali Vitaliev, in his review (in *Engineering and Technology Magazine*, 21 March 2018) of the 2018 reissue of Anthony Burton's *The Railway Empire* (Pen & Sword Transport, 2018; first edn, 1994, published by John Murray).

157 Curr's report of 1784 [SA ACM/S/221] noted that some £4,700 had been invested in the Wood Pits (Medlicott 1982, p. 55), with further investment at Manor Colliery. Such investment, though, was dwarfed by that expended on Attercliffe Colliery, work on which began in late 1796. Expenditure on the new colliery amounted to £13,823 (Medlicott 1982, p. 57) by mid-1790.

158 Agreement dated 5 March 1789 (noted in Medlicott 1982, p. 58): *'…in consideration of the great Costs and Charges amounting to many Thousand pounds which have already incurred and still remain to be incurred in the Opening Winning and Working the Seams or Beds of Coal … Hereby demised … in Consideration of the yearly Rent … made payable to … the said, Duke of Norfolk hath granted … unto the said Vincent Eyre … that Mine Seam or Bed of Coal lately opened and now called the Attercliffe or Darnall Coal.'* [SA ACM/SD/666].

159 Medlicott 1982, p. 58: A moiety on the property was given to Eyre as security on the capital invested: *'The several Collieries of the Duke of Norfolk being carried on at the joint & equal expense of his Grace, & his Agent Vin. Eyre, & in partnership betwixt them the Duke as Ground Landlord or Owner of the Coal, receiving from (the Partnership Fund a Rent for such coal).'* [Medlicott's source corrected to: SA ACM/SD/274].

160 Medlicott 1982, p. 188, quoting SA ACM/S/231, 226 & 185.

161 Medlicott 1999, p. 73.

162 Lee 1960–61, p. 1. The present author notes, for example, that Dendy Marshall commented on this: 'Curr's own book [*The Coal Viewer*] shows that the flanges were on the inside of the rails, but it may be remarked that they are shown incorrectly, on the outside, by Herbert in his *Engineers' and Mechanics' Cyclopedia* (1836) and his *Practical Treatise on Rail-Roads* (1837); also in all three editions of Wood [*Treatise on Railroads*, first published in 1825] (from Dendy Marshall 1938 (1971), *A history of British railways…*, pp. 146 & 158).

163 Lee 1960–61, p. 1.

164 Lee 1960–61, pp. 3, 5–8.

165 Rieuwerts 1981, *A Technological History of the Drainage of the Derbyshire Lead Mines* (PhD thesis). SA/OD/1152 (Sheffield Archives/Oakes' Deed) is a reckoning book (containing fortnightly production accounts) for Ladywash Lead Mine (located just north of the village of Eyam) covering the period 1745 to 1753.

166 Author's note: Lewis (1974, p. 170) concurs: *'Indeed, it seems clear that the invention of the cast-iron bar to lay on top of the wooden rails took place at*

Coalbrookdale in 1767, and that its use spread only to the Shropshire colonies.'

167 Bracegirdle 1973, *The Archaeology of the Industrial Revolution*, p. 72.
168 Lewis 1974, pp. 284–85.
169 Lewis 1974, pp. 314–15, citing Blue Books 1842, xvi.665.
   Author's comment: These government-published blue bound reports covered subjects as diverse as trade, agriculture, governance and demographics. They are a rich source of information.
   Although not clearly specified, it seems that the Parliamentary Commission referred to by Lewis is the same that produced the Children's Employment Commission report noted below.
170 *Children's Employment Commission* 1842, *First Report of the Commissioners: Mines…*
171 *Children's Employment Commission* 1842, Appendix 1; evidence of Geo. Lindsay, Part I, p. 363.
172 For example, Alexander Lock (2015) in *ODNB*: 'the new method of haulage was so effective that it encouraged cheaper child labour in place of animals and adult putters'; and Chadwick and Knights 1988 in *The Story of Tunnels*, p. 22: 'the system … incidentally led children into extended employment'.
173 *Condition and Treatment of the Children Employed* 1842, with a preface signed W. C. May 13th, 1842.
174 Flinn 1984, p. 96.
175 Lewis 1974, p. 319. Author's comment: Interviews conducted by the 1842 Commissioners established that no girls were employed in the pits belonging to the Duke of Norfolk's neighbour, Earl Fitzwilliam. Lists of employees show no girls for several decades previous to 1842. Boys were employed as pony drivers, as they were in the Duke's pits. It is hard to visualise children having the strength to haul or push a loaded corf weighing up to 10 cwt., let alone a train of corves, more than a short distance. Personal communication with John Hunter, 23 February 2023.
176 Ashton and Sykes 1929, p. 68.
177 Dunn 1844, *An Historical, Geological, and Descriptive View of the Coal Trade of the North of England*, pp. 107–8.
178 Medlicott 1983.
179 Flinn 1984, p. 95.
180 Author's note: See Wilson 1843, *The Pitman's Pay and Other Poems*, pp. 27–32. Lewis has included some parts of this poem in his 1974 book; I've included more verses from Wilson's poem here.
181 Lewis 1974, pp. 315–16, including parts of Wilson's poem. Inby, outby and rolley are noted in the Glossary.
182 Flinn 1984, p. 3. John Rees Harris (1934–2018) was professor of economics at Boston University. Harris's work, in a series of articles in journals such as *Scientific American* and *History*, highlights the critical role of coal technology in Great Britain's industrial growth. The identity of 'J.C.' is unknown.
183 Farey 1827 (1971), *A Treatise on the Steam Engine…*, vol. 1, p. 204.
184 From C. E. Lee's new introduction for the 1970 reprint of *The Coal Viewer*, p. 4.
185 As cited in Liffen's 2008 paper: Science Museum Library MSS 588 (now held at the Library site at Science Museum Swindon).
186 Liffen 2008, The discovery of a manuscript version dated 1795 of John Curr's *The Coal Viewer* (1797).
187 Correspondence, Bland 1930–31, p. 130.
188 Dendy Marshall 1938 (1971), p. 145. John Liffen (2008, n. 7): 'At the auction of the Dendy Marshall collection in November 1945 this copy was sold for £9 10s to

someone named Benger, but its present whereabouts are unknown.' Author's note: Almost certainly the bookdealer Frank B. Benger of Leatherhead.

189 English Short Title Catalogue (ESTC), print and probability search interface. ESTC does not record a 1796 edition.
190 *The Monthly Review; or Literary Journal*, vol. 29, August 1799, p. 451.
191 Galloway 1898, *Annals*, p. 327.
192 Smiles 1857, p. 61.
193 Cited in Lee 1960–61, p. 12. Original source: Printed as Appendix C in Mary Frances Outram, *Margaret Outram, 1778–1863*, London: John Murray, 1932. The author was the great-grand-daughter of Benjamin and Margaret Outram.
194 Mott 1969, pp. 14–15.
195 *The Coal Viewer*, pp. 23–29 & plate 2.
196 In the case of overland tramroads, as explained by Mott (1969–70, p. 14), 'The Merthyr tramroad had stone blocks and not cross-sleepers, this making it easier for the horse to move freely.'
197 Bland 1930–31, p. 126.
198 Lewis 1974, p. 316.
199 Galloway 1898, *Annals*, p. 323.
200 Notably, small wooden railways with wheeled rollies were used in the Newcastle area.
201 *The Coal Viewer*, p. 14.
202 *The Coal Viewer*, p. 8.
203 Galloway 1882, *History*, p. 211.
204 DRO NCB1/JB/354, 6 June 1800.
205 *The Coal Viewer*, pp. 15–22 & plate 1.
206 Medlicott 1999, p. 66.
207 *Hatchett Diary* (Raistrick 1967), p. 70.
208 Medlicott 1982, p. 119.
209 Medlicott 1982, p. 56, citing SA ACM/S/198.
210 SA ACM/S/214.
211 Galloway 1898, *Annals*, p. 305.
212 See SA ACM/S/215, 217, 218, 219, 221, 222.
213 See SA ACM/S/198.
214 Lewis 1974, pp. 317–18.
215 Ashton and Sykes 1929, p. 64.
216 SA ACM/S/223.
217 Medlicott 1982, p. 119.
218 Mott 1969, p. 16, 'Discussion', citing ACM/S/201.
219 Medlicott 1999, p. 67. No source given.
220 Lewis 1974, p. 318 (for the following quoted texts with included notes).
221 Author's note: Little is known of John Stephenson (?–1810). Almost certainly Roman Catholic (his will mentions Richard Rimmer, who had charge of the Sheffield Mission from 1787 to 1828), he worked a colliery, leased from the Earl of Effingham (another Howard), in Kimberworth in partnership with Micah Barber.
222 See TCR/NEIMME/Bud/14, cited in Lewis 1974, p. 318.
223 See SA ACM/S/201, cited in Lewis 1974, p. 318.
224 Leader 1901, p. 341, cited in Lewis 1974, p. 318.
225 As cited in Ashton and Sykes 1929: Barrowman pp. 274–75.
226 As cited in Ashton and Sykes 1929: see Charles Bond in Appendix B of H. F. Bulman and R. A. S. Redmayne, *Colliery Working and Management*, London: Crosby Lockwood and Son, 1906, 40.

227 As cited in Ashton and Sykes 1929: MSS in Bell and Watson Collection, Newcastle; Kenneth Vickers, *History of Northumberland*, xi. 426.
228 The whole quotation is from Ashton and Sykes 1929, p. 68.
229 Lewis 1974, p. 319. Author's note: Lewis's paragraph ends with a note referring to Ashton and Sykes (as noted above).
230 T. Y. Hall, another native of Co. Durham, of Greenside, near Ryton.
231 Galloway 1882, *History*, p. 218.
232 Author's note: Alexander Lock has explored the life and career of Sir Thomas Gascoigne in his 2016 book *Catholicism, identity and politics in the Age of Enlightenment: the life and career of Sir Thomas Gascoigne, 1745–1810*, and I draw on material from his text in this section.
233 Cited in Lock 2016, p. 202 (fn 101): Gascoigne Mss: WYL115/uncatalogued item awaiting repair, '*J. Curr's neat savings in working the intended colliery on Garforth Moor*', 16 December 1790. [A premium of £400 seems excessive.]
234 Cited in Lock 2016, p. 203 (fn 105): WYL115/Box 72, '*Sir Thomas Gascoigne to George Townsend*', 5 January 1791.
235 Cited in Lock 2016, p. 203 (fn 106): WYL115/GC/C4, '*Account book: Garforth and Barnbow collieries*', 1796.
236 Lock 2016, pp. 203–4.
237 Cited in Lock 2016, p. 202 (fn 103): WYL115/GC/C12, '*Statement of expence of drawing coals*', 18 September 1801.
238 The significance of 1801 is '*that James Watt's patent for the separate condenser had just expired, and it became possible for other engine builders to make new (condensing) rotary steam engines without fear of legal challenge. James Pickard's patent on the crank had expired in 1798 (this patent had forced Watt to devise the sun and planet gearing for his rotary engines). Therefore … around 1800 it suddenly became cheaper to erect a rotary steam engine (with a simple crank action). This was a viable alternative to a water-returning system with a water-wheel, which used a conventional (and cheaper) reciprocal pumping engine (instead of rotary). However, such choices were often site-specific – e.g. a reliable supply of fresh water was needed – for boilers, condensers etc. At a number of collieries engines were adapted to do double duty, i.e. both pumping and winding. Curr built a waterwheel at Attercliffe colliery, but there is little description of its use.*' John Hunter, personal communication with author, 15 August 2022.
239 Lock 2016, pp. 202–3.
240 Lewis 1974, p. 290.
241 Cited in Trinder 1973, p. 133, note 77: John Curr to Richard Dearman, 25 May 1793. The original is in Coalbrookdale Museum. Reprinted in Bland 1930–31, pp. 127–29. This letter is shown in full in Appendix 4.
242 Cited in Trinder 1973, p. 133, note 77: Science Museum, *William Reynold's Sketch Book*, f. 53.
243 Trinder 1973, pp. 133–34.
244 Trinder 1974, *The Darbys of Coalbrookdale*, p. 60.
245 Author's note: In 1801 Jessop was appointed engineer for the Surrey Iron Railway, which employed Curr's plate rails.
246 Information sourced from Mott 1969–70, pp. 10–11.
247 See, for example, Ashton 1963, *Iron & steel in the Industrial Revolution*, pp. 142ff (cited in Lewis 1974, p. 292, n. 37).
248 Lewis 1974, p. 292.
249 Farey 1817, *A General View of the Agriculture and Minerals of Derbyshire*, vol. 3, p. 288 (cited in Lewis 1974, ch. 17, p. 318, n. 57).

250 Leader 1901, p. 341 (cited in Lewis 1974, ch. 15, p. 293, n. 42).

251 National Library of Wales, Maybery Papers, Bundle 14 (cited in Lewis 1974, ch. 15, p. 293, n. 43).

252 Lewis 1974, p. 293.

253 Leeds Central Libraries, Archives Dept, MC51, MC203 (cited in Lewis 1974, ch. 17, p. 318, n. 58).

254 Bland 1930-31, p. 127; Labouchere Coll, particulars of the expence of a corve, 1794; *Hatchett Diary* (Raistrick 1967), p. 60 (both cited in Lewis 1974, ch. 17, p. 319, n. 59).

255 Ashton and Sykes 1929, p. 108. Hughes, ii. 203. Paymen, P. L., in *Business History*, iii. Liverpool (1961), 75ff. NRO, Blackett (Wylam) papers, colliery ledger. TCR NEIMME/Bud/15/7 (all cited in Lewis 1974, ch. 17, p. 319, n. 60).

256 Lewis 1974, ch. 17, p. 319, n. 61, citing *Blue Books* 1842, xvi.574f. This quoted text is from Lewis 1974, pp. 318-19.

257 SA ACM/S/214.

258 This list of the eleven rail roads is from Mott 1969-70, pp. 12-14.

259 Mott 1969-70, p. 13: The statement (The Coalbrookdale Co. 1709-1959, *Brief History*, July 1965) that the angled rails were developed at Coalbrookdale in 1767 is incorrect; this design was not adopted until 1793 and its design by John Curr is beyond question.

260 Baxter 1966, n.p. (cited in Mott 1969-70, p. 13).

261 G. G. Hopkinson, *Derby Arch. J.*, S.99 (1959), 28 (cited in Mott 1969-70, p. 13).

262 From Mott 1969-70, p. 13: Note 4: S. Mercer, *Newc. Soc. Trans.*, XXVI (1947-49), quoting C. Wilkins, *History of the Iron, Steel, Tinplate and other Industries*, 1903, 135.

Note 5: A similar error is shown in a letter from John Overton to Josiah John Guest, dated 1 August 1822 (Dowlais Ironworks Letters 175): 'I am of the opinion that as I go through Sheffield I had better stay some time and visit a few Founderies. I think my friend Mr. Carr is residing abroad, or he could give me some information, as he sent me down a plan of the wooden Dovetail'd Sleeper into Worcestershire 28 years ago' – which at least shows that Curr was in touch with Overton in 1794.

263 C. E. Lee, *Railway Mag.*, August 1967 (cited in Mott 1969-70, n. 6, p. 13).

264 C. E. Lee, *Newc. Soc. Trans.*, XXI (1940-41), 49-79 (cited in Mott 1969-70, n. 7, p. 13).

265 T. V. Simpson, *Trans. Inst. Min. Eng.*, LXXI (1930-31), 82 (cited in Mott 1969-70, n. 8, p. 13).

266 Baxter 1966, p. 185 (cited in Mott 1969-70, n. 9, p. 13).

267 Baxter 1966, p. 165 (cited in Mott 1969-70, n. 10, p. 13).

268 Baxter 1966, p. 192 (cited in Mott 1969-70, n. 11, p. 13).

269 G. G. Hopkinson, *Derby Arch. J.*, 99, 1959, 28 (cited in Mott 1969-70, n. 12, p. 13).

270 Baxter 1966, p. 180 (cited in Mott 1969-70, n. 13, p. 13).

271 Baxter 1966, p. 174 (cited in Mott 1969-70, n. 14, p. 13).

272 J. Aikin, *Description of the Country from thirty to forty miles round Manchester*, 1795, 582 (cited in Mott 1969-70, n. 15, p. 13).

273 C. E. Lee, *Newc. Soc. Trans.*, XXI, (1940-41), Plates XI, XX1 (cited in Mott 1969-70, n. 1, p. 14).

274 From Mott 1969-70, Appendix II, Discussion, p. 18.

275 Tomlinson 1915, *The North Eastern Railway…*, p. 33.

276 Trinder 1974, p. 60.

277 Lewis 1974, p. 289.

278 Dunn 1844, pp. 39–40 & 45.
279 Flinn 1984, p. 102: *'By 1800 it is estimated (Dr. J. Kanefsky) that at least 130 steam winding-engines had been installed in British collieries. All but twelve of these had been installed in the 1790s, so that there is little difficulty in establishing 1790 as the effective beginning of steam-winding generally in Britain. Forty-three of these engines were of Boulton and Watt design, whether licensed or pirated, fifty-two of basically Newcomen design, and thirty-seven of other types.'*
280 Dunn 1844, p. 45.
281 SA ACM/S/223.
282 Freely adapted from descriptions in Galloway 1882, *History* (p. 117) and Galloway 1898, *Annals* (pp. 314 & 323).
283 Phillips 1844, *Memoirs of William Smith*, p. 12: *'In the summer of 1794 Smith, who is credited with compiling the first geological map of England, made a brief tour of "the north" as a "fact-finding" exercise in preparation for surveying the route for a canal in Somerset.'* Author's note: White Lane Colliery is in Hesley Wood, which is in Chapeltown, just north of Sheffield.
284 *Hatchett Diary* (Raistrick 1967), p. 70.
285 From information provided by John Hunter, personal communication.
286 General Meeting. Held at the Queen's Hotel, Leeds, on Tuesday, 30th August 1887. The passage is extracted from the President's inaugural address, published in *Transactions of the Midland Institute of Mining, Civil & Mechanical Engineers*, vol. XL (1887–89), pp. 42–43.
287 TCR NEIMME/Bell/3/327.
288 DRO NCB1/JB/365. 29 December 1800.
289 Galloway 1882, *History*, pp. 212–13.
290 Buddle 1788, A life preserved after falling into a coal-pit… For the full account, see Appendix 1.
291 Flinn 1984, p. 103 and Wilson 1843, p. 25.
292 See note 230 above and the section Curr's railroad, III.
293 DRO NCB1/JB/366.
294 Dendy Marshall 1938 (1971), p. 145: He refers to this as *'the gravity method of conveyance'*.
295 Galloway 1882, *History*, p. 185.
296 Quotations from *The Coal Viewer*, p. 30.
297 DRO/NCB1/JB/357.
298 *Hatchett Diary* (Raistrick 1967), pp. 70–71.
299 SA ACM/S/223.
300 Shown in Figures 1–4 and 7–9 on Curr's plate 3 [see Figure 8.24].
301 SA ACM/S/215.
302 *Hatchett Diary* (Raistrick 1967), p. 71.
303 Medlicott 1982, pp. 117–18.
304 Mike Chrimes, entry for John Curr and *The Coal Viewer*, in Skempton (ed.) 2002, *A Biographical Dictionary of Civil Engineers* …
305 *The Coal Viewer*, p. 37.
306 Hunter 2021, p. 63.
307 Noted in Mott 1969–70, p. 7.
308 Farey 1827 (1971), vol. 1, pp. 205–9 and plate VIII.
309 See Hunter 2021, pp. 54, 58, 61–70 & 73.
310 SA ACM/S/221 (cited in Hunter 2021, p. 61).
311 Author's note: In March 1789 the duke entered into partnership with Vincent Eyre III, steward of the Norfolk's Hallamshire estate.

312 SA ACM/S/474/5 (cited in Hunter 2021, p. 62).
313 Author's note: this information is from Medlicott 1982, p. 57, n. 270.
314 SA ACM/S/201 (cited in Hunter 2021, p. 62).
315 SA ACM/S/223 (cited in Hunter 2021, p. 62).
316 TCR NEIMME/Bud/70 (cited in Hunter 2021, p. 62, n. 73).
317 For example, SA ACM/SheD/777 (1789 amendment); FC/SheD/141; WC/1752 (cited in Hunter 2021, p. 63, n. 74).
318 *The Coal Viewer*, Plate IV; Farey 1827 (1971), pp. 205–11 and Plate VIII. Both cited in Hunter 2021, p. 63, n. 75.
319 SA ACM/S/224/7 (cited in Hunter 2021, p. 65, n. 88).
320 SA WC/1752 (cited in Hunter 2021, p. 66, n. 90).
321 SA ACM/S/200 (cited in Hunter 2021, p. 66, n. 91).
322 SA FC/FB70, p. 29 (cited in Hunter 2021, p. 66, n. 92).
323 The elevation at the bottom of Staniforth's deep pit was similar to that at Darnall Colliery, but well above the shaft bottom at Attercliffe Common Colliery (cited in Hunter 2021, p. 67, n. 93).
324 SA ACM/65344/3 (cited in Hunter 2021, p. 67, n. 94).
325 SA MD/1741/2 (cited in Hunter 2021, p. 68, n. 96).
326 SA ACM/S/225 (cited in Hunter 2021, p. 68, n. 97).
327 SA MD/1741/2 (cited in Hunter 2021, p. 68, n. 98).
328 SA ACM/S/274 (cited in Hunter 2021, p. 68, n. 99).
329 SA ACM/S/225/1 (cited in Hunter 2021, p. 68, n. 104).
330 SA ACM/S/202 (cited in Hunter 2021, p. 68, n. 105).
331 SA ACM/S/274 (cited in Hunter 2021, p. 69, n. 107).
332 Holland 1837, *The tour of the Don…* (cited in Hunter 2021, p. 69, n. 108).
333 SA MD/1738/70 (cited in Hunter 2021, p. 69, n. 109).
334 SA MD/1738/94 (cited in Hunter 2021, p. 69, n. 110).
335 SA MD/1738/51 (cited in Hunter 2021, p. 70, n. 111).
336 SA SSC/675, MD/1738/110 (cited in Hunter 2021, p. 70, n. 113).
337 SA ACM/S/274, MD/1738/111 (cited in Hunter 2021, p. 70, n. 114).
338 SA ACM/S/274 (cited in Hunter 2021, p. 70, n. 115).
339 Author's note: The partners' costs were £665, the challenge cost a further £30.
340 Ashton and Sykes 1929, p. 68.
341 Medlicott 1982: Attercliffe and Ponds collieries paid 'tolerably well', and in 1793 they returned £4,600 after the deduction of rent and punchwood (p. 67). In 1800/01 a small profit of £1,182 14s 7d was returned (p. 67). At Sheffield Park and Manor collieries profits were £2,380 4s 6½d in 1796/97; £1,690 11s 4d in 1798/99; and, £200 12s 9½d in 1800/01 (p. 68).
342 SA ACM/S/205.
343 SA ACM/SD/18.
344 SA ACM/65344. Author's note: This partnership eventually became the Sheffield Coal Company, which was to last until nationalisation.
345 Lock 2015 (*ODNB*).
346 SA ACM/S/214. Unless otherwise noted, the quotations in this section (Curr's letter and the Duke's reply) are from this archive.
347 Mott 1969–70, p. 7.
348 Christine Hiskey (1978) noted the effects of inflation as they affected Buddle's operations:

'As causes of the falling profits, Buddle was chiefly concerned with *"the immense increase in price on every article of colliery consumption, as well as the price of labour"*, during the Napoleonic wars. … Increases in the price of timber, ropes, horses and wages were such that it was estimated in the trade that working charges had increased by 3/- to 3/6d. per chaldron [measure of coal, up to 53 cwt. on Tyneside] in the two or three years up to 1800. ... Oats for the horses in 1800 cost over twice the price per quarter at which Lambton had to meet the excess, ... In 1801 Lambton's subsidy on his workmen's bread-corn was costing him £5,728 – an increase of about £1,400 since the previous year. Hay for the horses increased in price by £2,000 in the same year ...'

349   SA ACM/S/214.

9. After dismissal

350   Dunn 1844, p. 146.
351   Hiskey 1978, p. 94.
352   Lock 2015 (*ODNB*).
353   Dendy Marshall 1938 (1971), p. 8.
354   Heesom 2004 (*ODNB*).
355   National Coal Board Records, first deposit: Papers of John Buddle. DRO NCB1/JB/345–369.
356   Buddle Atkinson Collection, TCR NEIMME/Bud/15.
357   DRO NCB1/JB/370–373.
358   TCR NEIMME/Bud/15/62 & 62b, 70 and 74, 74a & 74b.
359   TCR NEIMME/Bud/15/65.
360   SA ACM/S/214.
361   SA ACM/S/224.
362   John Curr III (1783–1850). In his book, *Railway Locomotion and Steam Navigation* (Curr [III] 1847, *Railway Locomotion*), pp. 1 & 9, he wrote *'although thirty-two years had passed since I relinquished that profession … at 18 years of age he was entrusted solely with the engineering department, both as to the plans and their execution, of what was then considered a rather extensive establishment in Sheffield Park, belonging to his father, and in which profession he continued about 16 years.'*

    From this it can be deduced that he was working at the foundry from, at the least, 1801, until 1811, when he left for Norwich.

363   *The Iris; or, The Sheffield Independent*, 22 May 1806 (cited in Flavell 1996, p. 120, n. 48).
364   Attercliffe Colliery accounts show that £370 was paid in August [1796] to Curr's foundry for a 27-inch-diameter cylinder for drawing coals.
365   Medlicott 1982, pp. 62, 187, 235, 256 & 273. SA ACM/S/199–202, 232 & 274.
366   DRO NCB1/JB/346 & 347.
367   DRO NCB1/JB/348.
368   DRO NCB1/JB/352.
369   TCR NEIMME/Bud/15/6, Walls-end, 15 September 1802.
370   Lewis 1974, p. 295: *'The only ones of any size were the Bedlington Iron-works, the Tyne Iron-works at Lemington, and a little later Losh's Walker Foundry.'*
371   DRO NCB1/JB/354 & 356.
372   DRO NCB1/JB/360.
373   DRO NCB1/JB/363 (17 October 1800).
374   TCR NEIMME/Bud/15/137 (23 February 1806).
375   SA OD/1500. (Sheffield Archives/Oakes' Deed). This is an account book/diary that belonged to a mine agent called Isaac Morton. The individual pages in the book are not numbered. A copy of Curr's invoice was written on a single page of the book.
376   DRO NCB1/JB/357.

377 TCR NEIMME/Bud/15/74a.
378 Killamarsh, in Derbyshire, is on the Chesterfield Canal, to the south-east of Sheffield.
379 Pipes were cast vertically and required a crane to manoeuvre. This suggests that the foundry had become well developed and quite extensive.
380 TCR NEIMME/Bud/15/47 & 67.
381 DRO NCB1/JB/346.
382 DRO NCB1/JB/348.
383 DRO NCB1/JB/354.
384 DRO NCB1/JB/360.
385 Improvement was initiated by a private Act of Parliament in 1726 and the river made navigable as far upstream as Aldwark by 1733, and to Tinsley by 1751. The principal method involved was to create a greater depth of water by means of a weir. This method involved the construction of a lock in order that vessels could bypass the weir. Because sail-power was unreliable, tow-paths were often constructed to enable hauling.
386 *Sheffield Register*, Weekly sailings to Hull, 30 May 1789.
387 The following excerpts, to the end of this section, can be found in the following documents: DRO NCB1/JB/353 (3 June 1800), 354 (6 June 1800), 357 (16 July 1800), 358 (13 August 1800), 362 (4 October 1800) & 363 (17 October 1800).
388 Edward Sargent (Chairman), 'A history of rope making'. Presented at the Docklands History Group Meeting, March 2015.
389 Galloway 1882, *History*, p. 117.
390 Flinn 1984, pp. 103–4.
391 Based on the entry for William Chapman, by Anita McConnell, *ODNB* (2004).
392 Chapman 1808, *A treatise on the progressive endeavours to improve the manufacture and duration of cordage*. The book contains descriptions of Curr's patents, nos. 2270, 2914 and 2960.
393 SA ACM/S/214.
394 Galloway 1882, *History*, p. 279.
395 Dunn 1844, p. 50.
396 DRO NCB1/JB/117.
397 The location of former roperies may often be found in names relating to rope. In this instance 'Ropery Row' can be discerned just above the overlaid ropery.
398 Dickinson 1942–43, A condensed history of rope-making, p. 71.
399 DRO NCB1/JB/345.
400 DRO NCB1/JB/347.
401 DRO NCB1/JB/348.
402 DRO NCB1/JB/354.
403 DRO NCB1/JB/355.
404 DRO NCB1/JB/356.
405 DRO NCB1/JB/357.
406 DRO NCB1/JB/358.
407 DRO NCB1/JB/359.
408 TCR NEIMME/Bud/15/6 & 7.
409 TCR NEIMME/Bud/15/8 & 9.
410 TCR NEIMME/Bud/15/17, 22, 26a, 62, 67, & 74.
411 TCR NEIMME/Wat/1/26/1.
412 TCR NEIMME/Bud/15/134–137.
413 As late as 1891, D. H. and G. Haggie of Wearmouth Patent Rope Works, Sunderland, released a catalogue listing both round and flat steel ropes.
414 National Maritime Museum: The Caird Library and Archive, ADM 354/233/98, 6

September 1808. Wm. Wellesley-Pole, an older brother of the Duke of Wellington, served as Secretary of the Admiralty 1807–9.

415 DRO NCB1/JB/117.
416 This information is from Dickinson 1942–43, p. 76.
417 DRO NCB1/JB/117.
418 Medlicott 1999, p. 76.
419 Galloway 1882, *History*, pp. 184–85.
420 TCR NEIMME/Bud/15/9, Walls-end, 28 September 1802.
421 Dendy Marshall 1938 (1971), p. 158, n. 11: *'Humble's waggonway, running from coal pits on Birtley Common to a staith at Fatfield was established in 1741. A branch to Black Fell was established before 1787 (it is shown on Gibson's map of 1787).'*
422 Galloway 1882, *History*, p. 186; Galloway 1898, *Annals*, p. 484.
423 Dunn 1844, p. 37.

10. Children

424 As were all her sisters.

The Convent of the Institute of the Blessed Virgin at Micklegate Bar, York, known as Bar Convent. The creation of the Convent was inspired, at least in part by Sir Thomas Gascoigne, a fervent Catholic, who declared 'we must have a school for our daughters'. Gascoigne even went as far as providing a gift of £450, part of which purchased a property on the Convent site and a boarding school, and day school, for Catholic girls was set up on the site by nuns in 1686 (Wikipedia).

Cousin Dorothy Gurner, a great great granddaughter of John Curr II, left a manuscript note: 'From information given her by Alice Perrot, a great granddaughter, in 1954. Great aunts Florence & Georgie at school there & 3 Florence Currs before them.'

425 For the text see: http://www.saintsbooks.net/books/Fr.%20John%20Furniss%20-%20The%20Sight%20of%20Hell.html.
426 EMC *Memoranda*.
427 EMC *Memoranda*.
428 Curr [III] 1847, *Railway Locomotion*.
429 TCR NEIMME/Bud/15/17–19, 21, 22, 26a, 47, 62 & 67.
430 From John Hunter (personal communication): 'the common type was by then outmoded and unsuited to the task'.
431 TCR NEIMME/Bud/15/6.
432 TCR NEIMME/Bud/15/17.
433 DRO NCB1/JB/366.
434 William Emerson, *The Principles of Mechanics*, first published 1754. For further information see note 56 above.
435 TCR NEIMME/Bud/15/21.
436 DRO NCB1/JB/367.
437 TCR NEIMME/Bud/15/26a.
438 TCR NEIMME/Bud/15/67.
439 TCR NEIMME/Bud/13/110 & 113.
440 Advert for sale of coals & cinders – Aggs & Curr – best quality for housekeepers, the small having been used at their iron foundry. Good malting cinders sell at 19s/chaldron.
441 Thurston 1895, *A History of the Growth of the Steam Engine*, p. 240.
442 DRO NCB1/JB/370, 19 July 1814.

443 Dunn 1844, p. 52.

444 Dawson 2001, PS *Orwell*, 1813–14, p. 214.

445 Malster 1971, *Wherries and Waterways &c.*, p. 61 (cited in Dawson 2001).

446 DRO NCB1/JB/373 (July) and DRO NCB1/JB/372 (August).

447 DRO NCB1/JB/371.

448 *Ipswich Journal*, 25 September 1813 and 2 April 1814.

449 Clark 1943, East Anglia's first steamboats, *The Model Engineer*, p. 143.

450 *Ipswich Journal*, 12 August 1815.

451 Malster 1971 (cited in Dawson 2001, note 20).

452 C. Dawson's article ascribes the sketch to John Constable; however, the caption for the sketch shows: 'The steamship *Orwell*, after Richard Constable.'

453 Author's note: In 1815 three steam vessels worked on the Thames. The first two were named *Thames* and *Regent*. Aggs & Curr's *Defiance* was the third.

454 The quoted text is extracted from a Letter to the Editor (from New York, June 1834): 'A sketch of the origin, progress, and improvement of Steam Navigation, in the Port of London, from 1814 to 1824', in *Journal of the Franklin Institute*, no. 2, vol. 14, August 1834, pp. 84–87.

455 See Appendix A in Dawson 2001, citing his previous article, 'PS *Defiance*, the first steamship in Holland', *The Mariner's Mirror*, vol. 84, no. 1 (February) & no. 3 (August) 1998.

456 Curr [III] 1847, *Railway Locomotion*, pp. 12–13.

457 Mackie 1801–50, *Norfolk Annals...*, p. 141: 'The *Courier* steam packet made its passage from Foundry Bridge, Norwich, to Yarmouth in three hours twenty-five minutes' (November 1816).

458 Cited in Mackie 1801–50, pp. 145–46.

459 George Dodd, discussed in Partington 1822, *An Historical and Descriptive Account of the Steam Engine*, chapter IV, p. 106. Abstract of evidence before a Select Committee of the House of Commons on steam navigation. The Committee commenced its sittings on 8 May 1817.

460 Partington 1822, chapter IV, p. 102.

461 From the *Morning Chronicle*, 22 April 1817: '... *A German Paper, in noticing the blowing up of the* Courier *Steam-boat, of Yarmouth, makes the following observations: – "Mr. TREVIRANUS, a mechanician in Bremen, with a ship carpenter of the name of LANGE, and Mr. SPILKER, ship captain, was last year commissioned by Mr. SCHRODER, the merchant, to visit England and Scotland, for the purpose of examining the Steam Boats on the different rivers, and if possible to purchase a good one for the Weser. The purchase, however, did not take place, as the English had only bad ones for sale. With respect to this very steam boat, the* Courier, *Mr. TREVIRANUS made a report to Mr. SCHRODER, so far back as 6th August, 1816, as follows: –* 'This boat could not be purchased, because in the first place, it is in all its parts too slightly built, and is violently shaken by the motion of the machine; and secondly, because it has a high pressure machine, which on any want of precaution in the management, may prove dangerous to the lives of the passengers.' *The two remaining steam boats, which still ply on the Yare between Yarmouth and Norwich, have also high pressure machines."'*

462 Partington 1822, chapter IV, p. 102. Evidence given by Mr Richard Wright of Blackfriar's Road, engineer.

463 Mr Bryan Donkin in Partington 1822, pp. 71–73.

464 Partington 1822, pp. 120–22.

465 *Journal of the Franklin Institute*, vol. 14, no. 2, August 1834, p. 84.

466 *Journal of the Franklin Institute*, vol. 14, no. 6, December 1834, pp. 361–62.

467 EMC *Memoranda*.

468 Curr [III] 1847, *Railway Locomotion*, p. 83.

469 Book Review. *Mechanics' Magazine*, Saturday, 2 October 1847, pp. 335–40.

470 Editorial, *Mechanics' Magazine*, Saturday, 30 October 1847, pp. 438–39.

471 Curr [III] 1847, *Railway Locomotion*, pp. 11–12.

472 Curr [III] 1847, *The Learned Donkeys of 1847*, being a Review of the Reviewers of *Railway Locomotion &c.*, by John Curr of New South Wales, p. 12.

473 *London Gazette*, 4 October 1817, issue 17291, p. 206: 'Notice of partnership dissolved – Henry Aggs and John Curr, doing business as iron founders and tobacco and snuff manufacturers, carried on in the city of Norwich under the name of Aggs & Curr, June 30 1817, and the tobacco and snuff manufactory will in future be carried on by John Curr only.'

474 Marriage Register, Norwich, St Peter Parmentergate. Norfolk Record Office PD 162/13:

'John Curr of this Parish, Bachelor, and Martha Melton of the Parish of Saint Swithin, Spinster were married in this Church by Licence this fourteenth Day of May in the Year One thousand eight hundred and eighteen By me James B Thompson, Off.t Curate.

This Marriage was solemnized between us, [signed] John Curr & Martha Melton X her mark

In the Presence of John Webb & John Busby(?)'

475 EMC *Memoranda*.

476 EMC *Memoranda*.

477 Norfolk Electoral Registers 1832–1915 (online at FindMyPast – held at Norfolk Record Office).

478 EMC *Memoranda*. The typewritten copy of the Memoranda does say 'elder', but the elder children were being educated in England at that time.

479 *Cornwall Chronicle* (Launceston), 24 October 1835, 'Oct. 19 – Arrived the barque *Ann*, 339 tons, John Virtue, Commander, from London. Passengers Mr. and Mrs. Curr and 4 children … with a general cargo assigned to … John Curr …'

480 'SUPREME COURT. [Charles] Curr v. Goodwin [for libel]'. *Launceston Courier*, Monday, 10 January 1842.

'Mr. Curr was a young gentleman of high respectability, the son of a respectable merchant at Sydney, nephew of Mr. Curr the agent for the VDL Co, and at present holding the office of reporter to the *Launceston Advertiser*.'

481 EMC *Memoranda*: 'Eventually he left for New South Wales and tried unsuccessfully to grow tobacco on a farm near Wollongong. He eventually went to Sydney and again became a tobacco merchant.'

482 The lists of unclaimed letters at the General Post Office, Sydney, find him in 1843 at Paramatta Road, and in 1845 at New Town (a suburb). In 1845 he is listed as 'Mr. John Curr, tobacconist'.

483 'ARRIVALS. From Launceston yesterday, whence she sailed 2nd instant, the brig *William*, Captain Le Grand, with wheat. Passengers, Mrs Curr and three children, Miss Curr …'. *The Sydney Gazette*, Thursday, 8 July 1841.

484 *Sydney Gazette and New South Wales Advertiser*, 15 February 1840 (134 words).

485 'The Rail-Road and Locomotive Conveyance', *Sydney Morning Herald*, 1 December 1842 (735 words).

486 'The Aerial Steam Carriage', *Sydney Morning Herald*, 11 July 1843 (948 words).

487 'REVIEW. Magazine day in Sydney', *Sydney Morning Herald*, Wednesday, 13 Sept 1843 (p. 4).

488 'Watering the Streets', *Morning Chronicle*, 11 October 1843.

489   Curr, John, *Statistics of Banking in the Colony of New South Wales*. Sydney: Reading, 1843.

490   *Weekly Register of Politics*, 21 October 1843.

491   'Statistics of Banking', *Morning Chronicle*, 21 October 1843. Author's note: The end of the pastoral boom in the early 1840s caused an economic depression. In 1843 this resulted in the failure of a number of colonial banks. These included the Bank of Australia and the Sydney Banking Company.

492   *Sydney Morning Herald*, Monday, 23 October 1843 (p. 2).

493   'To Correspondents', *Weekly Register of Politics*, 28 October 1843.

494   'Original Correspondence. The Circulation', *Australian*, 4 November 1843 (661 words).

495   'Original Correspondence. Report from the Select Committee on Monetary Confusion', *Australian*, 6 November 1843 (532 words).

496   'Railways', *Sydney Morning Herald*, 12 June (1942 words) & 15 July (517 words) 1846.

497   'SAILED.– The ship *ST GEORGE*, 605 tons, Jones, master, for London. Passengers … Mr. J. Curr …' *The Australian*, Saturday, 27 March 1847.

498   Reference number: AP/32/6. Date: 01 November 1849

      Description: In this paper, which is a continuation of a former paper bearing the same title [see AP/32/5], Curr applies his previously stated laws to new scenarios of hydraulics. He considers that the power of expansive engines has been greatly overrated, instancing those of the Great Britain, which were of the estimated power of 1200 horses, but which he states he can prove did not exceed in actual power that of 300 horses. This he attributes to the inapplicability of Mariotte's law without a particular limitation. He gives general laws for estimating the pressure of steam when cut off from its generating source.

      Subject: Hydraulics

      Received 30 May 1850. Read 13 June 1850. Communicated by John Scott Russell.

      Written by Curr at 7 Upper Penton Street, [London].

      Whilst the Royal Society declined to publish this paper in full, an abstract of the paper was published in volume 5 of *Abstracts of the Papers Printed in the Philosophical Transactions of the Royal society of London* [later *Proceedings of the Royal Society*] as 'On the temperature of steam, and its corresponding pressure'.

499   *Moreton Bay Courier* (Brisbane), Monday, 14 January 1850.

500   JOHN CURR, 19 Goulden Terrace, Islington, London; or at I. A. Beauvoisin's, Esq., Belmont Park Sheffield.

501   'ARRIVED. Sunday, September 15 – The ship *Stebonheath*, 1030 tons, Sargent, master, from London 1st June, and Plymouth 16th June [dates as printed]. Passengers – For Adelaide and port Phillip: … Mr. John Curr … in the cabin.', *Adelaide Observer*, Saturday, 21 September 1850.

502   EMC *Memoranda*.

503   See: https://www.findagrave.com/memorial/152894832/john-curr#

504   So named in the wills of both John Curr II and Joseph Curr.

505   DRO NCB1/JB/369.

506   See http://www.wakefieldfhs.org.uk/blog/2016/07/the-somme-remembered-6th-july-1916-2/

      Wakefield Family History Sharing: Private Louis D'Argenson Beauvoisin, Royal Warwickshire Regiment, died of wounds 6th July 1916.

507   1821: Beauvoisin Louis Amand, teacher of the French language, 22 Bridge St; 1828: PROFESSORS & TEACHERS, Beauvoisin, Louis Amand (French), 16

	Clarence Street.
508	1861 census (address as in advert).
509	Adrian's obituary is found at: http://www.plantata.org.uk/obits/smith/beauvoisin_a.htm
510	EMC *Memoranda*.
511	Kirkus (Sister Gregory) 2001, *An IBVM Biographical Dictionary ...*, p. 68.
512	Hudson 1874, *Life of John Holland of Sheffield Park...*: 'Mr Holland wrote a poem entitled 'Theresa's Tree', taking for a motto the words of Montgomery "whose spirit in the willow spoke".' Author's note: James Montgomery (1771–1854), was a poet and editor of the *Sheffield Iris*. He also published Holland's poetry.
513	John Holland, *Sheffield Park*, A descriptive poem, 1820. These stanzas are quoted in chapter 6.
514	Hudson 1874, pp. 104–6.
515	Thomas Penswick (1772–1836), Vicar Apostolic of the Northern District.
516	EMC *Memoranda*.
517	*Miscellanea VI*: Other papers VII. Catholic registers of Callaly Castle, Northumberland, 1769–1839, with historical notes by Joseph Gillow, p. 322.
518	Hadfield 1889, p. 59.
519	Hadfield 1889, p. 59.
520	Hadfield 1889, p. 59.
521	Gillow 1885–1902, *A literary and biographical history... of the English Catholics*, p. 609.
522	Gillow 1885–1902, pp. 610–11.
523	'A letter to the Rev. Joseph Curr occasioned by his letter to Sir Oswald Mosley, Bart., President of the Manchester and Salford Auxiliary Bible Society'. Printed by Joseph Pratt at Manchester, 1821. Of this there were at least four editions.
524	Gillow 1885–1902, p. 609.
525	Hadfield 1889, p. 59.
526	Hadfield 1889, p. 59. Author's note: the book ran to other editions, e.g. Dublin, 1834.
527	*Miscellanea VI*: Other papers VII, p. 322.
528	Gillow 1885–1902, p. 609.
529	Gillow 1885–1902, p. 611.
530	Saint Alphonsus Liguori (1696–1787).
531	Louis Bourdaloue (1632–1704).
532	Hadfield 1889, p. 60.
533	Gillow 1885–1902, pp. 611–12.
534	Hadfield 1889, p. 60.
535	Hadfield 1889, p. 60.
536	Initially in a Leeds newspaper, reprinted in *Lloyd's Weekly Newspaper*, 11 July 1847.
537	Hadfield 1889, p. 60. Gillow (1885–1902) and the *Catholic Encyclopedia* record his death as 29 June and *Miscellanea IV*, p. 367, as 30 June.
538	EMC, *Memoranda*. Author's note: Dorothy Gurner noted that her cousin Alice Perrott had told her in 1954 that 'some pictures there [Bar Convent, York] were left them by great great uncle Father Joseph Curr'. See also note 424.
539	Kirkus (Sister Gregory) 2001, p. 68.
540	EMC *Memoranda*.
541	*Miscellanea IV*: no. 9, 'The nuns of the Institute of Mary at York from 1677–1825', p. 367.
542	Kirkus (Sister Gregory) 2001, pp. 68–69.

543 Coleridge 1887, *St. Mary's Convent...*, pp. 361–62.
544 *Miscellanea IV*: no. 9, 'The nuns of the Institute of Mary ...', p. 367.
545 The preceding and succeeding entries are for 19 February.
546 'The late Mr. Edward Curr', *Argus*, Melbourne, 20 November 1850, p. 2.
547 Cheshire Marriage Licence Bonds and Allegations 1606–1905 – Winwick.
548 Derbyshire, Church of England Burials 1813–1991, Glossop Parish Register, p. 168.
549 *Yorkshire Gazette*, 23 June 1832.

Appendices

550 His eldest son.
551 His second-eldest son.
552 A son-in-law, married to Mary Ann, his second-eldest daughter.
553 Son of John Bernard Furniss & his wife, Elizabeth, née Curr.
554 Personal communication from John Hunter:
   'Fairly common for a time, these were cross-shaped in cross-section but must have been extremely heavy. Even two hefty colliers would struggle to set up a 6 foot long cast iron punch. Wooden props, being lighter, were much easier to handle and gave warning signals when under stress.'
555 Patent no. 3157, published 30 July 1808, 'Apparatus for towing vessels, catching or detaining whales, &c.' (applying flat ropes and flat bands or belts to capstans and windlasses of ships or vessels, for towing the same; applying flat or round ropes, lines, bands or belts, for catching and detaining whales).

Glossary

556 Gresley 1883, *A glossary of terms used in coal mining*, p. 225.
557 Galloway 1898, *Annals*, p. 368.
558 Cast iron is a group of iron–carbon alloys with a carbon content more than 2%. Its usefulness derives from its relatively low melting temperature. Iron alloys with lower carbon content are known as steel.
559 Lewis 1974, pp. 308–9.
560 Flinn 1984, pp. 82–83.
561 *Hatchett Diary* (Raistrick 1967), pp. 71–72.
562 Lewis 1974, Glossary, pp. 358–66.
563 SA ACM/S/201 (cited in Medlicott 1982, pp. 114–15).
564 Medlicott 1982, pp. 114–15.

INDEX

Page numbers for illustrations have the suffix fig, and for glossary entries have the suffix g. For endnotes, they are in the format 234n56.

Abercynon to Merthyr Tydfil tramroad 41, 53, 72, 77, 78, 78fig, 79
accidents
    collieries 16, 85, 171–72
    Courier steamboat explosion 149–51
Adkins, Joseph 112
Admiralty 139
advertisements
    for *The Coal Viewer* (Curr) 60–61, 177–78
    for Curr's iron foundry 61
    for ropes 135–36, 188–89, 193–98
Aggs, Henry 149, 150
Aggs & Curr (partnership) 147–54
Agricola, *De Re Metallica* 52, 54, 55fig
Aire (river) 119, 120
Aislaby Hall, Whitby 9
Ampleforth School 23
Anstice, William 135
anti-Catholicism 162–64
Ashton, T. S. 2, 19, 57, 67–68, 69–70
Ashton-in-Makerfield 168
Ashton-le-Willows 164
Attercliffe Common Colliery 50, 52, 93–104, 94fig, 100fig, 103fig
    plate rails at 75, 76
    steam engines at 41, 90, 96–99, 99fig
Australia 155–59
    Launceston, Van Diemen's Land (Tasmania) 155
    Sydney 155–59
    Van Diemen's Land (Tasmania) 155, 168–69
    Wollongong 155, 158–59

bankruptcy 159–60
Bar Convent, York 8, 23, 70, 143, 160, 167
Barnsley/hard coal 95, 101, 201g
barroways, wooden 56
bassets 98, 108, 201g
Beauvoisin, Henry (grandson) 26, 39fig, 159–60, 167
Beauvoisin, Louis Amand (son-in-law) 9, 39fig, 159, 167, 228fig
Beauvoisin, Mary Ann (daughter, formerly Curr) 9, 39fig, 159–60, 159fig, 228fig
Beauvoisin, Mary Ann (granddaughter) 159, 159fig, 167, 228fig
Belle Vue House, Sheffield Park 27–32, 27fig, 28fig, 29fig, 30fig, 31fig, 32fig, 45–46
Bewick, Thomas 16
Bible Society 162–64
Biddick, County Durham 14, 14fig
Binks, Booth & Hartop/Booth & Co (iron founders) 69, 96, 97, 98, 112
Blackburn 166
Blackwell, Henry 37
Bland, Fred 1, 63–65
Booth & Co/Binks, Booth & Hartop (iron founders) 69, 96, 97, 98, 112
Bourn, William 141
Bower, Robert 52
Boys, John 16, 171–72
Bracegirdle, Brian 55
The British Palladium (magazine) 14, 16
Broomhead, Rev Rowland 161
Buddle, Ann (formerly Reay) 16
Buddle, John Snr 13–20, 41, 80, 85, 104
    An account of an improved method of drawing coals and extracting ores &c. from mines 16, 82fig, 83, 83fig

colliery notebook 17, 203fig, 207fig
Instance of a life preserved 171–72
The Marquis of Worcester's Century of Inventions 16
'Observations on the Raising of Hedges' 16
A Representation of a Coal Waggon 15fig
Unitarianism 14
working relationships 229fig
—Sheffield Park Colliery reports 12, 49
1773 42–43, 42fig, 43fig, 44–45
1787 67–69, 68fig
Buddle, John Jnr 16, 17, 19–20, 83fig, 110–12
gives evidence to House of Commons Committee 12–13
John Curr III, correspondence with 112–19, 144–46, 147–48
Unitarianism 14
working relationships 229fig
Bugsworth to Peak Forest tramroad 77
buntons 81
Bushblades, County Durham 5, 6fig, 7fig, 14fig, 44
colliery 5–6, 6fig, 16

Callaly Castle, Northumberland 165, 166
canals 72–74, 76–77, 119
underground 89
cast iron 18–19, 52–53, 55–56, 79, 112, 117, 183, 203g
Castle Comer coal field, County Kilkenny, Ireland 109–10
Catholic Record Society 5
Catholicism 5–7, 9, 23–24, 33–39, 70, 143, 160–67
Catton, James 70
Chester-le-Street, County Durham 14–15, 14fig, 16
child labour 56–58
Clay, Joseph 28
Clay & Co (colliers) 100, 101
coal
demand for 99
hard/Barnsley 95, 101, 201g
profitability of 49–50
retail cost of 45
seams 6fig, 94fig
small/slack 49, 95, 97–98, 201g
coal mining 51, 207g
child labour in 56–58

conductors/guide rods for raising 66, 80–86, 83fig, 84fig
extraction methods 204g–05g
incline planes, self-acting 87–88, 88fig, 142
jinneys 86–88, 87fig
opencast 32
Royal Commission of Inquiry into Children's Employment in Mines and Manufactories 56–57
transport from coalface 56, 65–67
underground canals 89
winding 58, 79–86, 83fig, 84fig, 97fig, 203fig
winnings 18, 96, 118, 206g–07g
see also collieries; corves; railroads; ropes; steam engines; waggonways
The Coal Viewer and Engine Builder's Practical Companion (Curr) 2, 41, 53
on conductors 80, 85–86
on corves 65–66, 66fig
dedication 52
on jinneys 87–88, 87fig
prospectus 60–61, 177–78
publication history 59–61, 61fig
on rail plates 55–56, 62fig, 63, 74
reviews 61
on steam engines 90–93, 90fig, 91fig
coal viewer role 11–13, 16–17, 110–11
Coalbrookdale, Shropshire 41, 53, 72–74, 75, 76, 179–80
collieries 14fig, 111fig
accidents 16, 85, 171–72
Bushblades 5–6, 6fig, 16
Darnall 95, 98, 99fig, 100, 100fig, 101–02, 103fig
Dore House 101, 103–04, 103fig
on Duke of Norfolk's estate 17, 58, 95–104, 112
of Durham Cathedral 18
Elswick 111, 116
fire risks 81
Flatworth 116
Garforth 71–72
Gascoigne 70–72
Gatherick 70
Greenside, Ryton 17
Heaton 70, 111, 111fig
Hebburn 18, 111, 116, 133, 134
Hemingfield 81–82, 202fig
Hesley Wood 52, 76, 81, 105
innovations, cost of 96, 97, 98, 102–03, 104

Jarrow 111
Lambton 44, 44fig, 135, 136
Lings 76
Manor 49, 52, 104, 105, 112
Middleton 55, 70, 75, 88fig
Parlington 72
Pensher 135, 136, 146, 146fig
Percy Main 19, 110, 111fig, 144, 145–46
Ponds 52, 96, 207
profitability of 49–50, 95, 97–98, 102–03, 104, 106
props, cast-iron 117, 183
shafts 56, 66, 80–86
Sheriff Hill 111
Staniforth 95, 100–101, 102–03, 103fig, 104
Stanley Kinghill 70
Sturton 71
Tanfield Moor 110
ventilation 19, 81, 90
Washington 70, 111fig, 136
water breaches 101, 102–03, 104
water management 16, 17–19, 95–99, 100fig
Whitefield 116
Wingerworth 53, 74, 76, 77
workers, lowering/raising of 85
Worsley 89
Wykin 111, 116
see also Attercliffe Common Colliery; High Hazles colliery; Sheffield Park Colliery; steam engines; waggonways; Wallsend
Colliers Row, Sheffield Park 25, 25fig
common engines *see* steam engines
Consall Plateway tramroad, Stafford 77
corves 53, 56, 57, 118–19, 200fig, 202g
conductors for raising 80–86, 83fig, 84fig
for railroads 65–67, 66fig
County Durham 14fig, 43fig, 111fig
Biddick 14, 14fig
Bushblades 5, 6fig, 7fig, 14fig, 44
Chester-le-Street 14–15, 14fig, 16
Pontop Pike 5, 6fig, 44
Tanfield 44
West Kyo 14, 14fig, 15–16
Courier (paddle steamer), explosion of 149–51
Cromford Canal 76
Crook Hall, Durham 161
Curr family tree 228fig
Curr, Charles (grandson) 155

Curr, Charles (son) 167–68, 167fig
Curr, Edward Charles (son) 8, 9, 33, 155, 156, 168–69, 228fig
Curr, Edward Micklethwaite (grandson) 8, 23–24, 27–28, 33, 46–47, 161–62, 167, 228fig
Curr, Elizabeth (mother) 5, 7, 228fig
Curr, Elizabeth Mary (later Furniss, daughter) 9, 30, 33, 37–38, 39, 39fig, 143, 228fig
Curr, George (brother) 6–7, 50–51, 50fig, 228fig
Curr, Hannah (wife, formerly Wilson) 7–9, 33, 228fig
Curr, Harriet (*in religione* Gertrude, daughter) 9, 167, 228fig
Curr, John I (1712–77, father) 5–6, 17, 46–47, 228fig, 229fig
CURR, JOHN II (1756–1823) 228fig
birth 1, 5, 41
Catholicism 5, 23–24, 33–39
character 23–24
death 1, 9, 109
disinterment and reburial 37–38
education 11, 17, 20
in *The Gentleman's Diary* (magazine) 20, 21fig
grave and memorials 37–38, 37fig, 38fig
handwriting 23fig
lives at Belle Vue House, Sheffield Park 27–30
marriage 7–9
New Chapel, Norfolk Row, funds building of 35, 173–76
Paris, visit to 9, 109
wealth 25–26
will 25, 155, 174–75
—**career** 12, 49–51
and child labour 56–58
at Coalbrookdale 72–74, 179–80
Company of Cutlers, freeman of 24
consultancy work 109–10
criticism of 78–79, 103
dismissal by 11th Duke of Norfolk 41, 104–07
at Gascoigne collieries 70–72
income 52, 107
Ireland, visit to 109–10, 145
Literary and Philosophical Society, Newcastle-upon-Tyne, honorary member 24
management skills, limitations of 105

recognition and legacy 1–3, 69–70, 88–89
Report of Inventions 105–07
shipping of goods 119–21
steam engines, supply of 116–17, 117fig
Vincent Eyre III, relationship with 52, 104–05, 112
working relationships 229fig
—*The Coal Viewer and Engine Builder's Practical Companion* 2, 41, 53
   on conductors 80, 85–86
   on corves 65–66, 66fig
   dedication 52
   on jinneys 87–88, 87fig
   prospectus 60–61, 177–78
   publication history 59–61, 61fig
   on rail plates 55–56, 62fig, 63, 74
   reviews 61
   on steam engines 90–93, 90fig, 91fig
—**innovations** 51–53, 56–58, 79–93
   conductors 80–86, 83fig, 84fig
   corf development 65–67, 118–19
   cost of 105–07
   incline planes, self-acting 87–88, 88fig, 142
   jinneys 86–88, 87fig
   rail plates 55–56, 62fig, 63, 74
   railroad development 62–74
   ropes 123–27, 140, 186–87
   steam engines 90–93, 98–99, 141–42
   ventilation improvements 90
   waggonways 45–47, 46fig
—**iron foundry** 41, 107, 112–17, 114fig, 115fig, 144
   advertising and promotion 61
   permission to establish 52
   sale of 199–200
   shipping of goods 119–21
—**patents** 41, 53, 66, 107, 109, 136–41, 186–87
   Patent 1660 (drawing coals) 41, 82–83, 83fig, 84fig, 86
   Patent 1660 (raising coals) 16
   Patent 1924 (ropes) 41, 86, 124fig, 125
   Patent 2270 (ropes) 41, 86, 125–26, 125fig, 186
   Patent 2891 (ropes) 109, 136–37, 137fig
   Patent 2914 (ropes) 109, 137, 186
   Patent 2947 (ships' cables) 137–38, 138fig
   Patent 2960 (ropes) 109, 139fig, 140, 187
   Patent 3157 (towing vessels) 109, 140, 140fig
   Patent 3502 (ropes) 109, 140–41
   Patent 3711 (ropes) 109, 140fig, 141
—**ropery** 27fig, 52, 121, 127–41, 128fig
   advertisements for 135–36, 188–89, 193–98
   sale of 141
   success of 135–36, 191–92
   technical developments 127–34
—**at Sheffield Park Colliery**
   early work 42–43, 49–51
   Superintendent of Coal Works 12, 41, 50, 52
   waggonway 45–47, 46fig
Curr, John III (1783–1860, son) 86, 111, 143–59, 160, 167, 228fig
   in Australia 155–59
   death 158–59
   iron foundry works 112, 147
   The Learned Donkeys 153–54, 154fig
   Railway Locomotion 45–46, 149, 152–54, 158
   Statistics of Banking in the Colony of New South Wales 157–58
   steam engines supplied by 144–46
   steam power, press correspondence 156–57, 158
   steamboat manufacture 147–54
   tobacco and snuff manufacture 154, 154fig
   working relationships 229fig
Curr, John (IV, grandson) 155
Curr, Joseph Richard (son) 9, 161–67, 169, 228fig
   death and legacy 166–67
   and New Chapel 35, 38fig, 39, 39fig
   publications 162–66
Curr, Juliana Teresa (later Ellison, daughter) 9, 39fig, 169, 228fig
Curr, Martha (formerly Melton, daughter-in-law) 154, 158–59, 228fig
Curr, Mary Ann (daughter, later Beauvoisin) 9, 39fig, 159–60, 159fig, 228fig
Curr, Mary Anne (granddaughter) 160, 228fig
Curr, Teresa (daughter) 25, 160–61, 228fig
Curr, William (brother) 6–7, 50, 228fig

Dale Abbey tramroad 77
Darnall colliery 95, 98, 99fig, 100, 100fig, 101–02, 103fig
Deakin, Joseph 98, 101

debts 159–60
Defiance (paddle steamer) 148–49
Dendy Marshall, C. F. 60, 86–87, 239n162, 240n189
Derby Canal 76–77
Dickinson, H. W. 127, 184–85
Dixon, George 70
Dodd, George 150
Don (river) 94fig, 100fig, 102, 103fig, 119, 120
Donkin, Bryan 151, 153
Dore House colliery 101, 103–04, 103fig
Duchesne, Franciscus Nicholas 25–26
Dunn, Matthias 1–2, 16, 57–58, 79, 126, 142, 147
Dunn, William 101, 103
Durham *see* County Durham
Durham Cathedral, collieries of 18
Durham Record Office (DRO) 111

East India Company 139
Elliott, Martin 52
Ellison, Juliana Teresa (formerly Curr, daughter) 9, 39fig, 169, 228fig
Ellison, Thomas (son-in-law) 39fig, 168, 228fig
Elswick colliery 111, 116
Emerson, William 15, 234n56
Evans, Oliver 152
Eyre, Nathaniel 48
Eyre, Vincent Henry 104, 229fig
Eyre, Vincent II 48
Eyre, Vincent III 12, 41, 48, 97–98, 103–04
   John Curr III, relationship with 52, 104–05, 112
   death 105
   11th Duke of Norfolk, partnership with 51, 95
   John Buddle Snr consulted by 68, 80
   working relationships 229fig

Farey, John 74, 92–93, 92fig, 97
Fawcett, J. W. 5
fire risks 81
Flatworth colliery 116
Flinn, Michael W. 12, 17–18, 57, 59–60, 123
floods 101, 102–03, 104
Frank Cass & Co (publisher) 60
freight transport 119–21, 120fig
Froghall railway 52, 53
Fulford, York 166

Furniss, Elizabeth Mary (formerly Curr, daughter) 9, 30, 33, 37–38, 39, 39fig, 143, 228fig
Furniss, John Bernard (son-in-law) 30, 143, 228fig
   and New Chapel 36, 37–38, 39, 39fig, 173–76
Furniss, Rev John (grandson) 35, 39fig, 143, 228fig

Gainsford, Major A. J. 29, 31
Galloway, R. L. 2, 15, 17
   on railroads 58, 61, 65, 70, 87
   on rope 123, 126
   on winding 85
Garforth colliery 71–72
Gascoigne, Sir Thomas 41, 70–72
Gascoigne collieries 70–72
Gatherick colliery 70
The General Magazine 15, 15fig
The Gentleman and Lady's Palladium (magazine) 14–15
The Gentleman's Diary (magazine) 14, 20, 21fig
Gentleman's Magazine 16
Gillow, Joseph 161–66
gins 58, 79–80, 203fig, 203g
   horse-gins 58, 72, 79–80, 100
goods, transport of 119–21, 120fig
Green, Rev. Robert 165–66
Greenside colliery, Ryton 17
Gurner, Dorothy 1

Hadfield, Charles 34–36, 161, 163, 173–76
Hadrian's Wall 19
Hall, Thomas Young 5, 85
hard/Barnsley coal 95, 101, 201g
Harris, John Rees 59, 240n182
Harrison, David 19
Haslock, H. R. 47
Hatchett, Charles 50, 66–67, 81, 88–89
hauliers/hurriers 58, 203g
Heaton colliery 70, 111, 111fig
Hebburn colliery 18, 111, 116, 133, 134
Hemingfield colliery 81–82, 202fig
Hesley Wood colliery 52, 76, 81, 105
hewers 70, 203g
High Hazles colliery 100, 100fig, 103fig
   steam engines at 99fig, 101
   water management 102, 103, 104
High Main Coal seam 6fig
Hillyard, Jen 235n71

259

Hiskey, Christine E. 12, 14, 245n348
Holland, John
   Sheffield Park: A Descriptive Poem 29–30
   'Theresa's Tree' 160–61
Hopkinson, G. G. 2
Horne, Rev Melville 163
horse-gins 58, 72, 79–80, 100
horses 57–59, 65–68, 73, 85, 88–89, 112
   putting ponies 57, 58, 206g
   and rail/tramroads 75, 78, 79
   transport of goods by 119
House of Commons Select Committees
   on child labour 12–13
   Courier explosion inquiry 150–51
Howard, Charles (1720–1786) *see* Norfolk, 10th Duke, Charles Howard
Howard, Charles (1746–1815) *see* Norfolk, 11th Duke, Charles Howard
Howard, Edward *see* Norfolk, 9th Duke, Edward Howard (1685–1777)
Howard, Henry 41, 48, 49, 52
Huddart, Captain Joseph 139
Hudson, William 160–61
Hunter, John 42, 92, 93, 95–104, 126, 127, 181–82
hurriers (hauliers) 58, 203g
hurrying (hauling) 56, 63, 68, 203g
Hutton, Charles 15, 234n56

Idle (river) 119, 120
inby *see* rolley
incline planes, self-acting 87–88, 88fig, 142
inflation 106
investments 25–26
Ireland 109–10, 145
iron 61, 203g–04g
   cast 18–19, 52–53, 55–56, 79, 112, 117, 183
   cost of 74
   plate 113, 116
   props 117, 183
   wrought 53, 55, 79
iron foundries
   Booth & Co/Binks, Booth & Hartop 69, 96, 97, 98, 112
   of John Curr III 41, 52, 61, 107, 112–17, 114fig, 115fig, 144
   of John Curr III, sale of 199–200
   of John Curr III, shipping of goods 119–21
   Norwich Iron Foundry 147
   Walkers & Co 97

iron rails 52, 53, 113, 116
   Curr's development of 55–56, 62–69, 62fig, 64fig
   John Curr I's disapproval of 46–47

Jacob, Margaret 17
Jarrow colliery 111
Jeffcock, John 52, 103
Jessop, William 53, 62–63, 73–74, 77–78
jinneys 86–88, 87fig
Juliana (steamboat ferry) 152

Lambton, John George 18
Lambton colliery 44, 44fig, 135, 136
Launceston, Van Diemen's Land (Tasmania) 155
Lee, Charles E. 2, 51, 53–54, 60
Leeds 75, 161, 166
Leppingwell, James R. 147
Lewis, M. J. T.
   on barroways 56
   on child labour 57
   on mining working conditions 58–59
   on rail plates 78–79, 239n166
   on railroad/railway development 47, 55, 65, 67, 69, 70, 74–75
   on tramroads 74
Liffen, John 60
Lings colliery 76
Little Eaton, Derbyshire 53, 76–77
Lock, Alexander 71–72, 105
Locke, John, Snr 52
Locke, John, Jnr 52
Lord's House, Sheffield 33, 33fig, 35, 48

Major, J. K. 78
Manchester 161–66
Manor colliery 49, 52, 104, 105, 112
Mateaux, C. L. 47
Mechanics' Magazine 152–53, 154
Medlicott, Ian R. 2–3, 66fig, 67, 68, 69, 105, 141
Meinzies, Michael 86–87
Melton, Martha (later Curr, daughter-in-law) 154, 158–59, 228fig
Merthyr Tydfil to Abercynon tramroad 41, 53, 72, 77, 78, 78fig, 79
Middleton colliery 55, 70, 75, 88fig
Midland Institute of Mining, Civil & Mechanical Engineers 82–83
mines *see* collieries
mining *see* coal mining

Miscellaneous Correspondence (magazine) 15
Monmouthshire Canal 74
Mott, R. A. 38, 62–63, 75–76, 77, 78, 105

NEIMME *see* North of England Institute of Mining and Mechanical Engineers
New Chapel, Norfolk Row, Sheffield 35–36, 36fig, 37fig, 173–76
Newcomen, Thomas 92fig, 202
Newcomen engines 91, 99, 181–82
nonconformist religion 13, 14
Norfolk, 9th Duke, Edward Howard (1685–1777) 47, 49
Norfolk, 10th Duke, Charles Howard (1720–1786) 41, 45, 47–48, 229fig
Norfolk, 11th Duke, Charles Howard (1746–1815) 1, 35, 48, 52
   collieries, development of 95–104
   John Curr III, dismissal of 41, 104–07
   John Curr III, relationship with 106, 229fig
   Vincent Eyre III, partnership with 51
Norfolk Row, Sheffield
   New Chapel 35–36, 36fig, 37fig, 173–76
   Old Chapel 34, 35fig
   St Marie's Church 36fig, 37–39
North of England Institute of Mining and Mechanical Engineers (NEIMME) 17, 111, 235n71
Norwich Iron Foundry 147

Odell, Thomas 159–60
Odom, William 33
Ogden, Francis 152
Old Chapel, Norfolk Row, Sheffield 34, 35fig
Oldroyd, David 11–12
Orwell (paddle steamer) 147–48, 148fig
outby *see* rolley
Outram, Benjamin 53, 62, 73–75, 77
Outram & Co (railroad builders) 62, 73–74, 76
Overton, George 41, 72–74, 77
Overton, John 72, 243n262

Parlington colliery 72
patents 53, 66, 107, 136–41, 186–87, 205g
   Patent 1660 (drawing coals) 41, 82–83, 83fig, 84fig, 86
   Patent 1660 (raising coals) 16
   Patent 1924 (ropes) 41, 86, 124fig, 125
   Patent 2270 (ropes) 41, 86, 125–26, 125fig, 186
   Patent 2891 (ropes) 109, 136–37, 137fig
   Patent 2914 (ropes) 109, 137, 186
   Patent 2947 (ships' cables) 137–38, 138fig
   Patent 2960 (ropes) 109, 139fig, 140, 187
   Patent 3157 (towing vessels) 109, 140, 140fig
   Patent 3502 (ropes) 109, 140–41
   Patent 3711 (ropes) 109, 140fig, 141
Penniston, Isaac 127
Pensher colliery 135, 136, 146, 146fig
Percy Main colliery 19, 110, 111fig, 144, 145–46
pipes 117
Ponds colliery 52, 96, 207
Pontop Pike, County Durham 5, 6fig, 44
props (punches) 117, 183
putting ponies 57, 58, 206g
putting/putters 58, 70, 80, 206g

R. & W. Furley & Co. (ropemakers) 141
rail plates 55–56, 62fig, 63, 74, 78–79, 239n166
railroads 53, 62–79
   for canal transport 72–74
   cast-iron 56
   corves for 65–67, 66fig
   definition 56
   Sheffield Park Colliery 41
   spread of 74–78
   underground 67–70
   see also tramroads; waggonways
railways 51, 52–56, 54fig, 55fig
   above-ground 53
   cast-iron rails 52, 53
   flanged cast-iron rails (plates) 53, 56, 62–69, 62fig, 64fig, 72–78, 76fig
   flanged-wheel vehicles 52, 54, 55fig
   iron-edged boards 52, 53
Reay, Ann (later Buddle) 16
religion
   anti-Catholicism 162–64
   Catholicism 5–7, 9, 23–24, 33–39, 70, 143, 160–67
   nonconformist 13, 14
Reynolds, William 41, 72
Rieuwerts, Jim 55
Rimmer, Richard 33, 34, 36, 38fig, 167
riots 45, 46, 47
river transport 119, 120fig

Robinson, Thomas 14
Roby, Rev W. 163
rolleys 57, 59, 65–66, 70, 80, 206g
Roman Catholic Relief Act 1791 33
ropery 109, 122fig
ropes 41, 86, 109, 121–27, 140, 184–98
  double 125
  flat 41, 81, 123–36, 133fig, 186, 188–89
  flat, performance of 128–36
  flat, problems with 125–26, 190–91
  patents 41, 86, 124fig, 125fig, 136–37, 137fig, 139fig, 140–41, 140fig, 186–87
  problems with, in mining 123–24
  twisting 128–33
rope-walks 127
Rother (river) 94fig, 101, 103fig
Royal Commission of Inquiry into Children's Employment in Mines and Manufactories 56–57
Royal Navy 139
Russell, William 18

Sargent, Edward 121–23
Scarbrough, 4th Earl, Richard Lumley-Saunderson 48
Sedgley Park School 161
shaft conductors 66, 80–86, 83fig, 84fig
Sheffield
  Belle Mount 159
  Belle Vue House, Sheffield Park 27–32, 27fig, 28fig, 29fig, 30fig, 31fig, 32fig, 45–46
  Buckwell 30
  coal stage 45
  Colliers Row, Sheffield Park 25, 25fig
  Duke of Norfolk's Hallamshire estate 41, 48, 48fig
  Lord's House 33, 33fig, 35, 48
  New Chapel, Norfolk Row 35–36, 36fig, 37fig, 173–76
  Old Chapel, Norfolk Row 34, 35fig
  Ponds area 6–7, 8
  St Marie's Church, Norfolk Row 36fig, 37–39
  see also Attercliffe Common Colliery; Hesley Wood colliery; Manor colliery
Sheffield Park Colliery 41, 49–52, 95
  John Curr III's improvements at 88–89
  plans and reports 42–43, 42fig, 43fig
  profitability of 104
  railroads 66–69, 75, 76
  waggonway 42–43, 44–47, 46fig
  winding 81
Sheriff Hill colliery 111
ships' cables 137–39
Shropshire Canal, Brierley Hill terminus 72–74
slack/small coal 49, 95, 97–98, 201g
Slatcher, W. N. 69
Smeaton, John 70–71, 92–93
Smiles, Samuel 47
Smilter, John 143, 173, 174
Smith, Ellen 167
Smith, William 'Strata' 81
social unrest 45, 46, 47
Society for the Encouragement of Arts, Manufactures and Commerce 70
soughs 101, 102, 206g
St George's Convent, Menin 160
St Marie's Church, Norfolk Row, Sheffield 36fig, 37–39
Staniforth, Thomas 95, 100, 104
Staniforth colliery 95, 100–101, 102–03, 103fig, 104
Stanley Kinghill colliery 70
Stapehill Abbey, Dorset 160
steam engines 16, 41, 66, 92fig, 99fig, 181–82, 202g
  in *The Coal Viewer* 60, 90–93, 90fig, 91fig
  Curr, supplied by 116–17, 117fig
  Greenland Engines 99fig, 100fig, 101–02, 104
  problems with 144–46
  stationary 109, 141–42
  tenders for 146–47, 146fig
  for water management 96–104
  winding 72, 79–80
steamboats 147–54
  explosions 149–52
Stephenson, John 98
Stevens, John 152
Stockton and Darlington Railway 53, 75, 79
Stonyhurst College 155, 161
Strafford, 2nd Earl, William Wentworth (1722–1791) 48
Sturton colliery 71
Surrey, Earl of see Norfolk, 11th Duke, Charles Howard
Surrey Iron Railway 53, 76fig, 77, 78
Sydney, Australia 155–59
Sykes, J. 2, 57, 67–68, 69–70

Tanfield, County Durham 44
Tanfield Moor colliery 110
Tasmania (Van Diemen's Land) 155, 168–69
Tattersfield, Nigel 16
The Town and Country Magazine 20
Thurston, Robert 147
tobacco and snuff manufacture 154, 154fig, 155
Tomlinson, W. W. 78
Townsend and Furniss (lessees, Sheffield Park Colliery) 44, 49–50
tramroads 53, 74
   Bugsworth to Peak Forest 77
   Consall Plateway, Stafford 77
   Dale Abbey 77
   Merthyr Tydfil to Abercynon 41, 53, 72, 77, 78, 78fig, 79
   Warwick 77
   Wigan 77
transport
   from coalface 56, 65–67
   of goods 119–21, 120fig
   see also canals; railroads; railways
transportation, of convicts 168
Trent (river) 119, 120
Trevithick, Richard 53, 79
Trinder, Barrie 2, 73, 78
Tyne (river) 14fig, 43–44, 111fig, 120
Tyneside 14fig, 111fig

underground canals 89
underground railroads 67–70
Unitarianism 13, 14
United States of America 152
Ushaw College 23, 161, 164

Van Diemen's Land Company 155
Van Diemen's Land (Tasmania) 155, 168–69
Vane-Tempest, Sir Henry 18
ventilation 19, 81, 90
Vitaliev, Vitali 51

waggonways 6fig, 43–47, 43fig, 44fig, 52, 56
   Sheffield Park Colliery 41, 44–47, 46fig, 53
   underground 67–68
   wooden vs. iron 46–47
Walkers, Eyre and Stanley (bank) 48
Walkers & Co (iron founders) 97

Wallsend 18–19, 110, 111fig
   conductors at 84–85
   flat ropes at 128–29, 133–34
   railroads at 112–13, 116
Warwick tramroad 77
Washington colliery 70, 111fig, 136
water management 16, 17–19, 95–99, 100fig
   breaches/floods 101, 102–03, 104
   steam engines for 96–104
Watt, James 202, 242n242
Wear (river) 14fig, 111fig
Welford, Richard 14, 16–17
Wellesley-Pole, William 139
Wentworth, William, 2nd Earl Strafford (1722–1791) 48
West Kyo, County Durham 14, 14fig, 15–16
Whitby 166
Whitefield colliery 116
Wigan tramroad 77
Wilmot, George 48
Wilson, A. N. 13
Wilson, E. C. 5
Wilson, Hannah (wife, later Curr) 7–9, 33, 228fig
Wilson, Richard 7, 8
Wilson, Thomas, *The Pitman's Pay* 58–59, 85
winding 58, 79–86, 83fig, 84fig, 97fig, 203fig
Wingerworth colliery 53, 74, 76, 77
winnings 18, 96, 118, 206g–07g
Wollongong, Australia 155, 158–59
Worsley colliery 89
wrought iron 53, 55, 79
Wykin colliery 111, 116

www.ingramcontent.com/pod-product-compliance
Lightning Source LLC
Chambersburg PA
CBHW081102070526
44584CB00021B/3172